VHDL FOR LOGIC SYNTHESIS

Third Edition

VHDL FOR LOGIC SYNTHESIS

Third Edition

Andrew Rushton

A John Wiley and Sons, Ltd, Publication

Library of Congress Cataloging-in-Publication Data

Rushton, Andrew.
  VHDL for logic synthesis / Andrew Rushton. – 3rd ed.
    p. cm.
  Includes index.
  Summary: "Macrocycles: Construction, Chemistry and Nanotechnology Applications is an essential introduction this important class of molecules and describes how to synthesise them, their chemistry, how they can be used as nanotechnology building blocks, and their applications"– Provided by publisher.
  ISBN 978-0-470-68847-2 (hardback)
  1. VHDL (Computer hardware description language) 2. Logic design–Data processing. 3. Computer-aided design. I. Title.
  TK7885.7.R87 2011
  621.39′5–dc22

                                                                                                    2010045678

A catalogue record for this book is available from the British Library.

Print ISBN: 9780470688472
E-PDF ISBN: 9780470977927
O-book ISBN: 9781119995852
E-Pub ISBN: 9780470977972

Set in 10/12pt Times by Thomson Digfital, Noida, India.

Contents

Preface

The motivation for writing this book originally came from my own frustration at the lack of a synthesis-orientated book when I was learning VHDL. Not only was there a lack of information on the synthesis subset, but I found that most books on VHDL had a common problem: they described absolutely everything in an indiscriminate way, and left the reader to sort out which bits were relevant and useful. It was extremely difficult to deduce the synthesis subset from this approach.

In *this* book, I cover the features of VHDL that you need to know for logic synthesis, from a hardware designer's viewpoint. Each feature of the language is explained in hardware terms and the mapping from VHDL to hardware is shown. Furthermore, only the synthesisable features are presented and so there is no possibility of confusion between synthesisable and non-synthesisable features.

The exception to this rule is the chapter on test benches. Even hardware designers using the language exclusively for logic synthesis will have to write test benches and since these are not synthesised, the whole language becomes available (but not necessarily useful). So the test bench chapter introduces those parts of the language that are relevant and useful for writing test benches.

The reason that a book like this is necessary is that VHDL is a very large and clumsy language. It suffers from design-by-committee and as a result is difficult to learn, has many useless features, and I can say from my own experience, is extremely difficult to implement. I am not a champion of VHDL, but I recognise that it is still probably the best hardware description language for logic synthesis that we have. I hope that, by sharing what I have learnt of the language and how it is used for synthesis, I can help you avoid the many pitfalls that lie in wait.

I have this perspective on VHDL because I started my career as an Electronics Engineer, specialising in Digital Systems Design and gaining a BSc and PhD from the Department of Electronics at Southampton University, UK, in 1983 and 1987 respectively. However, I then moved into software engineering, but using my hardware background to develop software within the Electronics Design Automation industry. I have been working on VHDL and Electronic Design Automation using VHDL since 1988.

Initially I worked on logic synthesis systems, first for Plessey Research Roke Manor which is now a part of Siemens' UK operation. Then, in 1992 our then manager and CEO-to-be Jim Douglas arranged a management buyout of the synthesis technology that we had developed, supported by venture-capital funding from MTI Partners. Thus was born TransEDA Limited.

He took with him the key engineers for the project, and so I became one of the founder members of the new company. I was Research Manager for the new company and continued working on the logic synthesis project.

Our intention was to develop our in-house logic synthesis tool to commercial standard and sell it under the name TransGate. One of my first tasks was to help develop a VHDL front-end to the tool to replace the existing proprietary language front-end. I was very proud of the results that we achieved – TransGate had a very comprehensive support for the language, competitive with the best in the market at the time and considerably better than the majority of tools.

When we first released TransGate, we expected that engineers would take to VHDL easily, so we concentrated on the purely technical aspects of developing the synthesis algorithms. However, it gradually became apparent from feedback that users were experiencing problems with using VHDL for logic synthesis due to the learning curve associated with what was, at that time, a completely new hardware design paradigm.

As a consequence of this realisation, in 1992 I developed a new training course, offered as a public or on-site course and called 'VHDL for Hardware Design'. This course was based on my inside knowledge of how VHDL is interpreted by a synthesiser and also on the practical problem solving that I had been involved with as part of the company's customer support programme.

The first edition of this book, published in 1995 by McGraw-Hill, grew out of that training course. Much of the text and some of the examples were taken straight from the course. However, there is far more to a book than can be covered in a three-day long training course, so the book covered more material in far more detail than was possible in the training course.

Furthermore, at the time of writing the first edition, there was an international standardisation effort to define a standard set of arithmetic packages and common interpretation and subset for VHDL for logic synthesis. Although this standardisation was still some way from completion at the time, nevertheless there were some aspects of logic synthesis from VHDL that had a wide consensus and this was used to inform the writing of the book.

Back at TransEDA, we were finding that the logic synthesis market niche was not only already occupied but comprehensively filled by well-established companies and we made little progress in selling our synthesis tools.

Fortunately, we branched off into code coverage tools and created a niche for ourselves in this market instead. I became the lead systems developer for the VHDLCover system. Through this project, which involved a lot of collaboration with customers, I gained experience of scores of large synthesisable VHDL designs involving hundreds of designers working in many different styles.

This change in direction of our company had a strong influence on the second edition of this book that was published in 1998 by John Wiley and Sons. Three years had passed and the standards committee had at last ratified a standard for the synthesis packages. Furthermore, exposure to many other designers' work allowed me to take a broader view of the use of synthesis and its place in the design cycle. This made the book more user-orientated than the first edition, which did tend to dwell too much on the way that synthesisers worked. I think that the change in emphasis (slight though it was) improved the book significantly.

I left TransEDA in 1999, and since I left the company has gone bust, unfortunately disbanding the development team. However, the code coverage technology and the company name has been bought out and so TransEDA still sells VHDLCover but now under the name VN-Cover.

After TransEDA, I joined Southampton University and became a founding member of the university spin-off company Leaf-Mould Enterprises (LME). LME was formed with the intention of developing commercial behavioural synthesis systems using VHDL and based on a research programme within my old department, the Department of Electronics and Computer Science. I was responsible for the VHDL library manager, compiler and assembler which produced the concurrent assembly code from which behavioural synthesis was performed. Unfortunately, funding problems led to the demise of LME in 2001.

Since then I have become a self-employed consultant, working in a diversified range of fields: programmer, Web applications designer, systems engineer and counsellor.

It is 12 years since the publication of the second edition and it is interesting to see what has changed in the field of synthesis. The main change is that designers are moving on to system-level synthesis using C-like languages such as System Verilog, SystemC and Handel-C. However, there is clearly still a role for logic synthesis using VHDL for those who need more control over their design or, for that matter, as the synthesis engine for higher-level tools. There are now a plethora of logic synthesis tools available, for both ASIC and FPGA design.

However, VHDL itself has hardly changed at all for most of that time, with just minor tweaks to the language in 2000 and 2002. Then, in 2008, a major update was published to address a wide range of problems and to expand the range of pre-defined packages delivered with the language. Many of these changes affect synthesis. So, the time has come for a third edition of the book to reflect these changes. I have updated the whole book to reflect the current position, where the full VHDL-2008 standard is not yet available in any commercial tool, either for simulation or for synthesis, but some of the synthesis-specific features are gradually becoming available, either incorporated into the synthesis tools or as downloadable add-ons.

Andrew Rushton, London, 2010

List of Figures

List of Tables

1

Introduction

This chapter looks at the way in which VHDL is used in digital systems design, the historical reasons why VHDL was created and the international project to maintain and upgrade the language.

1.1 The VHDL Design Cycle

From its conception, VHDL was intended to support all levels of the hardware design cycle. This is clear from the preface of the Language Reference Manual (LRM) (IEEE-1076, 2008) which defines the language, from which the following quote has been taken:

> VHDL is a formal notation intended for use in all phases of the creation of electronic systems. Because it is both machine readable and human readable, it supports the development, verification, synthesis, and testing of hardware designs; the communication of hardware design data; and the maintenance, modification, and procurement of hardware.

The key phrase is 'all phases'. This means that VHDL is intended to cover every level of the design cycle from system specification to netlist. As a result, the language is rather large and cumbersome. However, this does not necessarily make it difficult to learn. It is best to think of VHDL as a hybrid language, containing features appropriate to one or more of the stages of the design cycle, so that each stage is in effect covered by a separate language that also happens to be a subset of the whole. Each subset is relatively easy to learn, provided there is guidance as to what is in, and what is not in, that subset.

In the idealised design process, there are three subsets in use – since there are three stages that use VHDL. These are: system modelling (specification phase), register-transfer level (RTL) modelling (design phase) and netlist (implementation phase).

In addition to these VHDL-based phases, there will be an initial requirements phase that is conventionally in plain (human) language. Thus, there are three stages of transformation of a design: from requirements to specification, from specification to design and from design to implementation. The first two phases are carried out by human designers, the last phase is now largely performed by synthesis.

Figure 1.1 illustrates this idealised design cycle.

VHDL for Logic Synthesis, Third Edition. Andrew Rushton.
© 2011 John Wiley & Sons, Ltd. Published 2011 by John Wiley & Sons, Ltd.

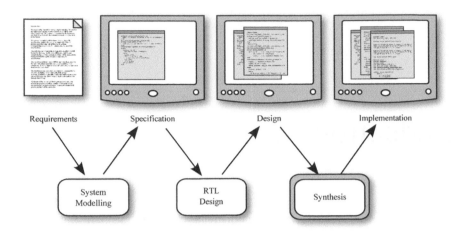

Figure 1.1   The VHDL-based hardware design cycle.

Typically, the system model will be a VHDL model that represents the algorithm to be performed without any hardware implementation in mind. The purpose is to create a simulation model that can be used as a formal specification of the design and that can be run in a simulator to check its functionality. This specification can also be used to confirm with a customer that the requirements have been fully understood.

The system model is then transformed into a register-transfer level (RTL) design in preparation for synthesis. The transformation is aimed at a particular hardware implementation but at this stage, at a coarse-grain level. In particular, the timing is specified at the clock cycle level at this stage of the design process. Also, the particular hardware resources to be used in the implementation are specified at the block level.

The final stage of the design cycle is to synthesise the RTL design to produce a netlist, which should meet the area constraints and timing requirements of the implementation. Of course, in practice, this may not be the case, so modifications will be required which will impact on the earlier stages of the design process. However, this process is the basic, idealised, design process using VHDL and logic synthesis.

1.2   The Origins of VHDL

VHDL originated from the American Department of Defense, who recognised that they had a problem in their hardware procurement programmes. The problem was that they were receiving designs in proprietary hardware description languages, which meant that, not only was it impossible to transfer design data to other companies for second sourcing, but also there was no guarantee that these languages would survive for the life expectancy of the hardware they described.

The solution was to have a single, standard hardware description language, with a guaranteed future. Specification of such a language went ahead as part of the Very-High Speed Integrated Circuits programme (VHSIC) in the early 1980s. For this reason, the language was later named the VHSIC Hardware Description Language (VHDL).

If the language had remained merely a requirement for military procurement, it would quite possibly have remained an obscure language of interest only to DoD contractors. However, the importance of the language development, and especially the importance of standardisation of the language, was recognised by the larger electronic engineering community and so the formative language was passed into the public domain by placing it in the hands of the IEEE in 1986. The IEEE proceeded to consolidate the language into a standard that was ratified as IEEE standard number 1076 in 1987. This standard is encapsulated in the VHDL Language Reference Manual (LRM).

1.3   The Standardisation Process

Part of the standardisation process was to define a standard way of upgrading the language periodically. Thus, there is a built-in requirement for the language to be re-standardised every five years. However, in practice updates have been irregular and driven by a desire to improve the language according to demand rather than this arbitrary 5-year cycle. Because the language has changed over the years, it is sometimes important to differentiate between versions. This is done in this book by referring to the year in which the standard was ratified by the IEEE. For example, the original standard, IEEE standard number 1076, ratified in 1987, is usually referred to as VHDL-1987. Subsequent revisions of the standard will be referred to in a similar way according to their year of ratification.

Here is a summary of the different versions and the features that affect the use of the language for synthesis:

VHDL-1987   The original standard.
VHDL-1993   Added extended identifiers, xnor and shift operators, direct instantiation of components, improved I/O for writing test benches.
Most of the synthesis subset of VHDL is based on VHDL-1993.
VHDL-2000 (minor revision)   Nothing of relevance to synthesis.
VHDL-2002 (minor revision)   Nothing of relevance to synthesis.
VHDL-2008   Added fixed-point and floating-point packages.
Added generic types and packages, enabling the use of generics to define reusable packages and subprograms. Enhanced versions of conditionals. Reading of out ports. Improved I/O for writing test benches.
Unification of VHDL standards.

As you can see, there are only three versions of VHDL relevant to synthesis: VHDL-1987, VHDL-1993 and VHDL-2008. VHDL-1993 was the last revision to add features useful for synthesis. So VHDL-2008 is the first significant change in 15 years. A lot has been added in VHDL-2008 (Ashenden and Lewis, 2008) and most of it has some relevance to synthesis.

However, synthesis tool vendors are historically slow to adopt new language features. This is for good reasons – the focus of synthesis is the quality of the synthesised circuit and effectiveness of the synthesis optimisations, not the list of language features supported. This means that it is expected that several years will pass before the more significant changes in VHDL-2008 are implemented by synthesis tools and many never will be. In effect, synthesis users are still using VHDL-1993 and will continue to do so for the foreseeable future.

As a consequence, this book is based mainly on VHDL-1993. However, the more recent extensions are discussed where relevant, particularly with regard to the new fixed-point and floating-point packages added in VHDL-2008 but that have been made available as VHDL-1993 compatibility packages so that they can be used immediately on synthesisers that do not yet support the rest of VHDL-2008.

1.4 Unification of VHDL Standards

One of the largest changes in the VHDL-2008 standard is the unification of the many standards that define parts of the language and its environment.

The management of the standardisation process is down to the VHDL Analysis and Standardisation Group (VASG), part of the IEEE standardisation structure. In addition to the main standardisation process of the language itself, there are a number of working-groups working on standardisation of the ways in which VHDL is used. In the past, these working-groups have published standards of their own. For example, there was a group working on using VHDL for analogue modelling (VHDL-AMS – VHDL Analogue and Mixed-Signal – standard 1076.1), a group working on standard synthesisable numeric packages (VHDL Synthesis Package – standard 1076.3 (1997)), a group working on accelerating gate-level simulation (VITAL – the VHDL Initiative Towards ASIC Libraries – standard 1076.4), and a group working on the standard interpretation of VHDL for logic synthesis (VHDL Synthesis Interoperability – standard 1076.6). In addition, the 9-value logic type std_logic that is almost universally used for synthesis was developed as a completely different IEEE standard (VHDL Multivalue Logic Packages – standard 1164).

This separation of the standardisation of the various application domains of VHDL was effective in the early days of language development, because it allowed the subgroups to get on with their work independently of the main VHDL standardisation process and furthermore meant that they could publish their standards when ready, rather than waiting for the next formal release of the VHDL standard. However, this separation has become a problem as the working-groups' work has become mature, stable and in common use. For example, a release of a new standard for VHDL could leave the subgroups' standards lagging behind, compatible with the previous version and lacking the new language features.

So, in VHDL-2008, those working group standards that are specific to synthesis have been *partly* merged into the VHDL standard itself. Standard 1076 now includes the standard logic types (1164), the standard numeric types (1076.3) and some parts of the standard synthesis interpretation (1076.6). This doesn't make any difference to the user, but it does formalise these parts of the language as an integral part of VHDL and ensures that they stay in step with language developments in the future.

As you can probably imagine, this makes the Language Reference Manual (IEEE-1076, 2008) quite massive.

1.5 Portability

Synthesisable RTL designs can have a long life span due to their technology independence. The same design can be targeted at different technologies, revised and targeted at a newer technology and so on for many years after the original design was written. It is a wise designer who plans for the long-term support of their designs. It is therefore good practice to write using a safe, common style of VHDL that can be expected to be supported for years to come, rather than use 'clever' tool-specific tricks that might not continue to be supported.

Also, it is not unusual for a company to change their preferred tools, or for a designer to be obliged to use a different synthesis tool because a different technology is being targeted. So it is good practice to write using a portable subset of synthesisable VHDL that will work across many different tools.

The problem with this principle is that synthesis relies on an interpretation of VHDL according to a set of templates, and historically each synthesis vendor has developed their own set of templates. This means that in practice, each synthesis tool supports a slightly different subset of VHDL. However, there has always been a lot of overlap between these subsets and this book attempts to identify the common denominator.

To make life more complicated, the IEEE Design Automation Standards Committee have specified a synthesis standard for VHDL (IEEE- 1076.6, 2004) that seems to be a superset rather than a subset of the VHDL supported by commercial tools. Therefore, adhering to the standard does not mean that a design will be synthesisable with any specific synthesis tool. It also seems unlikely that any single tool will implement every detail of this standard.

It is recommended that a subset is used that is common to all synthesis tools. As a consequence, this book focuses on the common subset and avoids the more obscure tool-specific features of VHDL, even if those obscure features are in the synthesis standard.

2

Register-Transfer Level Design

Logic synthesis works on register-transfer level (RTL) designs. What logic synthesis offers is an automated route from an RTL design to a gate-level design.

For this reason, it is important that the user of logic synthesis is familiar with RTL design to the extent that it is second nature. This chapter has been included because many designers have never used RTL design *formally*. This chapter serves as a simple introduction to RTL design for those readers not familiar with it. It is not meant to be a comprehensive study but it does touch on all the main issues that a designer encounters when using the method.

RTL is a medium-level design methodology that can be used for any digital system. Its use is not restricted to logic synthesis: it is equally useful for hand-crafted designs. It is an essential part of the top-down digital design process.

Register-transfer level design is a grand name for a simple concept. In RTL design, a circuit is described as a set of registers and a set of transfer functions describing the flow of data between the registers. The registers are implemented directly as flip-flops, whilst the transfer functions are implemented as blocks of combinational logic.

This division of the design into registers and transfer functions is an important part of the design process and is the main objective of the hardware designer using synthesis. The synthesis style of VHDL has a direct one-to-one relationship with the registers and transfer functions in the design.

RTL is inherently a synchronous design methodology, and this is apparent in the design of all synthesis tools.

This chapter outlines the basic steps in the RTL methodology. It is recommended that these basic steps are used when designing for logic synthesis. To illustrate the connection between RTL and logic synthesis, the examples will be written in VHDL. You are not expected to understand the full details of the VHDL at this stage, but all the VHDL used will be covered in later chapters.

VHDL for Logic Synthesis, Third Edition. Andrew Rushton.
© 2011 John Wiley & Sons, Ltd. Published 2011 by John Wiley & Sons, Ltd.

2.1 The RTL Design Stages

The basis of RTL design is that circuits can be thought of as a set of registers and a set of transfer functions defining the datapaths between registers. The method gives a clear way of thinking about these datapaths and trying different circuit architectures while still at an abstract level.

The first stage of the design is to specify at a system level (i.e. *not* RTL) what is to be achieved by the circuit. Typically this will be a set of arithmetic and logic operations on data coming in at the primary inputs of the circuit. At this stage there is no hardware implementation in mind; the purpose is just to create a simulation model that can then be used as the formal specification of the design. At this stage the system-level model looks more like software than hardware. The system-level model can also be used to confirm with a customer that their design requirements have been understood. Even at this early stage in the design, long before the RTL design process is complete, it is possible to write a VHDL model for simulation purposes only (not intended to be synthesisable). This is a worthwhile exercise since it tests the understanding of the problem and allows the algorithm to be checked for correctness. Later, this VHDL model can be used for comparison with the completed RTL design to verify the correctness of the design procedure. This ability to cross-check different representations of a design in the same design language using the same simulator is a powerful feature of VHDL.

The second stage of the design is to transform the system level design into an RTL design. It is rare for a design to be directly implemented in exactly the same form as the system-level model. For example, if the design performs a number of multiplications or divisions, the circuit area of the direct implementation would be excessive.

The basic design steps in using RTL are:

- identify the data operations;
- determine the type and precision of the operations;
- decide what data processing resources to provide;
- allocate operations to resources;
- allocate registers for intermediate results;
- design the controller;
- design the reset mechanism.

The VHDL model of the RTL design can be simulated and checked against the system design.

The third stage of the design is to synthesise the RTL design. The resulting gate-level netlist or schematic can be (and should be) simulated against the RTL design to confirm that the synthesised circuit has the same behaviour.

Finally, the netlist or schematic produced by synthesis is supplied to the placement and routing tools for circuit layout.

Needless to say, the design will probably need to go through the design/synthesise/layout cycle several times with minor or even major modifications before all the design constraints are met. Synthesis does not eliminate the need to re-iterate designs, but it does speed up the iteration time considerably.

2.2 Example Circuit

The best way to illustrate the RTL design method is with an example. In this case, the example will be a quite artificial circuit for calculating the dot product of two vectors.

The dot product of two vectors is defined by:

$$\mathbf{a} \cdot \mathbf{b} = \sum_{i=0}^{n-1} a_i * b_i$$

For the purpose of this example, to keep it simple, the size of the vectors will be fixed at 8 elements.

The system-level model in VHDL is:

```
package dot_product_types is
  type int_vector is array (0 to 7) of integer;
end;
use work.dot_product_types.all;
entity dot_product is
  port (a, b : in int_vector; z : out integer);
end;
architecture system of dot_product is
begin
  process (a, b)
    variable accumulator : integer;
  begin
    accumulator := 0;
    for i in 0 to 7 loop
      accumulator := accumulator + a(i)*b(i);
    end loop;
    z <= accumulator;
  end process;
end;
```

This VHDL model is generally referred to as the system model. It is the simplest possible statement of the algorithm to be carried out, with no regard for data precision, timing or data storage.

In fact, since this is a very simple example, it *is* possible to synthesise this system model. This would not normally be the case and it should be assumed during the system modelling phase that the full range of VHDL can be used since the result is never going to be synthesised. In this example, synthesising the system model is of interest because it will give a means of comparison so that the effect of the RTL design process can be measured.

The system model was synthesised using a commercial synthesis system and targeted at a commercial ASIC library. It is not relevant which system and which library because the purpose of performing the synthesis is just to compare this direct implementation of the algorithm with the RTL model that will be developed over the rest of the chapter.

The results of synthesis were

- area – 40 000 NAND gate equivalents;
- I/O – 546 ports;
- storage – 0 registers.

It can be seen from the lack of registers that the system model synthesises to a purely combinational circuit. This circuit contains eight multipliers and seven adders. One of the reasons why this is such a large circuit is that the standard interpretation of integers is a 32-bit 2's complement representation. This means that the multipliers and adders are all 32-bit circuits.

Clearly the direct implementation of the system model is unacceptable and a better solution should be sought. This is where RTL design comes in.

2.3   Identify the Data Operations

The first stage in the design process is to identify what data operations are being performed in the problem. This can be seen more clearly in the form of a data-flow diagram showing the relationship between the datapaths and the operations performed on them. This is illustrated in Figure 2.1.

It can be seen from this diagram that the dot-product calculation requires eight 2-way multiplications and one 8-way addition. These are the basic data operations required to perform the calculation.

At this stage the type of the operation should also be considered. Are the calculations acting on integers, fixed-point or floating-point types? Will a transformation be needed? For example, performing floating-point calculations is very expensive in hardware and time, so significant speed and area improvements could be made by recasting the problem onto fixed-point or even integer types.

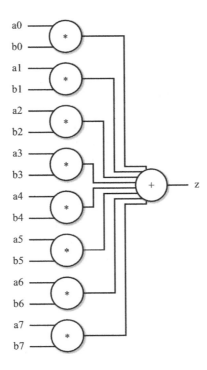

Figure 2.1   Cross-product calculator – data-flow diagram.

For this example, all the operations are assumed to be 2's-complement integer arithmetic.

The diagram also shows the dependencies on the data operations. The multiplications can be performed in any order or even all simultaneously since they are independent of each other. However, the additions must be carried out after the multiplications.

The additions have been lumped together as one operation. In practice, the additions will be performed as a series of two-way additions. They are lumped together in the figure because the ordering of the additions is irrelevant and can be chosen by the designer at a later stage in the design process so as to simplify the circuit design. This means that there are a number of structures for the data-flow diagram depending on the chosen ordering of the additions. The optimum ordering of these two-way additions will often become obvious as a design progresses. The two most likely candidates for the ordering of the additions are shown in Figures 2.2 and 2.3.

The different orderings of adders place different requirements on the ordering of the multiplications. The balanced tree for example allows an addition to be performed when any two adjacent multiplications have been performed. The multiplication pairs can be performed in any order or simultaneously. The skewed tree on the other hand places a stricter ordering on the multiplications but allows an addition after every multiplication except the first.

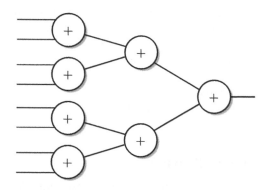

Figure 2.2   Adder – balanced tree.

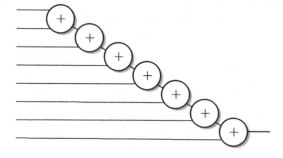

Figure 2.3   Adder – skewed tree.

No decision will be made at this stage of the design process, but it will become clear later in the design process that the skewed tree data-flow turns out to be the ordering for the chosen solution for this design.

Note that the two orderings of the additions illustrated here, and indeed all of the possible orderings, require seven 2-way additions.

In conclusion then, the data operations required to perform the dot-product calculation are:

- 8 multiplications;
- 7 additions.

2.4 Determine the Data Precision

In a real design, the specification would place requirements on the design, such as the expected data range, the required overflow behaviour and the maximum allowable cumulative error (for example when sampling real-world data). These factors will vary from design to design, but the key step in the design process will always be the same: to assign a precision to every data-flow such that the design meets the requirements.

This example is for illustration only, so the precision of the calculations will be chosen arbitrarily. In this case overflow during the addition will be allowed but will be ignored to keep the example simple.

In this example the following will be assumed:

- data inputs 8-bit 2's-complement;
- all other datapaths 16-bit 2's-complement.

2.5 Choose Resources to Provide

Having determined the data operations to be performed and the precision of those operations, it is now possible to decide what hardware resources will be provided in the circuit design to implement the algorithm.

In the simplest case, there would be a one-to-one mapping of operations onto resources. This would be a direct implementation of the algorithm in hardware. In this example, a direct implementation would require eight 8-bit multipliers (with 16-bit outputs) plus seven 16-bit adders. This is the same circuit as the system specification but with reduced precision on the datapaths.

Since this is just an example, there are no design constraints as such. However, for the purposes of the exercise, it will be assumed that there are design constraints that effectively restrict the hardware resources to one multiplier. The system will be clocked and the result accumulated over several clock cycles. No limit is placed on the number of clock cycles that can be used or on the length of the clock cycle, but it will also be assumed that a complete multiply and add can be performed in one clock cycle. This means that, since there is only one multiplier, the design also only needs one adder.

So, in summary, the hardware resources available are:

- one, 8-bit input, 16-bit output, multiplier;
- one, 16-bit input, 16-bit output, adder.

2.6 Allocate Operations to Resources

The next stage in the RTL design cycle is commonly referred to as Allocation and Scheduling. Allocation refers to the mapping of data operations onto hardware resources. Scheduling refers to the choice of clock cycle on which an operation will be performed in a multi-cycle operation. Registers must also be allocated to all values that cross over from one clock cycle to a later one. Allocation and Scheduling are interlinked and normally must be carried out simultaneously. The aim is to maximise the resource usage and simultaneously to minimise the registers required to store intermediate results.

Due to the simplicity of this example, the allocation stage is trivial since all multiplications must be allocated to the one multiplier and all the additions to the one adder.

The scheduling operation means choosing which clock cycle each multiplication and addition is to be performed. This is confused slightly by the fact that all the additions are interchangeable. Since the specification allows a multiplication and an addition in one clock cycle, the schedule can allow the product of a multiplication to be fed directly to the adder in the same clock cycle, therefore avoiding an intermediate register.

The scheduling and allocation scheme is illustrated by Table 2.1.

The whole operation of calculating the dot-product takes eight clock cycles. The algorithm has been simplified slightly by adding an eighth addition in the first cycle that effectively resets the accumulated result by adding 0 to `product0` instead of adding the result so far. This saves the need for a reset cycle.

Only one register is required by this scheduling since the only value that needs to be saved from one clock cycle to another is the result that is accumulated over the eight clock cycles.

It is now possible to design the datapath part of the circuit minus its controller. The datapath consists of a multiplier with two inputs, one multiplexed from the set of `a0` to `a7`, the other multiplexed from the set of `b0` to `b7`. The product is then added to either the accumulated `result` or 0. Finally, the accumulated `result` is saved in a register. The circuit is shown in Figure 2.4.

Table 2.1 Scheduling and allocation for cross-product calculator

Cycle	* Operator	+ Operator
1	a0*b0 $\Rightarrow$ product0	0 + product0 $\Rightarrow$ result
2	a1*b1 $\Rightarrow$ product1	result + product1 $\Rightarrow$ result
3	a2*b2 $\Rightarrow$ product2	result + product2 $\Rightarrow$ result
4	a3*b3 $\Rightarrow$ product3	result + product3 $\Rightarrow$ result
5	a4*b4 $\Rightarrow$ product4	result + product4 $\Rightarrow$ result
6	a5*b5 $\Rightarrow$ product5	result + product5 $\Rightarrow$ result
7	a6*b6 $\Rightarrow$ product6	result + product6 $\Rightarrow$ result
8	a7*b7 $\Rightarrow$ product7	result + product7 $\Rightarrow$ result

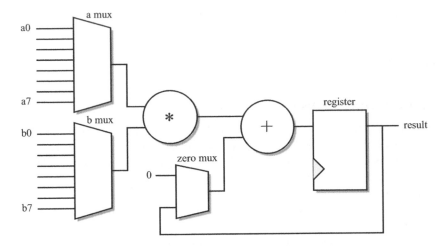

Figure 2.4   Cross-product calculator – datapath.

2.7   Design the Controller

The penultimate stage in the design of the dot-product calculator is to design a controller to sequence the operations over the eight clock cycles. There are three multiplexers and a register to control in this circuit. Their operation for each of the eight clock cycles is shown in Table 2.2.

It can be seen that the multiplexers selecting between the a and b vector elements have identical operation; the `zero` multiplexer selects the zero input on clock 1 and the `result` input all the rest of the time; the register is permanently in `load` mode and so needs no control.

Normally, the controller would be implemented as a state machine. However, in this case, the state machine can be simplified to a counter that counts from 0 to 7 repeatedly. The output of the counter controls the a and b multiplexers directly. A zero detector on the counter output controls the `zero` multiplexer. The circuit for the controller is illustrated by Figure 2.5.

Table 2.2   Controller operations per clock cycle

Cycle	a mux	b mux	Zero mux	Register
1	select 0	select 0	select 0	load
2	select 1	select 1	select 1	load
3	select 2	select 2	select 1	load
4	select 3	select 3	select 1	load
5	select 4	select 4	select 1	load
6	select 5	select 5	select 1	load
7	select 6	select 6	select 1	load
8	select 7	select 7	select 1	load

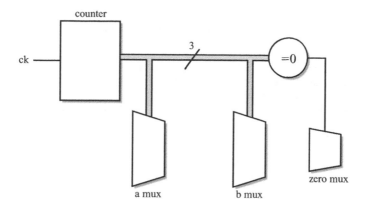

Figure 2.5   Cross-product calculator – controller.

2.8   Design the Reset Mechanism

The final stage of the RTL design is to design the reset mechanism. This is a simple, but essential stage of the design process. The design of a reset mechanism is an essential part of the design of the RTL system, although it is often the case that only the controller needs a reset control. If the reset mechanism is not designed into the RTL model, then there is no guarantee that the circuit will start up in a known state.

In this case, it is sufficient to reset the controller. The datapath will be cleared by the design of the controller, which resets the accumulator anyway at the start of the calculation. The controller's reset will be incorporated as a synchronous reset.

2.9   VHDL Description of the RTL Design

Now that the RTL design process has been completed, a VHDL model can be written. This model can be simulated to verify correct behaviour by comparison with the system model that we started with. The difference is that the RTL model is clocked and needs eight clock cycles to form a result, whilst the system model was combinational and formed the result instantaneously.

```
library ieee;
use ieee.std_logic_1164.all, ieee.numeric_std.all;
package dot_product_types is
   subtype sig8 is signed (7 downto 0);
   type sig8_vector is array (natural range <>) of sig8;
end;

library ieee;
use ieee.std_logic_1164.all, ieee.numeric_std.all;
use work.dot_product_types.all;
entity dot_product is
   port (a, b     : in  sig8_vector(7 downto 0);
```

```
              ck, reset: in  std_logic;
              result    : out signed(15 downto 0));
    end;

    architecture behaviour of dot_product is
      signal i : unsigned(2 downto 0);
      signal ai, bi : signed (7 downto 0);
      signal product, add_in, sum, accumulator : signed(15 downto 0);
    begin
      control: process
      begin
        wait until rising_edge(ck);
        if reset = '1' then
          i <= (others => '0');
        else
          i <= i + 1;
        end if;
      end process;
      a_mux: ai <= a(to_integer(i));
      b_mux: bi <= b(to_integer(i));
      multiply: product <= ai * bi;
      z_mux: add_in <= X"0000" when i = 0 else accumulator;
      add: sum <= product + add_in;
      accumulate: process
      begin
        wait until rising_edge(ck);
        accumulator <= sum;
      end process;
      output: result <= accumulator;
    end;
```

This design depends on an existing package called `numeric_std` that defines a set of numeric types. This will be examined in more detail in Chapter 6. For now it is sufficient to say that type `unsigned` represents unsigned (magnitude-only) numbers, and type `signed` represents signed (2's-complement) numbers. All the VHDL used in this circuit is explained in subsequent chapters and fits the common subset of VHDL that can be synthesised by current VHDL synthesis tools.

2.10   Synthesis Results

The RTL design exercise just completed was an area constrained design. It was assumed that there would only be sufficient logic gates available to this circuit to allow a single multiplier and a single adder. It is interesting at this stage to do a comparison with the unconstrained design based on the system specification at the start of the chapter.

   The RTL design was synthesised using the same synthesis system and the same target ASIC library as for the system specification.

   The results of synthesis were:

- area – 1200 NAND gate equivalents;
- I/O – 146 ports;
- storage – 19 registers.

Table 2.3 Comparison of synthesis results

	System model	RTL model
NAND equivalents	40 000	1200
ports	546	146
clock cycles	—	8
registers	0	19

The only strange result here is the number of ports – 146 I/O pins is clearly a large overhead. However, this is simply a result of the use of an artificial example that assumes that the two vectors being used to form the dot-product are primary inputs. In practice they would probably be time-multiplexed onto either one or two input buses.

For comparison, Table 2.3 compares the synthesised RTL results with the results from synthesising the system specification. This illustrates the importance of the RTL design process.

3

Combinational Logic

This chapter will describe the basics of VHDL required to describe combinational logic using basic types to create boolean equations and simple arithmetic circuits.

It will also introduce the simulation model of VHDL, with an introduction to modelling concurrency, how this is done using the event model and the concepts of simulation time and delta time.

This chapter will then show how this model is used by synthesis tools to control the mapping of VHDL descriptions to circuits, and introduces synthesis templates.

3.1 Design Units

Design Units are the basic building blocks of VHDL. They are indivisible in that a design unit must be completely contained in a single file. A file may contain any number of design units.

When a file is analysed using a VHDL simulator or synthesiser, the file is, in effect, broken up into its individual design units and each design unit is analysed separately as if they had been in separate files.

There are six kinds of design units in VHDL. These are:

- entity;
  - architecture;
- package;
  - package body;
- configuration declaration;
- context declaration.

The six kinds of design unit are further classified as primary or secondary units. A primary design unit can exist on its own. A secondary design unit cannot exist without its corresponding primary unit. In other words, it is not possible to analyse a secondary unit before its primary unit is analysed. The secondary units are shown above indented and immediately below their corresponding primary units.

VHDL for Logic Synthesis, Third Edition. Andrew Rushton.
© 2011 John Wiley & Sons, Ltd. Published 2011 by John Wiley & Sons, Ltd.

The entity is a primary design unit that defines the interface to a circuit. Its corresponding secondary unit is the architecture that defines the contents of the circuit. There can be many architectures associated with a particular entity, but this feature is rarely, if ever, used in synthesis and so will not be covered here.

The package is also a primary design unit. A package declares types, subprograms, operations, components and other objects that can then be used in the description of a circuit. The package body is the corresponding secondary design unit that contains the implementations of subprograms and operations declared in its package. This will not be covered yet, but the usage of packages supplied with the synthesiser is covered throughout the book and how to declare your own is covered in Chapters 10 and 11.

The configuration declaration is a primary design unit with no corresponding secondary. It is used to define the way in which a hierarchical design is to be built from a range of subcomponents. However, it is not generally used for logic synthesis and will not be covered in this book.

The context declaration is a new primary unit with no corresponding secondary, and was added in VHDL-2008. It allows multiple context clauses (i.e. library and use clauses) to be grouped together. However, because it is not in common use it will not be used in this book, except in Chapter 6 where other VHDL-2008 features are discussed.

3.2   Entities and Architectures

An entity defines the interface to a circuit and the name of the circuit. An architecture defines the contents of the circuit itself. Entities and architectures therefore exist in pairs – a complete circuit description will generally have both an entity and an architecture. It is possible to have an entity without an architecture, but such examples are generally trivial and of no real use. Also, it is possible to have multiple architectures for a single entity, each one representing a different implementation of the same circuit. This can be useful when comparing different levels of model, such as comparing the RTL model with the gate-level model. It is not possible to have an architecture without an entity.

An example of an entity is:

```
entity adder_tree is
   port (a, b, c, d : in integer; sum : out integer);
end entity adder_tree;
```

In this case, the circuit adder_tree has five ports: four input ports and one output port. Note that the repeat of the keyword entity and the circuit name adder_tree after the end are both optional and in practice are usually omitted.

The structure of an architecture is illustrated by the following example:

```
architecture behaviour of adder_tree is
   signal sum1, sum2 : integer;
begin
   sum1 <= a + b;
   sum2 <= c + d;
   sum <= sum1 + sum2;
end architecture behaviour;
```

The architecture has the name `behaviour` and belongs to the entity `adder_tree`. It is common practice to use the architecture name `behaviour` for all synthesisable architectures. As with the entity, the repeat of the `architecture` keyword and name `behaviour` after the `end` is optional and usually omitted. Common alternatives to architecture `behaviour` are architecture `RTL` or architecture `synthesis`. Architecture names do not need to be unique, indeed the consistent use of the same architecture name throughout a VHDL design is considered best-practice because it makes it easy to tell at a glance whether a VHDL description is system level (architecture `system`), RTL (architecture `behaviour`) or gate-level (architecture `netlist`). It does not matter what naming convention is used for architectures but it is recommended that a consistent naming convention is adhered to.

The architecture has two parts.

The declarative part is the part before the keyword `begin`. In this example, additional internal signals have been declared here. Signals are similar to ports but are internal to the circuit.

A signal declaration looks like:

```
signal sum1, sum2 : integer;
```

This declares two signals called `sum1` and `sum2` that have a type called `integer`. Basic types will be dealt with in Chapter 4, and a set of synthesis-specific types are covered in Chapter 6, so for now it is sufficient to say that `integer` is a numeric type that can be used for calculations.

The statement part is the part after the `begin`. This is the description of the circuit itself. In this example the statement part only contains signal assignments describing the adder tree as three adders described by equations.

The simple signal assignment looks like:

```
sum1 <= a + b;
```

The left-hand side of the assignment is known as the target of the assignment (in this case `sum1`). The assignment itself has the symbol "$<=$" that is usually read 'gets', as in 'signal sum1 gets a plus b'.

The right-hand side of the assignment is known as the source of the assignment. The source expression can be as complex as you like. For example, the circuit of the `adder_tree` example could have been written using just one signal assignment:

```
sum <= (a + b) + (c + d);
```

The example was written with three assignments so that the relationship between assignments, ports and signal declarations could be explained.

In this example, the statements have been written in sequence so that the data flow is from top to bottom. However, this is done for readability only; the ordering of the statements is irrelevant. This is because each statement simply defines a relationship between its inputs (the source, on the right-hand side of the assignment) and its output (the target, on the left-hand side).

For example, the following architecture is functionally equivalent to the previous version:

```
architecture behaviour of adder_tree is
  signal sum1, sum2 : integer;
begin
  sum <= sum1 + sum2;
  sum2 <= c + d;
  sum1 <= a + b;
end;
```

3.3   Simulation Model

In order to really understand how VHDL works, it is useful to have a basic knowledge of
the underlying mechanisms of the language. This will help to explain many VHDL features
introduced in this and subsequent chapters.

VHDL has been designed from the start as a simulation language, so an understanding of the
language must come from examining the behaviour of a VHDL simulator. The definition of
VHDL contained in the Language Reference Manual includes a definition of how a simulator
should implement the language, so this behaviour must be common to all VHDL simulators.

The basis of VHDL simulation is event processing. All VHDL simulators are event-driven
simulators.

There are three essential concepts to event-driven simulation. These are: simulation time,
delta time and event processing.

During a simulation, the simulator keeps track of the current time that has been simulated,
that is, the circuit time that has been *modelled* by the simulator, not the time the simulation has
actually taken. This time is known as the *simulation time* and is usually measured as an integral
multiple of a basic unit of time known as the resolution limit. The simulator cannot measure
time delays less than the resolution limit. For gate-level simulations the resolution limit may be
quite fine, possibly 1 fs or less. For RTL simulations, there is no need to specify a fine resolution
since we are only interested in clock-cycle by clock-cycle behaviour and the transfer functions
are described with zero or unit time delay. In this case, a resolution limit of 1 ps is often used.
It is important to note that the resolution limit is a characteristic of the simulator, not of the
VHDL model. It is usually controlled by a simulator configuration setting.

The simulation cycle alternates between *event processing* and *process execution*. Put another
way, signals are updated as a batch in the event processing part of the cycle, then processes are
run as a batch in the process execution part. The signal updating and process execution are kept
completely separate. This is how VHDL models concurrency such that it can be modelled on a
sequential computer processor without having to use multiple processors or threads.

When a signal assignment (a simplified process) is performed, the signal that is the target of
the assignment is not updated immediately by the assignment; in fact it keeps its old value for
the remainder of the process execution phase. Instead, the assignment causes a *transaction* to
be added to a queue of transactions associated with the *driver* of the signal.

For example:

```
a <= '0' after 1 ns, '1' after 2 ns;
```

This signal assignment queues two transactions in the driver for signal a. The first transaction
has the value '0' and a time delay of 1 ns; the second transaction has the value '1' and a time
delay of 2 ns. It is also possible to have a zero-delay assignment:

```
a <= '0';
```

This contains one transaction with the value of '0' and no time delay. Even when there is no time delay the signal is not updated immediately, since the transaction will be scheduled for the next *delta cycle*.

When simulation time moves on to the point where a transaction becomes due on a signal, then during the event-processing phase that signal becomes *active*. The new value is then compared with the old value and, if the value has changed, then an *event* is generated for that signal. This event causes processes sensitive to the signal to be triggered. Note that, if the signal is assigned a value that is the same as its current value, it will become active but will not have an event and so will not trigger any processes.

An event is processed by updating the signal value, then working out which statements have that signal as an input (in VHDL-speak, all the statements that are *sensitive* to that signal). All signals are processed as a batch, that is, all signals that have an event in the current simulation cycle are updated in this way. The set of processes triggered by these signal updates are scheduled for execution during a later process execution phase. Each process can only be triggered once per simulation cycle, no matter how many of its inputs change.

During the process execution phase, each process is executed until it pauses. The simulator works its way through all the triggered processes in no particular order executing them until they pause. Only after all the triggered processes have paused will the simulator switch back to the event-processing phase.

Any signal assignments in the executed processes cause more transactions to be generated. These new transactions are processed in later simulation cycles. Zero-delay assignments will be processed in the next delta cycle.

The distinction between an active signal and a signal event is very important. Processes are sensitive to events, so will only be activated by a signal changing its value. This is generally what is wanted. For example, consider the following RS latch model:

```
P1: process (R, Qbar)
begin
  Q <= R nor Qbar;
end process;

P2: process (S, Q)
begin
  Qbar <= S nor Q;
end process;
```

This example has been written using processes to show the sensitivity list. This is the list of signals in parentheses after the keyword process, which represents the set of signals that will *trigger* the process. For combinational logic, the sensitivity list should include all of the inputs of the process, in this case all of the signals on the source (right-hand) side of the signal assignments.

The example could have been written without processes, using just simple signal assignments:

```
P1: Q <= R nor Qbar;
P2: Qbar <= S nor Q;
```

This is exactly equivalent since the VHDL standard states that a signal assignment has an implied sensitivity list containing all the signals on its right-hand side. In other words, assignment P1 will trigger on changes to R or Qbar, whilst assignment P2 will trigger on changes to S or Q.

Consider the case when R and S are '0', with Q at '1' and Qbar at '0'. Consider what then happens when R changes due to a transaction of value '1' at the current simulation time. The model will go through the following sequence:

delta 1, event processing
    The transaction makes R active and, since it is a change in value for R (from '0' to '1'), it causes an event on R. The event on R triggers process P1 that is sensitive to it.
delta 1, process execution
    P1 recalculates the value of Q, creating a transaction of value '0' (since '1' nor '0' is '0') at the current time. This transaction is added to the transaction queue for Q.
delta 2, event processing
    The transaction on Q makes Q active and, since it is a change in value for Q (from '1' to '0'), it causes an event on Q. The event on Q triggers process P2 that is sensitive to it.
delta 2, process execution
    P2 recalculates the value of Qbar, creating a transaction of value '1' (since '0' nor '0' is '1') at the current time. This transaction is added to the transaction queue for Qbar.
delta 3, event processing
    The transaction on Qbar makes Qbar active and, since it is a change in value for Qbar (from '0' to '1'), it causes an event on Qbar. The event on Qbar triggers process P1 for the second time.
delta 3, process execution
    P1 recalculates the value of Q, creating a transaction of value '0' (since '1' nor '1' is '0') at the current time. This transaction is added to the transaction queue for Q.
delta 4, event processing
    The transaction on Q makes Q active but, since it is not a change in value for Q (from '0' to '0'), it does not cause an event on Q.

Since there are no more transactions to process in the model, the model reaches a stable state at this point. The simulation time can now be moved on to the next scheduled transaction on R or S and a similar series of delta cycles will be carried out.

The important thing about the way VHDL models the circuit is that the signal/process delta cycles stopped because a transaction did not result in a change in a signal, so no events were generated, even though a signal did become active in the last cycle. As you can see from this example, this means that VHDL models asynchronous feedback simply and naturally. It also means that the order in which the processes or signal assignments are listed in the architecture has no effect on the simulation, since the decisions determining which processes to execute are based purely on the events and process sensitivity lists, not on the order of the statements. Swapping the two processes would result in exactly the same sequence. VHDL is a *concurrent* language.

Note: you would never model a latch like this in RTL, it was just used to illustrate how VHDL models feedback correctly.

To further illustrate the action of an event-driven simulator to show how values propagate through a circuit, consider the behaviour of the `adder_tree` example introduced earlier.

For this example, it will be assumed that the circuit is initially in a stable state with all the inputs set to 0. It can be seen from the description of the adder tree that all the internal signals and the outputs will also be 0. These values are set up during the initialisation (or elaboration) phase of simulation, which happens at time zero.

Consider what happens if the input b changes from 0 to 1 at a simulation time of 20 ns. This means that a transaction is generated for signal b and this transaction is posted at the first delta cycle of the 20 ns simulation time. When this transaction is processed, it is tested to see if it is a change in value, which it is, so this causes an event on b. The event processing causes the equations that are sensitive to b to be triggered. These are:

```
sum1 <= a + b;
```

So this equation is executed during the process execution phase. As a result of recalculating this equation, a transaction is generated for `sum1` at the current simulation time (20 ns), but at the next delta cycle. At this stage, signal `sum1` has not changed value; the only outcome of the process execution is that a transaction is posted for `sum1` specifying a new value of 1 (i.e. `0 + 1`) for signal `sum1`.

The next stage of the simulation is transaction processing of the second delta cycle. First, the transaction on signal `sum1` is tested to see if it changes its value, which it does, so the transaction is transformed into an event.

Then, all equations sensitive to `sum1` are triggered. The sensitive equations are:

```
sum <= sum1 + sum2;
```

Process execution is carried out on this equation, generating transactions for the next delta cycle. One new transaction is generated in this case and posted at the third delta cycle at the current simulation time. The transaction is a new value for `sum2` of 1. Once again, this value is not yet assigned to the signal.

Transaction processing of the third delta cycle causes the transaction on `sum` to be tested to see if it represents a change of value. Once again it is a change, so the transaction is transformed into an event, triggering equations sensitive to it. No equations are sensitive to the output signals, so there are no further transactions to process and simulation of the current simulation time has now completed. Simulation time can now be moved on.

The whole simulation cycle is summarised in Table 3.1.

Note how the change on the input propagated through to the `sum` output over three delta cycles. The result is that a minimum set of processes was re-executed as a result of the input change and that some processes were not re-executed at all.

3.4   Synthesis Templates

Since VHDL has been designed as a simulation language without regard to the needs of synthesis or any other application area, synthesisers must make an *interpretation* of the

Table 3.1   Event processing of adder tree

	20 ns	20 ns + δ	20 ns + 2δ	20 ns + 3δ
a	0	0	0	0
b	0	1	1	1
c	0	0	0	0
d	0	0	0	0
sum1	0	0	1	1
sum2	0	0	0	0
sum	0	0	0	1
transactions	b $\Rightarrow$ 1	sum1 $\Rightarrow$ 1	sum $\Rightarrow$ 1	
events	b	sum1	sum	

language. This interpretation is based on mappings of special VHDL constructs onto hardware with equivalent behaviour.

These special constructs are known as *templates*.

The mapping is not always straightforward. Some VHDL constructs have direct one-to-one mappings to hardware equivalents. Many VHDL constructs have no possible hardware equivalents, at least within the confines of logic synthesis, and these will cause errors during synthesis. Other constructs have to meet specific constraints in order to be mappable. Synthesisers must impose these constraints on the use of the language so that only VHDL constructs that have hardware equivalents can be used. In other words, your VHDL must conform to the appropriate template for the hardware structure you wish to build.

There are templates for combinational logic, simple registers, registers with asynchronous reset, registers with synchronous reset, latches, RAMs, ROMs, tristate drivers and finite-state machines, all of which will be covered later in this book.

Note: you might come across the synthesis subset of VHDL expressed as a restricted *syntax*. This is unhelpful since the synthesis subset is really a *semantic* subset. That is, most VHDL constructs are synthesisable provided that they are used in a particular, constrained way that fits one of the synthesis templates.

It is extremely important to conform to these templates since they dictate how VHDL must be written in order to be synthesisable. VHDL models must be written for synthesis from the start; it is not possible to take just any VHDL that simulates correctly and expect it to be synthesisable. Many person-years of work have been wasted by engineers who failed to realise this and wasted their time perfecting simulation models before considering the synthesis constraints.

Fortunately, in the case of the adder_tree example, the circuit interpretation is a simple and direct mapping to hardware. VHDL signal assignments map directly onto blocks of combinational logic. This can be seen by considering the event processing cycle described earlier. At each stage, every equation that is sensitive to a changing input is recalculated. This is the behaviour of combinational logic in which the output is re-evaluated whenever an input changes. The expressions used (+ operators) have direct equivalents in hardware too. These equivalences together give us the mapping from simulation behaviour to circuit structure.

In later chapters, similar parallels will be drawn to show how the simulation model of other constructs can be mimicked by certain hardware structures. It is this mimicry that gives us the

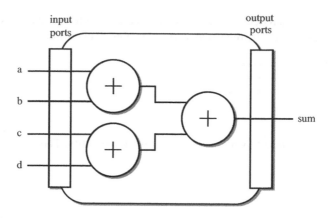

Figure 3.1   Adder tree circuit.

hardware mapping. It should always be remembered that VHDL is a simulation language and that not all simulation constructs have hardware equivalents. This is why all synthesisers must work on subsets of the language.

Figure 3.1 illustrates the circuit representation of the `adder_tree` entity/architecture pair.

In this figure, the operations have been represented by simple circles rather than as gates to highlight the fact that, at this stage, there has been no mapping to gates. Instead the circuit has been shown as a network of abstract arithmetic functions. A synthesiser will usually restructure these arithmetic functions to match actual gates in the target technology library at a late stage in the synthesis process known as the *technology mapping* stage. In this case the functions would be restructured into a full-adder circuit, but the exact type of adder will depend on the target technology and the speed and area constraints being use for the synthesis. For clarity in this discussion, the synthesis process has been frozen prior to this technology mapping phase so that the intermediate structure can be seen. All synthesisers perform synthesis in stages, starting with the interpretation of the source VHDL to form a functional network, followed by optimisation of the functional network and then finally technology mapping.

3.5   Signals and Ports

Signals are the carriers of data values around an architecture. Ports are the same as signals but also provide an interface through the entity so that the entity can be used as a subcircuit in a hierarchical design.

A signal is declared in the declarative part of an *architecture* (between the keywords `is` and `begin`) and the declaration has two parts:

```
architecture behaviour of adder_tree is
   signal sum1, sum2 : integer;
begin
   ...
```

The first part of the declaration is the keyword `signal` and a list of signal names: in this case there are two signals `sum1` and `sum2`. The second part, after the colon, is the type of the signals: in this case `integer`.

There can be many signal declarations in an architecture, each terminated by a semi-colon. The above declaration could be rewritten as two separate declarations:

```
architecture behaviour of adder_tree is
   signal sum1 : integer;
   signal sum2 : integer;
begin
   ...
```

Port declarations are enclosed by a *port specification* in the entity. The port specification has the following structure:

```
entity adder_tree is
   port ( port specification );
end;
```

Note that the port specification is enclosed in parentheses and always terminated by a semi-colon that is outside the parentheses.

A port is declared within the port specification. A port declaration has three parts:

```
entity adder_tree is
   port (a, b, c, d : in integer; sum : out integer);
end;
```

Looking at the first declaration in the port list, the first part is a list of port names: in this case a, b, c and d. The second part is the mode of the port: in this case the mode is in. The third part is the type as in the signal declaration: in this case the type is integer.

Each port declaration within the specification is separated by semi-colons from the others. Note that, unlike the signal declarations, which are each terminated by a semi-colon, port declarations are separated (not terminated) by semi-colons, so there is no semi-colon after the last declaration before the closing parenthesis.

The mode of a port determines the direction of data flow through the port. There are five port modes in VHDL: in, out, inout, buffer and linkage. If a mode is not given, then mode in will be assumed.

The meanings of the modes as they are used for logic synthesis are:

in	input port – cannot be assigned to in the circuit, can be read
out	output port – can be assigned to in the circuit, cannot be read
inout	bidirectional port – can only be used for tristate buses
buffer	output port – like mode out but can also be read
linkage	not used by synthesis

There is often confusion between mode out and mode buffer. Mode buffer is an anachronism and the reason for its existence in the language is obscure. The full behaviour of a buffer port is a restricted form of mode inout. However, to make the mode usable for synthesis, the rules for buffer ports are constrained so that they act like mode out ports with the added convenience that it is possible to read from the port within the architecture. There really is no reason to have two output modes, so it is recommended that only out mode is used.

So the port modes as recommended for use in logic synthesis are:

`in`	input port
`out`	output port
`inout`	bidirectional port for tristate buses

There is a problem if you need to read from an `out` mode port. This problem is illustrated by the following example of an `and` gate with true and inverted outputs. The first version shows an *illegal* description, because the `out` port z is read:

```
entity and_nand is
  port (a, b : in bit;
        z, zbar : out bit);
end;
architecture illegal of and_nand is
begin
  z <= a and b;
  zbar <= not z;
end;
```

The solution is to use an intermediate internal signal and read from that. The intermediate signal can then be assigned to the `out` mode ports.

The corrected example is:

```
entity and_nand is
  port (a, b : in bit;
        z, zbar : out bit);
end;
architecture behaviour of and_nand is
  signal result : bit;
begin
  result <= a and b;
  z <= result;
  zbar <= not result;
end;
```

It is good practice to always use intermediate signals for outputs and to assign them to the `out` ports at the end of the architecture. By doing this consistently, the pitfall of trying to read an `out` port is always avoided.

Note: in VHDL 2008, it is possible to read from `out` ports, thus making it unnecessary to have the intermediate signal. Therefore, in a few years time when VHDL-2008 becomes the norm, this problem will go away.

3.6 Initial Values

All signals have an initial value when simulation begins. This initial value can either be user-supplied in the signal (or port) declaration or will be given by default. The simulator will assign these initial values to signals during the elaboration phase of simulation.

Initial values given in the signal declaration look like this:

```
signal a : bit := '1';
```

This means that, at the start of simulation, signal a will take the value '1'.

If a signal does not have an explicit initial value given in the signal declaration, the signal will still have an initial value in simulation. This value will be the first value (referred to as the *left* value because it will appear on the left if you list the values in order) of the type. For type bit (see Chapter 4), the left value is '0', so all signals of type bit will be initialised with the value '0' unless an explicit initial value is used to override it. Signals of type std_logic (see Chapter 5) will initialise to the value 'U' that signifies an undefined value.

The rules for initial values ensure that all simulations start from the same, known, state. This means that identical simulations will give identical results even on different simulators.

For synthesis, there is no hardware interpretation of an initial value. It is not possible to initialise all signals in a circuit with a known value on power-up. Even though it is possible to do a power-on reset in some logic technologies, it would not be desirable to do this for every wire in the circuit. So, synthesis *must* ignore initial values.

This opens up a potential pitfall that is a common cause of circuit failure for a synthesis-generated design. That is, the synthesised circuit will not start up in a known state and will not therefore behave in the same way as the simulation predicted. It may even get permanently stuck in an unknown state that doesn't exist in the simulation model.

The point is that this is not a failure of the synthesiser, since the synthesiser must ignore initial values – it is the designer's responsibility to provide a mechanism for putting the circuit into a known state, either in the form of a global reset input to the system or by designing circuits so that they can be put into a known state in some other way such as by making the controller state machine re-entrant. The design of a reset or initialisation mechanism is an essential part of the RTL design process and must not be skipped. Generally it is registers that need to be reset or initialised, so the details of how to implement register resets will be deferred to Chapter 9.

One way of testing a reset mechanism is to give the internal signals undefined initial values (i.e. use std_logic for all signals) and then ensure that the simulation gives expected, defined results. This doesn't guarantee that the design is correct, but it is impractical to try all possible initial values, so the only real way of ensuring correct initialisation behaviour is by structured and rigorous design.

3.7 Simple Signal Assignments

The simple signal assignment statement has already been used in the adder_tree example. This section examines the signal assignment statement in more detail.

The simple signal assignment looks like this:

```
sum1 <= a + b;
```

The left-hand side of the assignment is known as the *target* of the assignment (in this case sum1). The right-hand side of the assignment is known as the *source* expression of the assignment (in this case a + b). The assignment itself is the symbol <= that is formed by combining the less-than and the equals symbols to form an arrow. There must be no space between the two characters in the symbol. Beware of confusion with the less-than-or-equal operator <= that looks exactly the same. Fortunately, there is no real chance of confusion because there are no situations where both meanings would be allowed by the language. Nevertheless, it takes some getting used to.

The rules of VHDL insist that the source of an assignment is the same *type* as the target. This example uses type `integer`. The addition of two signals of type `integer` gives a result that is also an `integer`. Therefore, the source and target are of the same type.

The source expression can be as complex as you like. For example, the circuit of the `adder_tree` example could have been written using just one signal assignments:

```
sum <= (a + b) + (c + d);
```

3.8   Conditional Signal Assignments

Conditional signal assignments are signal assignments with more than one source expression, where one of the source expressions is chosen by a control condition. The simplest form of the conditional signal assignment has only one condition:

```
sum <= a + b when sel = '1' else a - b;
```

In this case, the condition is the test for $sel = '1'$. The test can be any boolean expression, which generally means a test for equality or inequality, or a comparison such as less-than when using integer types.

The rule for the source expressions is that they must all be the same type as the target of the assignment. In this example, the sources and the target are of type integer.

The hardware mapping of the conditional signal assignment is a multiplexer that selects between the source expressions and that is controlled by the condition expression. Figure 3.2 shows the circuit of this example.

In the figure, the condition has been shown explicitly as a test for the equality of `sel` and the value `'1'`. This is the circuit that a synthesiser will map the VHDL onto initially, but the equality operator will be easily minimised to a single piece of wire such that `sel` controls the multiplexer directly. This kind of minimisation is routine in synthesis and it means that you don't have to worry about the exact implementation of the circuit.

This example is the simplest form of conditional signal assignment, with only one condition. It is possible to have any number of branches; each branch but the last has a different condition. The last branch must be an unconditional `else` so that one of the source expressions

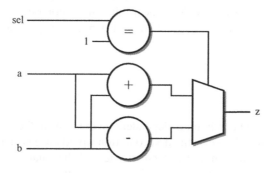

Figure 3.2   Hardware mapping of conditional signal assignment.

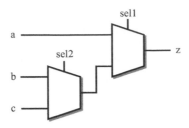

Figure 3.3   Multi-way conditional signal assignment.

will always be assigned to the target. An example of a multiple-branch conditional signal assignment is:

```
z <= a when sel1 = '1' else
     b when sel2 = '1' else
     c;
```

The conditions are evaluated in the order that they are written and the first condition that is true will be selected. This is equivalent in hardware terms to a series of two-way multiplexers, with the first condition controlling the multiplexer nearest to the output so that that condition overrides any of the later conditions.

The circuit for this example is illustrated in Figure 3.3.

In this example, the conditions have already been optimised so that sel1 and sel2 control the multiplexers directly. It can be seen from this circuit that when sel1 is '1', then a is selected regardless of the value of sel2. If sel1 is '0', then sel2 selects between b and c inputs.

With more branches, extra multiplexers are added to the structure so the result is a long chain of multiplexers. The chain will be skewed, not symmetrical, because of the precedence rules just described. This should be taken into account in the design; the further down the list of selections a source is, the more multiplexers it will pass through when synthesised.

A similar structure can be created by the if statement (see Section 8.4) which allows more complex structures; in particular it allows more than one signal to be assigned in each branch of the conditional.

Each condition in a conditional signal assignment is assumed to be independent of the others when mapping onto hardware. This means that, if the conditions are dependent (based on the same signal, for example), then there won't necessarily be any optimisation. For example:

```
z <= a when sel = '1' else
     b when sel = '0' else
     c;
```

In this example, the second condition is dependent on the first. In fact, in the second branch, sel can only possibly be '0', since the first condition would have been true if sel was '1'. Therefore, the second condition is a redundant one and the final else branch cannot be reached. In synthesis, this conditional signal assignment would still be mapped onto two multiplexers as shown in Figure 3.4.

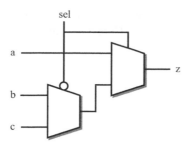

Figure 3.4   Redundant branch in conditional signal assignment.

With a simple example such as this, the synthesiser may identify and eliminate the redundant extra multiplexer during the logic optimisation phase of the synthesis process, but with more complex examples it could not be guaranteed to do so. Detecting redundant conditions is a very difficult task in the general case. This example was made simple to illustrate the problem, but in practice, these redundancies can be subtle and beyond the capability of logic minimisation. It is important to realise that you tend to get what you ask for in the mapping from VHDL to hardware, so if you ask for two multiplexers, that is probably what you will get.

The cause of this pitfall is the incorrect use of an extra branch in the conditional signal assignment in the first place. The point of the conditional signal assignment is that the individual conditions are independent of each other and the synthesis interpretation assumes this independence. In this example, the conditions are dependent on each other because they are repeated tests of the same signal. This should have been implemented as a two-input conditional signal assignment.

3.9   Selected Signal Assignment

Selected signal assignments, like conditional signal assignments, allow one of a number of source expressions to be selected based on a condition. The difference is that a selected signal assignment uses a single condition to select one from many branches. The condition can be of any type and therefore can select between any number of branches.

A simple example is:

```
with sel select
   z <= a when '1',
        b when '0';
```

The main difference between the selected signal assignment and the conditional signal assignment is that there is only one condition; all the tests have equal priority and are mutually exclusive.

This simple example could just as easily have been done with a conditional assignment, because it only has two branches, but it does illustrate the form of the selected signal assignment.

The condition in this case is the signal sel. In this example it is of type bit, but could be any type. For example, it could be a bus of signals, with a branch for each possible bit pattern on the bus. Then, the assignment itself contains a number of source expressions, each of which is selected by a given value of the condition. Here is a similar example where the selection signal is a 2-bit bus:

```
with sel select
  z <= a when "00",
       b when "01",
       c when "10",
       d when "11";
```

All values of the condition's type must be covered by the selections. A similar structure can be created by the case statement (see Section 8.5) that allows more complex structures; in particular it allows more than one signal to be assigned in each branch of the conditional.

Selected signal assignments come into their own when used with enumeration types, especially multi-valued logic types and integer types. The discussion of selected signal assignments with other types will be deferred until after the types have been covered, in Section 4.12.

3.10   Worked Example

3.10.1   Parity Generator

The example creates a simple 8-bit parity generator using concurrent signal assignments. The parity circuit will be switchable between two different modes depending on the values of a control input, as shown in Table 3.2.

In odd parity mode, the output is 0 if the input has an odd number of ones. In even parity mode, the output is 0 if the input has an even number of ones.

The block diagram for the circuit is shown in Figure 3.5.

The first stage is to write the entity, based on the block diagram. This looks like:

```
entity parity is
  port (d7, d6, d5, d4, d3, d2, d1, d0 : in bit;
        mode : in bit;
        result : out bit);
end;
```

The architecture for the solution is:

```
architecture behaviour of parity is
  signal sum : bit;
begin
  sum <= d0 xor d1 xor d2 xor d3 xor d4 xor d5 xor d6 xor d7;
  result <= sum when mode = '1' else not sum;
end;
```

Table 3.2   Parity-generator functions

Mode	Function
0	odd parity
1	even parity

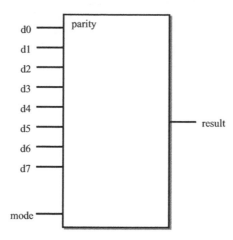

Figure 3.5   Parity generator interface.

The architecture is in two parts, each a separate signal assignment. One is a simple signal assignment that calculates the parity function and assigns it to the internal signal sum. The second part is a conditional signal assignment that selects the output appropriate to the function to be performed according to the values of the control input mode.

4

Basic Types

The use of types is fundamental to the understanding of VHDL, especially when used for logic synthesis. Furthermore, an understanding of types is necessary to understand the later chapters. For these reasons, this chapter has been placed very close to the beginning of the book.

VHDL is referred to as a strongly typed language. This means that every data flow (input, output, internal signal, etc.) has a type associated with it and that there can never be any ambiguity about what type a data flow is.

A feature of VHDL that newcomers sometimes struggle with is that the specific implementation of an operator is selected by the type of the signals used. So to add two numbers together you always use the + operator, but the implementation will be different depending on the type of the operands: an integer adder will be used if the operands are integers, a fixed-point adder will be used if the operands are fixed-point and so on. So, if you get the types right, you will get the operations you want. If you confuse your types, you will get into trouble.

There are a few built-in types that are part of the language, but the language also has the capability of defining additional types. Many of the types in common use in logic synthesis are additional types.

This chapter will cover the built-in types and the basic type handling of VHDL. Chapter 6 will cover the additional logical and numerical types added specifically for use in synthesis and which will be used for nearly all signals in a real design.

4.1 Synthesisable Types

There are eight classes of types in VHDL, but not all of them are synthesisable. Table 4.1 shows which of the classes of type are commonly synthesisable.

The four classes of type that are not synthesisable will not be covered in this book. The four classes that are synthesisable will be covered in the following sections.

4.2 Standard Types

A number of types are predefined in the language. The predefined types are to be found in a package called `standard` that must be a part of every VHDL system. Package `standard` is listed in Appendix A.1 for reference.

VHDL for Logic Synthesis, Third Edition. Andrew Rushton.
© 2011 John Wiley & Sons, Ltd. Published 2011 by John Wiley & Sons, Ltd.

Table 4.1 Synthesisable types

Class	Synthesisable
enumeration types	yes
integer types	yes
floating-point types	no
physical types	no
array types	yes
record types	yes
access types	no
file types	no

Table 4.2 Standard types

Type	Class	Synthesisable
boolean	enumeration type	yes
bit	enumeration type	yes
character	enumeration type	yes
severity_level	enumeration type	no
integer	integer type	yes
natural	subtype of integer	yes
positive	subtype of integer	yes
real	floating-point type	no
time	physical type	no
string	array of character	yes
bit_vector	array of bit	yes

Table 4.2 lists the type definitions to be found in package `standard`, the class each type belongs to and whether it is supported by synthesis.

The type `severity_level` has been marked as unsynthesisable because, although as an enumeration type it is technically synthesisable, it should never be used in this way.

The usage and interpretation of all of the synthesisable types will be covered in the rest of this chapter.

4.3 Standard Operators

Values, signals and variables (which will be covered in Section 8.3) of a type can be combined in expressions using operators. For example, to find the logical `and` of two signals of type `bit`, the `and` operator would be used. There is a comprehensive set of operators in VHDL, which will be introduced here briefly, but covered in much more detail in Chapter 5.

For the purposes of this book, the operators will be divided into 5 groups. These groups, and the relevant detail section of Chapter 5 are: boolean (Section 5.3), comparison (Section 5.4), shifting (Section 5.5), arithmetic (Section 5.6) and concatenation (Section 5.7):;

```
boolean:     not, and, or, nand, nor, xor, xnor
comparison:  =, /=, <, <=, >, >=
shifting:    sll, srl, sla, sra, rol, ror
```

arithmetic: sign + , sign -, abs, + , -, *,/, mod, rem, **
concatenation: &

Each group of operators can only be applied to a particular set of types. As each type is introduced in the following sections, the groups of operators applicable to it will be listed.

4.4 Type Bit

Type bit is the built-in logical type. Bit has two values, represented by the characters '0' and '1'. In other words the type definition is:

```
type bit is ('0', '1');
```

This is a kind of type known as an enumeration type. The quotes (notice that they are single quotes) are essential. This is because the values are characters, not numbers and in VHDL, characters are distinguished by enclosing them in single quotes. Enumeration types with character values are also known as character types.

The operators that apply to type bit are:

boolean: not, and, or, nand, nor, xor, xnor
comparison: =,/=, <, <=, > , >=

Bit has the full set of boolean operators. The boolean operators give a result that is also of type bit. This means that boolean operators can be combined in complex expressions and all the intermediate results are of type bit.

Bit has the full set of comparison operators, all of which give a boolean result. It is possible to test a bit signal for a value, but the result of the test is boolean and not bit. For example:

```
if a = b then
```

This tests if the value of bit signal a is the same as the value of bit signal b. The result will be true or false: boolean values. The boolean type will be covered in detail in the next section.

4.4.1 Synthesis Interpretation

Type bit is represented by a single wire, with the value '0' represented by logic 0 and the value '1' by logic 1. Logical operators are implemented directly, with each signal one bit wide.

In practice, bit is very rarely used since the two logic levels are insufficient for any but the most trivial of examples. Typically, at least four logic levels are required so that unknown and tristate values can be modelled. In practice, the nine-value type std_logic is usually used instead of bit. Type std_logic will be covered in Chapter 6.

4.5 Type Boolean

Type boolean is the built-in comparison type for VHDL. That is, the results of comparisons are of type boolean. It is very rarely used directly as a logical type, since type std_logic fills this role.

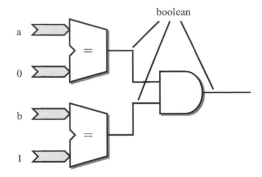

Figure 4.1   Using boolean as a comparison result.

The boolean type is predefined and has the following definition:

```
type boolean is (false, true);
```

This means that boolean is an enumeration type that has only two values, false and true.
The operators that apply to type boolean are:

boolean: not, and, or, nand, nor, xor, xnor
comparison: =, /=, <, <=, >, >=

Boolean is usually used indirectly whenever a comparison between two values of any other
type are made.

To illustrate this, consider the following example:

```
if a = 0 and b = 1 then
```

For this example, assume that a and b are 4-bit integers. Each of the comparisons is
implemented as a comparator. The result of each comparison is a boolean, which is
implemented as a single wire. The results are then combined using the and operator for type
boolean. The resultant circuit is shown in Figure 4.1.

The result of the test, in this case a test for equality, is a boolean value. Boolean will be the
result regardless of the types of a and b.

Boolean itself has the full set of six comparisons. That is, two boolean signals can be
compared using any one of the six comparisons listed earlier. However, these aren't of much
use and usually only the boolean operators are used.

4.5.1   Synthesis Interpretation

Boolean has two values: false and true. When boolean signals are mapped onto hardware
by a synthesiser, they are represented by a single wire. The boolean value false is represented
by a logic 0 and the value true by a logic 1.

Note that it is not possible to use any other type than `boolean` in a comparison. In particular, you cannot use a signal of type `std_logic` or `bit` in a conditional test. Since most logical signals are modelled by `std_logic`, it is therefore necessary to convert the signal to type `boolean`. This conversion is most easily carried out by using the equality or inequality tests. For example:

```
if s = '1' then
```

In this case, the `std_logic` signal `s` is being converted to boolean by simply testing for the `'1'` value. The result will be `true` when `s` is equal to `'1'` and `false` otherwise. This is a fairly crude type conversion, because there are eight other values in the type `std_logic`, all of which are treated as `false`, including the weak high value `'H'`.

In synthesis, this is not important, since only three of the nine values of `std_logic` have synthesis interpretations anyway. These three values are the logical values `'0'` and `'1'`, and the high impedance value `'Z'`. The last of these is only used in describing tristate drivers, so in all other situations there are only two valid values.

4.6 Integer Types

It is important when dealing with integers to distinguish between the type called `integer` and defined in the standard, and any other integer types that are user-defined. Throughout the text, the phrase 'type `integer`' refers to the built-in type called `integer`, whereas the phrase 'integer type' refers to any type that has integral values.

4.6.1 Type Integer

Type `integer` is the built-in numeric type that, as its name suggests, represents integral values.

The range of values covered by `integer` is not exactly defined by the VHDL standard, but must be at least the range -2147483647 to $+2147483647$. This is the range for 32-bit 1's-complement or sign-magnitude representations of numbers. It is a slightly smaller range than that covered by 2's-complement representation. The reason for specifying the reduced 1's-complement range as the standard is not clear but is presumably to allow VHDL tool vendors some freedom in the choice of representation since it does allow for a sign-magnitude implementation. In practice, all implementations of the language use 32-bit 2's-complement integers with a lowest value of -2147483648. It is safe to assume this slightly broader range.

The definition of type `integer` is therefore:

```
type integer is range -2147483648 to +2147483647;
```

The set of operators usable with type `integer` are the full set of comparison operators and the full set of arithmetic operators.

comparison:	`=,/=,<,<=,>,>=`
arithmetic:	`sign +`, `sign -`, `abs`, `+`, `-`, `*`,`/`, `mod`, `rem`, `**`

There are restrictions on the use of the `**` operator for synthesis – this will be covered in Section 5.6.

4.6.2   User-Defined Integers

In addition to the built-in type `integer`, it is possible to define other integer types. For example, if it is known that all calculations are to be performed in 8-bit arithmetic, an 8-bit integer type could be defined for those calculations. However, the use of user-defined integer types is not recommended, indeed it is strongly discouraged, and this section describes them for reference in case they are used in an existing design.

The only limitation on user-defined integer types is that they can have ranges n*o greater than the range of the built-in type `integer`. The built-in `integer` effectively defines the implementation limit for numeric types. This means that user-defined integers are limited to 32 bits.

An example of an integer type definition that defines a new type with an 8-bit 2's-complement range is:

```
type short is range -128 to 127;
```

In integer expressions, it is not possible to mix different integer types. For example, it is not possible to mix type `integer` and type `short` in the same expression. This clear division between different types is what is referred to as strong typing. It is claimed to be a very useful feature of the language since it allows a lot of errors to be discovered at an early stage in a design cycle. If an attempt is made to mix types inappropriately, it will result in an error. Strong typing promotes careful use of types with a clear understanding of which type is used to convey which information.

Having said this, it is *not* good practice to define a lot of unique types for each signal in a design. This is one of the most common pitfalls with new users of VHDL, who think that, because it is possible to define many distinct types, that this is a good idea. Inevitably, signals of different types will meet somewhere in the design. This then requires type conversions between types that tend to confuse and obscure the meaning. In practice there are very few errors that are trapped by the use of user-defined integer types, yet they can cause enormous frustration due to the strong typing. This is why it is strongly recommended that the built-in type `integer` is regarded as the only integer type.

Furthermore, the synthesis types described in Chapter 6 are the preferred types for defining numeric data, so in fact it is recommended to only use integer types to control for loops (Section 8.7) or index arrays (Section 4.10). They should not be used as datapath types.

The rules of VHDL insist that the result of an integer calculation must be within the range of the type. Thus, if you were using the type `short`, all expressions using `short` would have to give values in the range −128 to 127. If a calculation exceeds the range of the type, an error will occur during simulation. If this happens the result of the calculation is undefined and most simulators will stop the simulation at this point. The choice of integer range must be made with this in mind – VHDL integers do not wrap round on overflow. Again, the synthesis types described in Chapter 6 do not have this problem.

When an integer type is defined, VHDL automatically provides the following operators for the new type:

comparison   `=,/=,<,<=,>,>=`
arithmetic   `sign +, sign −, abs, +, −,*,/, mod, rem, **`

These operators will be interpreted for synthesis in exactly the same way and with the same limitations as for the standard type `integer`.

4.6.3   Integer Subtypes

A subtype is a restricted range of a type. The type a subtype is based on is known as its basetype.

For example, there is a predefined subtype of `integer` that is called `natural` and a second called `positive`. The definitions of these types are:

```
subtype natural is integer range 0 to integer'high;
subtype positive is integer range 1 to integer'high;
```

The value `integer'high` represents the highest value of type `integer`. Bear in mind that type `integer` is implementation defined but that this value is almost certainly +2147483647.

A signal declared to be of subtype `natural` is in fact an `integer`, but with restricted usage. All the operators that the type `integer` has are inherited by its subtypes. Furthermore, when type `natural` is used in calculations, the calculations are carried out using the basetype, `integer`, and then checked to ensure that they fit the subtype range of `natural`. This check is not carried out until an assignment is made.

The significance of this interpretation of subtypes is that intermediate values in an expression can exceed the subtype range, provided that they do not exceed the basetype range and provided the final value of the source expression is within the subtype range of the assignment target.

For example, start with the following definition of a 4-bit subtype of `integer`:

```
subtype nat4 is natural range 0 to 15;
```

Note how this creates a subtype of `natural`, itself a subtype of `integer`. There is no special significance in this subtype of a subtype: `nat4` is still just a subtype of `integer`.

Take four signals `w`, `x`, `y` and `z` of subtype `nat4`, related by the assignment:

```
w <= x - y + z;
```

Consider the case where the three source signals have the following values:

```
x = 3
y = 4
z = 5
```

The subtraction 3–4 is carried out first, giving an intermediate value of −1. Since the intermediate values of the expression are calculated using the basetype `integer`, this is a valid value. This value is then added to 5 to give the final result of 4. This value is then assigned to the target `w`, and is checked to ensure that it is within the range of `nat4`, which it is.

This contrasts with the same expression using a user-defined integer basetype with no negative values. In that case, the initial subtraction would exceed the range of the basetype and therefore cause an error.

4.6.4   Synthesis Interpretation

The synthesis interpretation of integers relies on the assumption that simulation has been carried out and that there are no simulation errors. This, completely reasonable, assumption allows simplifications to be made that give smallest-possible circuits for each operator. The particular rules that are used to optimise the mapping onto hardware are:

- intermediate values are within the range of the basetype of the expression;
- the value assigned to the target is in the range of the subtype of the target.

In addition, knowledge of the behaviour of arithmetic also provides some optimisations.

An integer type or subtype is represented by a bus of wires, with the number of wires in the bus depending on the range of the subtype. The number of wires will be the number of bits required to represent all the values of the subtype range.

Furthermore, the representation will be a 2's-complement if the range of the calculation includes negative numbers, but will be simply an unsigned magnitude if the calculation range does not include negative numbers.

It is not always made exactly clear where unsigned arithmetic will be used by the synthesiser, another reason to use the synthesis types instead, but generally if a calculation is made using an unsigned subtype and the result is immediately assigned to an unsigned target, then a synthesiser can infer that an unsigned circuit can be used. If the calculation is more complicated and there are intermediate values in the calculation (as in the example above), then the synthesiser must use signed representation for the intermediate values to allow for the fact that those intermediate values can be negative.

Consider again the following assignment that does contain intermediate values:

```
w <= x - y + z;
```

This can be written with parentheses:

```
w <= (x - y) + z;
```

This means that the subtraction takes place first, resulting in an intermediate value. Since this intermediate value can be negative, a signed representation must be used, even though an unsigned subtype is being used throughout. The addition must also be signed because one of its inputs (the output of the subtracter) is signed. Finally, the assignment to the unsigned signal w causes a conversion from the signed intermediate value to the unsigned result. This conversion is done by simply dropping the sign bit.

The representation of integer types always includes zero, even if the range itself does not. There is no optimisation of the implementation to suit an offset range.

For example, consider the following type:

```
type offset is range 14 to 15;
```

Even though this type has only two values, it will be represented by a 4-bit bus because this is the number of bits required to represent the maximum value, 15, as an unsigned integer.

Similarly, the 2's-complement representation of signed numbers means that the implementation will always be (almost) symmetrical around zero. There are no optimisations for offset ranges, nor for entirely negative ranges. The number of bits used to represent the type is the maximum of the number of bits required to represent the most negative and the most positive values of the type using a 2's-complement representation.

For example:

```
type negative is range −2147483648 to −1;
```

This type is a 32-bit type, even though it only has half the values of the full 32-bit range.

In practice, most integer subtypes used in circuit design will either be zero-based unsigned types or symmetrical signed types.

Another issue is the number of bits used to represent the intermediate values in an expression. Looking again at the assignment:

```
w <= x - y + z;
```

All four signals w, x, y and z are 4-bit unsigned numbers. But what size is the intermediate value from the calculation x − y? In principle, since the intermediate value is calculated using the basetype, the intermediate value should be the size of the basetype. For subtypes of type integer this would make all intermediate values 32-bits. However, this clumsy solution is optimised in practice by using knowledge of the behaviour of computer arithmetic.

First, the earlier argument showed that the calculation will use signed representation, so the two unsigned 4-bit values x and y will have to be converted to 5-bit signed values by adding a sign bit. We know that the largest result from subtracting two 5-bit signed numbers is a 6-bit value, so in fact the intermediate value is represented as a 6-bit signed number. This is then added to signal z, which must first be converted from 4-bit unsigned to 5-bit signed. The addition gives a 7-bit signed intermediate result. The assignment to w results in truncation of the result and conversion back to an unsigned representation by discarding the sign. This truncation reduces the result to 4-bits since w is a 4-bit number. The truncation is done by discarding the most significant bits. This data flow is illustrated by Figure 4.2.

The gradual expansion of the word length in an expression like this allows for the fact that intermediate values are allowed to exceed the range of the subtype, provided that they stay within the range of the basetype. This also puts an upper limit on the expansion: the largest intermediate value representation will be 32-bits when using integer subtypes. The truncation

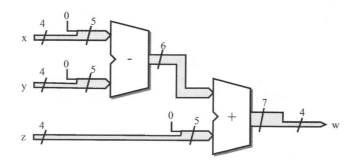

Figure 4.2   Intermediate value precisions.

by discarding the most significant bits is valid because this corresponds to the point where a simulator would check the value to ensure that it was within the target subtype. If there are no simulation errors, then allowing the synthesiser to discard the extra bits cannot affect the value.

4.7   Enumeration Types

An enumeration type is a type composed of a set of literal values. The most obvious example that illustrates an enumeration type is the state variable of a state machine. The literal values are names, so the type can be thought of as a set of names.

For example, here is an enumeration type that might be used for a traffic-light controller:

```
type state is (main_green, main_yellow, farm_green, farm_yellow);
```

This type has four literals, identified by the names `main_green`, `main_yellow`, `farm_green` and `farm_yellow`. Any names can be used as literals except the reserved words of the VHDL language (for example, `type` would be illegal as a literal).

Enumeration types can also be defined using character literals. Such enumeration types are referred to as character types. There are advantages in using character types, especially when using arrays, because VHDL provides some short cuts for defining arrays of characters. For this reason, most logic types are defined as character types. The specific details of the special characteristics of character arrays will be covered in Section 4.10.

As an example, a four-value logic type could be defined as follows:

```
type mvl4 is ('X', '0', '1', 'Z');
```

As always in VHDL, character literals are enclosed in single quotes. The characters that can be used as literals are all those that are found in the standard 8-bit ISO 8859-1 or Latin-1 character set, which contains 256 characters and incorporates ASCII as its first 128 values.

There is also a predefined character type called `character` (don't get confused between the general term character type and the specific type that happens to be called `character`). Type `character` contains the complete 8-bit Latin-1 character set, in other words all of the possible character literals. This can be used for synthesising hardware that is intended to manipulate text provided that the given encoding is the required one – in other words provided that ASCII or Latin-1 encoding is required.

In theory, an enumeration type can contain a mixture of character and name literals. In practice, this is very rarely done and most enumeration types will be pure character types or pure named literal types. One exception is the type `character` itself, which uses named literals to represent the control characters which would otherwise be impossible to represent. For example, the literal `HT` (without quotes) is the horizontal tab character.

Two predefined enumeration types have already been described. These are the types `boolean` and `bit`. Their type definitions are:

```
type boolean is (false, true);
type bit is ('0', '1');
```

`Boolean` has named literals, whilst `bit` has character literals.

The literals of an enumeration type are said to have a position number associated with them, with the first literal having the position 0 and subsequent literals having subsequent numbers. For example, the position numbers of the state type defined above are:

```
main_green  = 0
main_yellow = 1
farm_green  = 2
farm_yellow = 3
```

The position numbers of type `character` are the numeric codes for the character set. For example:

```
'A' = 65
'a' = 97
'0' = 48
```

The position numbers are used by synthesis to implement the type as an unsigned integer.

However – and anyone who has ever used the programming language 'C' take special note – enumeration types are not just integer types and the position values cannot be substituted for the literal values. Furthermore, it is not possible to perform arithmetic on enumeration types. Finally, the position values are predetermined by the language definition and therefore cannot be user-defined.

The best way to think of enumeration types is as sets of abstract values, the appropriate use of which makes any VHDL more readable and easy to understand.

The only operators predefined for an enumeration type are the six comparison operators:

comparison: $=, /=, <, <=, >, >=$

The comparison operators are defined in terms of the position values of the literals. This means that the first (or *left*) literal in the type is regarded as the smallest – zero – value and the last (or *right*) as the largest.

The enumeration types `boolean` and `bit` are special in that they also have logical operators defined, but this is not generally true and user-defined enumeration types will not have predefined boolean operators. However, it is possible to define your own logical operators using a mechanism called operator overloading. This will be covered in Section 11.3.

4.8   Multi-Valued Logic Types

A multi-valued logic type is a logic type that includes what are known as *metalogical* values. These are values that do not exist in the real world but that are a useful concept for simulation. A classic example is the high-impedance `'Z'` value. It would not be possible to attach a voltage meter to a circuit and measure a `'Z'`, but nevertheless the value is well understood and commonly used in the modelling of tristate buses. In a sense, all logic values are metalogical values. The values `'0'` and `'1'` do not exist, but they do have a real interpretation as voltages. It is the other values of a multi-valued logic type that are commonly referred to as the metalogical values of the type.

A multi-valued logic type could in principle be represented by any type, such as an integer type, with enough values to cover the full set of values being simulated. However, synthesisers restrict logic types to be one of a predefined set. In fact, most synthesisers will only allow `std_ulogic` (Chapter 6) to be used as a multi-value logic type, although some still have support for other types for historical reasons, which should all be considered obsolete. There really is no reason to use any other multi-valued logic type.

The concept of a multi-valued logic type is specific to synthesis, since in simulation terms a multi-valued logic type is just an enumeration type like any other.

A synthesiser needs to be able to identify multi-valued logic types so that they can be represented by a single wire, rather than a bus of wires representing the enumeration encoding of all the metalogical values as unsigned integers. A synthesiser usually recognises a multi-valued logic type by knowledge built-in to the synthesiser.

The use of multi-valued logic types for synthesis is full of potential pitfalls. The majority of them fall into the category of using metalogical values as if they were real values, for example, assigning a metalogical value such as the weak driving value `'L'` or an unknown `'X'` to a signal. Synthesis will either treat these as errors (the safest interpretation) or map them onto one of the two real values (the most dangerous interpretation). This quite arbitrary mapping may result in a subtle change in the behaviour of the circuit.

The safest rule in using multi-valued logic types for synthesis is to use them as if they were type `bit`. That is, not to refer to the metalogical values at all. There is no legitimate use for the metalogical values in synthesisable models, with the sole exception of the high-impedance value `'Z'` that is used in tristates (see Section 12.1).

4.9   Records

A record is a collection of *elements*, each of which can be of any constrained type or subtype. The only unconstrained types, which cannot be used in a record, are unconstrained arrays, which will be covered in more detail in the next section. Unfortunately, unconstrained arrays are the most commonly used and indeed the most useful types, so this constraint makes records very limited in their use. However, they are covered here in case they prove useful.

A record is declared as follows:

```
type pair is record
   first : integer;
   second : integer;
end record;
```

Once a record type has been declared, signals can be declared to be of that type in the same way as with any other type:

```
signal a, b, c : pair;
```

A signal of a record type is effectively a collection of signals, one for each element. The rules for the interpretation of each element are the rules for that element's type. For the example of type `pair` above, the elements are two 32-bit integer types.

The only operations that can be performed on the whole of records are equality and inequality:

comparison: $=, /=$

These are implemented as element by element comparisons, using the appropriate comparison for the element type. The element equalities are then anded together to form the overall equality. In other words, two record signals are equal if their corresponding elements are all equal.

To access an element of a record, a dot notation is used. For example, to assign the value 0 to the `first` part of signal `a`:

```
a.first <= 0;
```

The type of `a.first` is `integer`, because `integer` is the type of element `first` of type `pair`. All the `integer` operators can be used on the `first` element.

To assign values to all the elements of a record requires a notation called an *aggregate*. An aggregate is a collection of values.

An example of an aggregate is:

```
a <= (first => 0, second => 0);
```

This is the full form of the aggregate, using named association, in which each element is explicitly named and associated with a value. The symbol "=>" is known as *finger* and associates the following value with the named element of the record.

A shorter form of the aggregate uses positional notation. That is, the values of the elements are simply listed in the same order as the elements are defined in the type definition:

```
a <= (0, 0);
```

The aggregate can also be used to combine signals (rather than values) together to assign them to a record. For example, suppose there are two signals c and d that are of type `integer` and they are to be assigned to the `pair` signal a. This is done as a single aggregate assignment:

```
a <= (first => c, second => d);
```

The reverse can also be done, by using an aggregate as the target of the assignment.

For example, the integer signals c and d can be assigned from `pair` signal a in one assignment:

```
(c, d) <= a;
```

In this example, the signals c and d have been bundled together in an aggregate. This aggregated value is than assigned the value of the signal a using a whole-record assignment. The overall effect is the same as the following two separate assignments:

```
c <= a.first;
d <= a.second;
```

4.10 Arrays

An array is a collection of elements, all of which are of the same type. The elements are accessed by an *index* as opposed to a name as with records. The index can be of any integer or enumeration type. It is usually an integer type; in fact it is usually subtype `natural` so that indices cannot be negative.

An array type can be *unconstrained* or *constrained*. An unconstrained array type is one in which the size of the array is as yet unspecified. The type of the index is given but its range is not. It effectively defines a family of array subtypes with the same element type but with a variety of ranges. A constrained array type is one in which both the index type and its range are given. All signals of a constrained array type have the same range.

In fact, constrained arrays are implemented in VHDL as subtypes of an *anonymous* unconstrained array type. An anonymous type is one with no name, which cannot therefore be used by the user, but exists as a convenience for the analyser. In that sense, all array types in VHDL are unconstrained. However, because the anonymous unconstrained basetype cannot be referred to, it is impossible to use it directly and so only the constrained subtype can be used. Apart from this restriction, there is no other difference in the handling of unconstrained and constrained array types. The following discussion therefore only covers the use of unconstrained array types.

For synthesis, all datapaths must be constrained when they are declared. The reason they must be constrained is that they must be mapped onto hardware. A type maps onto a bus of wires. The datapath must be constrained so that the number of wires in the bus can be calculated.

The requirement that all datapaths are constrained means that either a constrained array type or subtype must be used, or an unconstrained array type combined with a range constraint in the declaration of the signal must be used. In practice, most array types used in synthesis are unconstrained and the signal is constrained in the signal declaration itself.

For example, there is a built-in type called `bit_vector`, defined in package `standard`, which defines an unconstrained array of type `bit`. The definition of this type is:

```
type bit_vector is array (natural range <>) of bit;
```

The symbol "`<>`" is known as *box* and signifies an unconstrained array range. The rest of the index type declaration shows that the range must be within the range of subtype `natural`, so no negative indices will be allowed.

In use, signals of this type are constrained in the signal declaration:

```
signal a : bit_vector(3 downto 0);
```

This defines a signal with four elements, indexed by the range 3 `downto` 0. This is called a *descending* range; it means that the first (*left*) element is element number 3, the second element 2 and so on down to the last (*right*) element that is numbered 0.

Arrays may be constrained with either an ascending range or a descending range for the index. The index values are simply values of subtype natural.

A descending range is the common convention for arrays representing buses, especially if they are bitwise representations of integers. The convention makes the m.s.b the leftmost bit and gives it the highest index. It is also part of the convention to make the l.s.b, which is the rightmost bit, have the index zero.

Note that this is just a convention, and *ascending* ranges could be used without changing the meaning of the model. However, most engineers already use this convention, so conforming with it will make the model much easier to understand and so is recommended best-practice.

An alternative way of achieving the same effect is to declare a constrained subtype of the unconstrained array, and then to declare the signal to be of this subtype. The equivalent declarations to give the signal the same characteristics as the signal declaration above are:

```
subtype bv4 is bit_vector (3 downto 0);
signal a : bv4;
```

Every signal that is defined in terms of `bit_vector` or any of its subtypes is of basetype `bit_vector`. These subtypes can be of any size. Therefore, since the basetype is always the same, signals of different sizes can be mixed in expressions. However, it is required that when an array is assigned to another array, the two arrays are of the same size, although they can have different ranges.

When a signal assignment is made with array signals, the assignment is made element by element from left to right without reference to the ranges of the arrays.

For example, consider the following two signals with different ranges:

```
signal up : bit_vector (1 to 4);
signal down : bit_vector (4 downto 1);
```

The assignment of one signal to the other is legal, since they are the same basetype and the same length, even though their actual ranges are different:

```
down <= up;
```

Since the assignments are made element by element from left to right, this is equivalent to:

```
up(1) <= down(4);
up(2) <= down(3);
up(3) <= down(2);
up(4) <= down(1);
```

Notice that it is the position of the element in an array, not its index that determines the outcome of the assignment. There is plenty of scope for pitfalls here.

This example also shows that the range doesn't have to begin or end at zero just because the index subtype is `natural`. Any range in which all the possible indices fall within the range of `natural` can be used.

Elements of the array can either be accessed using actual index values (static indexing), or by using a signal of the appropriate index type (dynamic indexing).

An example of static indexing is:

```
a(0) <= '1';
```

Since signal a is a `bit_vector` – that is, an array of type `bit` – the elements are of type `bit` and have all the operators appropriate to the type. This is generally true for all arrays – the elements can be manipulated individually using the operators for the element type.

For example:

```
z(0) <= (a(0) and b(0)) or (c(0) and d(0));
```

If a signal is to be used to access an array by dynamic indexing, then the indexing signal can be defined with either an ascending or a descending range regardless of the range of the array. The array is accessed by the value of the index, not its relative position in the type definition.

A descending range integer type could be used for the indexing signal. However, it is strongly recommended that only ascending ranges are used with integer types and that, furthermore, only subtypes of `integer` are used for array indices.

Thus, to define a signal to index the array:

```
signal item : integer range 0 to 3;
```

The array can be dynamically indexed with this signal:

```
a(item) <= '0';
```

Indexing allows access to the elements of an array one at a time. It is also possible to access a subrange of an array as a whole. This is done using a *slice*. For example:

```
b(1 downto 0) <= a(3 downto 2);
```

This example demonstrates the use of slices as both the source of the assignment and as the target. Both slices must be of the same size for the assignment to be legal. To be synthesisable, the slice must have a constant range; dynamic slices are not allowed. The type of a slice is the same as the type of the signal being sliced. For example, a slice of a `bit_vector` is a `bit_vector`.

4.10.1   Array Operators

The only operators available for all array types are the comparison operators and the concatenation operator.

comparison: $=, /=, <, <=, >, >=$
concatenation: &

However, for any array of type `boolean` or of type `bit`, the logical operators and the shifting operators are also predefined. Furthermore these operators have been added to `std_logic_vector` (see Chapter 6):

boolean: `not, and, or, nand, nor, xor, xnor`
shifting: `sll, srl, sla, sra, rol, ror`

The details of how these operators work is left until the relevant sections of the next chapter (Chapter 5). This section will just deal with the types used in the operators.

The comparison operators take two arrays of the same type and return a result of type `boolean`. The standard interpretation of these comparisons is unusual and not as would be expected for a bus representing an integer value (see Section 5.4). Fortunately, the standard synthesis packages (see Chapter 6) provide more sensible comparisons that correspond to a numerical interpretation of the type.

The concatenation operators allow an array to be built up out of smaller arrays and elements. For example, a 16-bit `bit_vector` can be built from two 8-bit `bit_vector` signals by concatenating them:

```
signal a, b : bit_vector(7 downto 0);
signal z : bit_vector(15 downto 0);
...
z <= a & b;
```

The result of the concatenation is the same type as the arguments – in this case bit_vector.

Similarly, single elements can be concatenated with an array to make a larger array. A common requirement is to convert an unsigned representation into a signed representation by adding a zero sign bit to the left end of the bus. In this case the buses are represented by the types in numeric_std, even though that type has not been covered yet (see Chapter 6). There are two tasks – adding the sign bit and converting the type. The example shows the conversion of a 7-bit unsigned to an 8-bit signed:

```
signal a : unsigned (6 downto 0);
signal z : signed (7 downto 0);
...
z <= signed('0' & a);
```

The concatenation creates an array of the same type as the array argument – in this case the signal a is of type unsigned, so the result of the concatenation is the same type. The whole expression is therefore wrapped up in a type conversion to convert it into type signed. Without the type conversion, the assignment would be illegal because it is not possible to assign an expression of one type to a signal of another type.

The boolean operators are similar to their corresponding single-bit operators. When a boolean operator is applied to an array, each element of the array is processed separately from left to right. So for example, to form the and of two buses, each element of one bus is combined with the corresponding element of the other bus, in left to right (not numerical) order to form a result of the same size. Both arrays must be the same size:

```
signal a, b, z : bit_vector (7 downto 0);
...
z <= a and b;
```

Finally, the shift operators shift an array by a distance specified by an integer argument, the *shift distance*. For example, to perform a logical shift left by two bits, the following example would be used. The result of the shift is an array of the same size as the array being shifted:

```
signal a, z : bit_vector (7 downto 0);
...
z <= a sll 2;
```

4.11   Aggregates, Strings and Bit-Strings

Array values can be created from a set of element values using aggregates:

```
a <= (3 => '1', 2 => '0', 1 => '0', 0 => '0');
```

The type of the aggregate is deduced by the analyser from the type of the target. In this case, the aggregate is of type bit_vector. The range constraint is worked out from the set of indices used within the aggregate except where the context gives some clues as is the case here where the assignment to signal a gives the expected range. For the assignment to be legal, the indices must fit within the index type (`natural` in this case), they must be contiguous (no gaps in the indices) and the lengths of the arrays must be the same.

To make this absolutely clear, this assignment is equivalent to the following four assignments:

```
a(3) <= '1';
a(2) <= '0';
a(1) <= '0';
a(0) <= '0';
```

The rules for calculating the range of aggregates can be confusing and has scope for pitfalls in the use of aggregates for arrays. To avoid problems, it is strongly recommended that aggregates are given exactly the same range as the target and in the same direction as was done in the example above.

It is possible to use aggregates to bundle together signals, rather than values, of the element type to create an array and to assign them all together in an array assignment. For example, given four signals of type `bit` called `elem0` - `elem3`, then they can be assigned to signal a in one assignment:

```
a <= (3 => elem3, 2 => elem2, 1 => elem1, 0 => elem0);
```

The reverse can also be done; that is, aggregates can be used as the target of an assignment:

```
(3 => elem3, 2 => elem2, 1 => elem1, 0 => elem0) <= a;
```

There are a number of alternative ways to represent array aggregates. All of the following assignments are exactly equivalent to the first aggregate example above.

The first alternative notation is to group indices together in a multiple-choice selection:

```
a <= (3 => '1', 2 | 1 | 0 => '0');
```

Once again, it is good practice to preserve the ordering of the target to avoid ambiguity.

The second example uses subrange selection to assign the same value to a subrange of the array:

```
a <= (3 => '1', 2 downto 0 => '0');
```

Here, also, the range of the target has been preserved by using a descending range in the subrange selection.

The final example uses the *others* selector:

```
a <= (3 => '1', others => '0');
```

The `others` choice must be the last choice in the aggregate and selects all the remaining elements of the target.

All of the notations so far have been *named* notations. This means that the indices have been explicitly named and associated using the finger "=> ") with a value. In addition to the named notations, array aggregates can use *positional* notation.

The same example in positional notation is:

```
a <= ('1', '0', '0', '0');
```

Because array assignment is carried out from left to right according to position, the value '1' will be assigned to the leftmost element of a, in this case element 3, and so on. This is usually clearer than the named association and is the preferred form. The only exception to this is the special case of assigning the same value to all elements using only an others clause:

```
a <= (others => '0');
```

This can only be done by named association, but remember that the range of the others clause is taken from the range of the target.

This notation cannot be used where the range is unknown. A common example is in a conditional test:

```
if a = (others => '0') then
```

This is illegal, because the range of the aggregate cannot be deduced from the context. The equality operator can take arrays of any size for each argument, so the size of a on the left cannot be used to deduce the size of the right operand. The others clause can only generally be used in assignments where the size can be deduced.

This can be rewritten to use a range attribute:

```
if a = (a'range => '0') then
```

For arrays of integers and named enumeration types, this is the full set of notations available. However, arrays of character types have two further notations that are extremely useful short-cuts and are used almost universally for assigning values to character arrays such as bit_vector and std_logic_vector (Chapter 6). The first of these notations is the *string literal*:

```
a <= "1000";
```

This is not only simpler, but much clearer than the other notations. It is a positional notation, so the elements are assigned from left to right, just as with the positional aggregate.

The alternative string notation is known as a *bit-string literal*:. This notation allows more flexibility in the representation of values, such as using underscores to divide values into groups and the options of binary, octal or hexadecimal notation. A bit-string literal is distinguished from a string literal by a prefix of B, O or X, which can be lowercase or uppercase, representing the use of binary, octal or hexadecimal within the bit-string. The extended values 10–15 in the hexadecimal notation are represented by the characters A–F, which may also be lowercase or uppercase. Examples are:

```
x <= B"0000_0000_1111";
x <= O"00_17";
x <= X"00F";
```

These all assign the same value (15) to signal x.

One limitation with the octal and hexadecimal notations is that they can only be used to assign to arrays that are a multiple of 3 or 4 bits long, respectively. The binary representation is equivalent to the ordinary string literal except that it may also include underscores to separate the bits into convenient groups. This is not possible for ordinary string literals.

Bit-string literals are available to all character array types that contain the literal values of '0' and '1'. The octal and hexadecimal values are converted into their binary string literal equivalents using the character values '0' and '1' and then the resulting string is assigned to the target.

4.12   Attributes

Attributes are a mechanism for eliciting information from a type or from the values of a type. They are useful, for example, in finding the left and right values in a type or in finding the positional value of an enumeration literal. In some circumstances it is good practice to use an attribute to refer to a value instead of the value itself, so that if the values of the type change later due to a design change, the reference will change in line with the redesign.

There are a number of attributes that apply to integer and enumeration types and another set, often with the same names, which apply to array values. To keep their meanings clear, they will be discussed as two completely separate sets of attributes.

4.12.1   Integer and Enumeration Types

This section will discuss the following attributes and their application to scalar types – that is, integer and enumeration types. All these attributes are predefined for all such scalar types:

```
type'left
type'right
type'high
type'low
type'pred(value)
type'succ(value)
type'leftof(value)
type'rightof(value)
type'pos(value)
type'val(value)
```

To illustrate this section, the following three types will be used:

```
type state is (main_green, main_yellow, farm_green, farm_yellow);
type short is range -128 to 127;
type backward is range 127 downto -128;
```

The reverse range integer type backward has been included for illustration, even though it is strongly recommended that descending range integers are never used.

The leftmost and rightmost values of a type can be found using the left and right attributes:

```
state'left = main_green
state'right = farm_yellow
```

```
short'left = -128
short'right = 127
backward'left = 127
backward'right = -128
```

It is also possible to find the lowest and highest values of a type using the `low` and `high` attributes. These are subtly different from the `left` and `right` values of a type, as can be seen from the results on the reverse range type:

```
state'low = main_green
state'high = farm_yellow
short'low = -128
short'high = 127
backward'low = -128
backward'high = 127
```

In other words, the `low` value is the `left` value for an ascending range and the `right` value for a descending range.

The two attributes `pos` and `val` convert an enumeration value into the integer representing its position number and vice versa. These attributes also work on an integer type, where the position value of the integer is the same as its actual value, so the feature is not very useful. The attributes take a single argument, which is the value to be converted. The value can be either a constant or a signal, in the latter case the attributes effectively perform type-conversion functions between integer types and enumeration types.

```
state'pos(main_green) = 0
state'val(3) = farm_yellow
```

The integer value returned from the `pos` attribute is of a type known as universal integer. Universal integer is not a type that can be explicitly used but is a convenience for the analyser, since universal integers can be assigned to any integer type. This means that the value returned from the `pos` attribute can be assigned to any integer type. Similarly, the argument of the `val` attribute is universal integer, so can come from any integer type.

For example:

```
signal short1, short2 : short;
signal state1, state2 : state;
...
short1 <= state'pos(state1);
state2 <= state'val(short2);
```

In this example, the two types in use are `short`, which is an enumeration type, and `state`, which is an enumeration type. Type conversion is not possible for this combination of types. The `pos` and `val` attributes give us a way of performing an equivalent operation. In the first assignment, the value of signal `state1`, of type `state`, is being converted into its positional value. Since this positional value is a universal integer, it can then be assigned to any integer type, in this case to type `short`. In the second assignment, the reverse process is taking place. The value of signal `short2`, of integer type `short`, is being used as the argument to the `val` attribute. This attribute can take any integer type as an argument and converts the value into the equivalent enumeration value of type `state`. This value can then be assigned to the signal `state2`.

Finally, there is a set of four attributes that can be used to increment or to decrement a value. These are the `succ`, `pred`, `leftof` and `rightof` attributes. The `succ` attribute finds the successor of its argument; that is, the next highest value of the type, regardless of whether the type is ascending or descending. The `pred` attribute finds the predecessor value; the opposite of `succ`. The `leftof` attribute finds the next value to the left of its argument; this will be the next lowest value for an ascending range and the next highest for a descending range. Finally, the `rightof` attribute finds the next value to the right of its argument; the opposite of `leftof`.

For example:

```
state'succ(main_green) = main_yellow
short'pred(0)  = -1
short'leftof(0) = -1
backward'pred(0) = -1
backward'leftof(0) = 1
```

These attributes can be used as incrementers and decrementers if used with a signal rather than a literal value as their argument. However, beware that they do not wrap around on overflow. In other words:

```
state'succ(farm_yellow) = error
```

4.12.2   Array Attributes

Array attributes are used to elicit information on the size, range and indexing of an array value. These attributes are far more widely used than the other type attributes and many applications for them will be seen in the rest of the book. It is generally considered good practice to use attributes to refer to the size or range of an array signal. In this way, if the size of the array is changed due to a design change, the VHDL statements that access the array will automatically adjust to the new size. Furthermore, in the case of `for loop` statements (Section 8.7) and `for generate` statements (Section 10.6), attributes are the only way to specify that elements are visited in a left to right order, regardless of whether they have an ascending or a descending index range: a common requirement. Further discussion of these applications will be left for the relevant sections.

This section will discuss the following attributes:

```
signal'left
signal'right
signal'low
signal'high
signal'range
signal'reverse_range
signal'length
```

Notice that these attributes act on an array signal, not a type like the attributes in the previous section. In principle, they can also be used on a constrained array subtype. This usage is less common in practice.

To illustrate this section, the following two signal declarations will be used:

```
signal up : bit_vector (0 to 3);
signal down : bit_vector (3 downto 0);
```

The `left` attribute returns the index of the leftmost element of the array, whilst the `right` attribute returns the index of the rightmost element. The `low` attribute returns the index of the lowest numbered element of the array; this is the leftmost for an ascending range and the rightmost for a descending range. Similarly, the `high` attribute returns the index of the highest numbered element.

For example:

```
up'left = 0
down'left = 3
up'right = 3
down'right = 0
up'low = 0
down'low = 0
up'high = 3
down'high = 3
```

All of these attributes return a value of the index type of the array. This means that they can be used to access the array directly. For example, suppose that the signal `down` was representing a signed number, using the common interpretation that the leftmost element is the m.s.b. and therefore represents the sign bit. If there is a signal of type `bit` that is to be assigned the value of the sign of `down`, then it would be done using the following signal assignment:

```
sign <= down(down'left);
```

The `range` and `reverse_range` attributes are mainly used in controlling `for loops` and `for generate` statements and will be discussed in more detail in Sections 8.7 and 10.6 respectively. They can also be used to define the subtype of a signal. They return the range constraint of an array signal. This is not a value that can be assigned to a signal because there is no way of declaring a signal that takes a range value. However, they can be used wherever an array range is used.

For example:

```
signal a : bit_vector (3 downto 0);
signal b : bit_vector (a'range);
```

In this example, the second signal b has been defined to have the same range as signal a. This can be a safe way of working since, if the design changes such that the size of signal a changes, then signal b will automatically adjust to the same size.

In the above example, the `range` attribute gave the value 3 downto 0. The `reverse_range` attribute literally reverses the range, so that the value in this example would be 0 to 3.

Finally, the `length` attribute returns the number of elements in an array. This value is a universal integer and so can be assigned to any integer signal. A common use of this attribute is in the normalisation of a signal. For example, given a signal with an unconventional index range:

```
signal c : bit_vector (13 to 24);
```

Then it is possible to create another signal of the same size as c, but normalised to the common convention of a descending range ending in zero:

```
signal d : bit_vector (c'length-1 downto 0);
```

It is now possible to assign the first signal to the second since they are the same size:

```
d <= c;
```

This normalisation is used extensively in VHDL, especially when writing subprograms, as discussed in Chapter 11.

4.13   More on Selected Signal Assignments

This section is effectively a continuation of Section 3.9. It covers the use of selected signal assignments with the types covered in this chapter.

Consider the problem of converting the traffic-light controller type, introduced earlier, into control signals for individual lights.

The traffic-light controller enumeration type was:

```
type state is (main_green, main_yellow, farm_green, farm_yellow);
```

The control signals for the individual lights will be represented by a std_logic_vector with three elements, the first for the red light, the second for the yellow and the third for the green. The encoding will reflect the UK traffic-light conventions.

The signal declarations are:

```
signal current_state : state;
signal main_lights : bit_vector (0 to 2);
```

The selected signal assignment to decode the current state is:

```
with current_state select
  main_lights <= "001" when main_green,
                 "010" when main_yellow,
                 "100" when farm_green,
                 "110" when farm_yellow;
```

The synthesis interpretation of this is a single multi-way multiplexer, as shown in Figure 4.3. The exact implementation of this multiplexer may vary slightly from synthesiser to synthesiser but the behaviour will always be the same. In this case, the control condition is an enumeration type, so this will also be converted into its synthesis representation to give a two-bit control bus.

As well as using a single value to match with, known as the *choice*, for each branch of the assignment, it is possible to specify multiple choices, range choices and to use an others

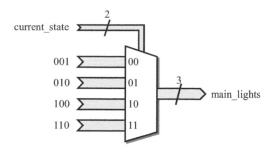

Figure 4.3   Multi-way selected signal assignment.

choice to mop-up any remaining selections not already covered. The formats for these multiple choices are the same as the multiple choices used with aggregates.

The following example shows how these various multiple choices are used. It shows how the control signals of the lights example above could be individually encoded in separate selected signal assignments.

```
with current_state select
  main_lights(0) <= '1' when farm_green to farm_yellow,
                    '0' when main_green to main_yellow;
with current_state select
  main_lights(1) <= '1' when main_yellow | farm_yellow,
                    '0' when main_green | farm_green;
with current_state select
  main_lights(2) <= '1' when main_green,
                    '0' when others;
```

It is generally good practice to end with a when others choice to mop up the remainder. It is a requirement of the VHDL language that all possible values are covered and this is the easiest way to guarantee it. However, as the first two assignments in the example shows, it is not essential to finish with an others choice.

The selection condition can be an array signal. For example, to do the reverse conversion from the light controls back to the state type, the following selected signal assignment can be used:

```
with main_lights select
  current_state <= main_green when "001",
                   main_yellow when "010",
                   farm_green when "100",
                   farm_yellow when others;
```

Note that the others choice in this selected assignment is not equivalent to a choice of '110'. The 3-bit bit_vector has 8 possible values, since each element can have any one of the two values of bit, but only 4 of them have legal interpretations in this example, since there are only four combinations of lights. The remaining 4 selections are unused and should never occur in this design. Nevertheless, all of these must be covered by the choices to make the selected signal assignment complete. In this example, all the unused encodings of the light controls have been mopped up into the selection for farm_yellow by using the others choice.

5

Operators

The VHDL language has a set of standard operators that can be used to perform comparisons, form boolean equations and perform arithmetic. This set of operators is the toolkit that is used to build up RTL models.

This chapter introduces the built-in operators and what they do, with an explanation of the rules that VHDL uses to decide the order of precedence when calculating a complicated expression.

5.1 The Standard Operators

The full set of operators in VHDL is listed here:

`not`	inversion
`and`	and function
`nand`	not-and function
`or`	or function
`nor`	not-or function
`xor`	exclusive-or function (bitwise inequality)
`xnor`	exclusive-nor function (bitwise equality)
`=`	equality
`/=`	inequality
`>=`	greater-than or equal
`>`	greater-than
`<=`	less-than or equal
`<`	less-than
`sll`	shift-left logical
`srl`	shift-right logical
`sla`	shift-left arithmetic
`sra`	shift-right arithmetic
`rol`	rotate left
`ror`	rotate right

VHDL for Logic Synthesis, Third Edition. Andrew Rushton.
© 2011 John Wiley & Sons, Ltd. Published 2011 by John Wiley & Sons, Ltd.

+	addition
−	subtraction
+	plus sign
−	minus sign
*	multiplication
/	division
mod	modulo arithmetic
rem	remainder after division
**	exponentiation
abs	absolute value
&	concatenation.

5.2 Operator Precedence

Operators are classified by the standard as logical, relational, adding, sign, multiplying and miscellaneous. The reason for classifying the operators is to allow for operator precedence. This is not the same classification I have used in the rest of the book, which is simpler and groups operators by what they do, rather than their precedence order. However, this section is about the precedence rules and so is based on the formal classification given by the VHDL standard.

The precedence of operators is the order in which they will be processed in an expression. For example, in the expression:

```
3 + 4 * 5
```

This is interpreted as:

```
3 + (4 * 5)
```

and not as:

```
(3 + 4) * 5
```

The result is 23, not 35. The multiplication is carried out first because multiplication in arithmetic has a higher precedence than addition.

If an expression has operators of the same precedence, they are processed from left to right. For example:

```
3 - 4 + 5
```

The result is 4, because subtraction and addition have the same precedence and so the expression is interpreted from left to right as:

```
(3 - 4) + 5
```

and not as:

```
3 - (4 + 5)
```

The classification of the standard operators defines their precedence. The classifications are listed here in order of precedence, with the highest first:

miscellaneous:	`**, abs, not`
multiplying:	`*, /, mod, rem`
sign:	`+, -`
adding:	`+, -, &`
shift:	`sll, srl, sla, sra, rol, ror`
relational:	`=, /=, <, <=, >, >=`
logical:	`and, or, nand, nor, xor, xnor`

The rules for operators in VHDL are a bit odd. Generally, they are more restrictive than you would expect from using other languages, so you end up with more parentheses than are strictly necessary, but within those restrictions VHDL on the whole follows the rules of arithmetic precedence. Unfortunately, there are some exceptions to this which may become pitfalls if you are not careful.

This description will be based on the syntax rules defined in the VHDL Language Reference Manual [IEEE-1076, 2008].

I have generally avoided any mention of syntax in this book, but this is really the best way to describe the operator precedence rules. If you cannot bring yourself to read syntax, skip to the next section!

Here then, is the syntax for operators from the VHDL LRM, starting with the overall definition, an expression:

```
expression ::=
   relation {  and relation }  | relation {  or relation }  |
   relation {  xor relation }  | relation {  xnor relation }  |
   relation [  nand relation ]  | relation [  nor relation ]
```

These rules have a number of consequences:

logical operators are the lowest precedence
    Logical operators are evaluated after the rest of the expression (the relations).
logical operators have the same precedence
    All the binary logical operators are at the same level in the syntax – the bottom, so therefore they must all have the same precedence. This is unusual in that in some other languages `and` is considered the same precedence as multiplication and `or` the same as addition. Not in VHDL!
it is impossible to mix logical operators
    Once you have used an operator, all subsequent relations must be combined using the same operator – for example:

```
a and b or c
```

is illegal and would result in an error. This restriction is included to resolve confusion caused by giving all logical operators the same precedence.

you must use parentheses a lot

Since logical operators cannot be mixed at the same level of an expression, parentheses must be used. For example:

```
a and b or c
```

is illegal, so it must be written:

```
(a and b) or c
```

nand and nor cannot be chained

By this I mean that the expression:

```
a nand b nand c
```

is illegal (note the use of square [] brackets rather than curly { } braces in the syntax). This restriction was incorporated into VHDL because it was felt that people might not realise that this example is not the result of nanding a, b and c together because the nand operator is not associative. In other words:

```
y <= (a nand b) nand c
```

is not the same as:

```
y <= a nand (b nand c)
```

To avoid the potential for error due to this non-associative behaviour, VHDL insists on the use of brackets to break up the inverting operators into two-way functions only.

The non-inverting operators (including xor) are associative and so can be chained together to form expressions of any length.

As you can see, VHDL logical operators are not as you would expect. Note that the unary operator not has not been dealt with yet – it is the exception to the above rules and will be dealt with later since it is classed as a miscellaneous operator.

The next level of syntax is the relation:

```
relation ::=
  shift_expression [ relational_operator shift_expression ]
relational_operator ::=
  = | /= | < | <= | >| >=
```

This appears more normal, although it is again restrictive compared with other languages because you cannot chain comparisons. For example, you cannot say

```
if a > b = false then ...
```

since this is an error. This is bad style anyway, but most languages would at least allow it. This also shows that logical operators are evaluated after comparisons, so the test

```
if a = '1' and b = '0' then ...
```

is evaluated exactly as expected, as

```
if (a = '1') and (b = '0') then ...
```

So there are no surprises there.

The next level of syntax is the shift_expression:

```
shift_expression ::=
  simple_expression [ shift_operator simple_expression ]
shift_operator ::=
  sll | srl | sla | sra | rol | ror
```

The rules here are very similar to the above relation operators. You cannot chain shift expressions. For example, you cannot say:

```
a sll 2 srl 2
```

to strip off the leftmost two bits of a bus, since this is an error. You would have to use parentheses again to achieve this effect.

Note that shifts are evaluated before relations, so the expression

```
if a sll 2 > 5 then ...
```

is evaluated as

```
if (a sll 2) > 5 then ...
```

as expected.

The next level of syntax is the simple_expression:

```
simple_expression ::=
  [ sign ] term { adding_operator term }
sign ::=
  + | -
adding_operator ::=
  + | - | &
```

These rules are again somewhat unusual. In particular, you cannot have a sign operator before a term except for the first one. In other words

```
a + -b
```

is an error and parentheses are needed again. Also, strangely, the abs operator is not classed as a sign operator and will be dealt with later. It will be seen that this creates some peculiar side effects.

The sign operator is evaluated before the adding operators, so only the first term is negated by the negation operator. In other words,

```
-a + b
```

is interpreted as

```
(-a) + b
```

The next level of syntax is the term:

```
term ::=
   factor { multiplying_operator factor }
multiplying_operator ::=
   * | / | mod | rem
```

This is what would be expected for normal arithmetic. Multiplying operators have a higher precedence than adding operators so are evaluated first. This means that

```
a + b * c
```

is evaluated as

```
a + (b * c)
```

as you would expect.

The interaction between multiplication and sign operators bears closer examination. Since multiplication operators have higher precedence than sign operators,

```
-a * b
```

is interpreted as

```
-(a * b)
```

This is quite different to the way that sign operators work with addition, which was shown earlier. Of course, this makes no difference to multiplication, since the negation of the result is the same as negating one of the arguments before multiplying, but it can make a difference with the other operators such as mod that do not have this symmetry.

The most likely problems with modulo and negation are due to the fact that

```
-a mod b
```

is actually interpreted as

```
-(a mod b)
```

and not, as you might expect

```
(-a) mod b
```

Modulo is not symmetrical around zero, so this could give a different result to that expected. For example, given:

```
a = 3
b = 4
```

then the expected result might be:

```
(-a) mod b = 1
```

whereas the actual result will be:

```
-(a mod b) = -3
```

The next level of syntax is the factor:

```
factor ::=
    primary [ ** primary ] | abs primary | not primary
```

This is the highest level of precedence in VHDL, since the primary is any object (signal, variable, array element, etc.) that is appropriate for the operators being used (not all objects have all the operators). Note that the logical operator not is defined at this level and so will be evaluated before other logical operators. Note also that the abs operator is included here and so will be evaluated before any sign operators.

There is one part of the primary that is significant to complete the picture so that is included here:

```
primary ::=
    ( expression ) | ... lots else
```

This means that a parenthesised expression is the highest precedence of all and so will be calculated before anything else.

Looking back at the definition of a factor, you cannot chain exponentiation. In conventional arithmetic, exponentiation has a right-to-left precedence instead of the left-to-right precedence of all the other binary operators. In other words,

```
a ** b ** c
```

should be evaluated as:

```
a ** (b ** c)
```

To avoid any possible confusion that this might have caused, the VHDL designers disallowed it.

It should by now be clear that the rules only allow chaining of a limited set of logical operators (and, or, xor, xnor), all the adding operators (+, -, &) and all the multiplying operators (*,/, mod, rem), all of which have left-to-right ordering in conventional arithmetic too.

The precedence rules of VHDL are generally what you would expect for most of the operators, it is the restrictions and the equal precedence of all binary logical operators that are very strange. Also, handling exponentiation in a right-to-left way would be trivial. However, to give the designers credit, it is true to say that most potentially confusing expressions are disallowed by the syntax rules so at least you can be sure of what you are getting. The result tends to be a lot of parentheses.

There are some other interesting restrictions and irregularities that come from this strange syntax. For example, in the use of negation.

The following are *illegal*:

```
a + -b
a ** -b
a / -b
```

However, the following are *legal*:

```
a < -b
a rol -b
```

interestingly, the following are also *legal*:

```
a + abs b
a / abs b
```

but the following is *illegal*

```
a ** abs b
```

It seems incredible that the sign operators are given a different precedence from the abs operator.

Now that the syntax has been dealt with, it is easier to understand the operators if they are divided into just the five kinds introduced in Chapter 4, which are called boolean, comparison, shifting, arithmetic and concatenation operators:

boolean:	not, and, or, nand, nor, xor, xnor
comparison:	=, /=, <, <=, >, >=
shifting:	sll, srl, sla, sra, rol, ror
arithmetic:	sign +, sign −, abs, +, -, *, /, mod, rem, **
concatenation:	&

These terms are not a part of the definition of the language but have been chosen to simplify this explanation. The meaning and the synthesis interpretation of these five kinds of operators are explained in the following sections. This terminology will be used throughout the rest of the book. The following sections explain the operation and the synthesis interpretation of the individual operators in detail.

5.3   Boolean Operators

There are three different families of boolean operators provided as built-in operators for bit and boolean as well as arrays of those types:

Basic boolean operators. They perform bitwise logic on each element of two parameters of the same size to produce a result of the same size.

Selecting boolean operators. They combine a single-bit input with each element of an array to produce an array of the same size.

Reducing boolean operators are only available in VHDL-2008. In earlier versions of VHDL they are occasionally provided as functions with names like and_reduce and so on. They combine all the elements of an array to produce a single-bit output.

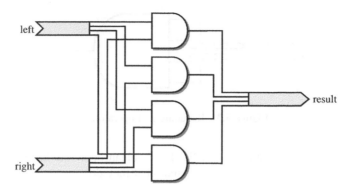

Figure 5.1   Basic and operator.

To illustrate the different boolean operators, consider the three types of and operator.

The basic and operator is defined for one-bit types and arrays. It has two arguments and as a result the same type and size. Each bit of the result is created by combining the corresponding bit from each argument, as illustrated in Figure 5.1.

The selecting and operator is defined for an array type so that one argument is the array type, the other is the element type. The result is also the array type and has the same size as the array argument. Each bit of the result array is created by combining the corresponding input array element single-bit input. This can be thought of as the one-bit signal acting as a select line controlling the connection from input to output. This is illustrated in Figure 5.2.

The reducing and operator is defined for an array type and only has one argument (referred to as a unary and operator). The result is the element type of the array. All elements of the array are combined by the logic operator to form the output. This is illustrated in Figure 5.3.

The reducing operators were introduced in VHDL-2008. Earlier versions used a function called and_reduce to do this job. These and_reduce functions are not available for the built-in types of VHDL, but are overloaded for the synthesis types described in Chapter 6. A typical reducing function is:

```
function and_reduce(l : <array_type>) return <element_type>;
```

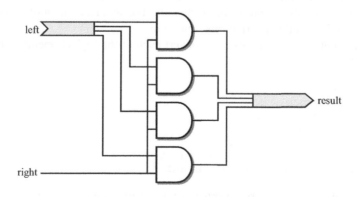

Figure 5.2   Selecting and operator.

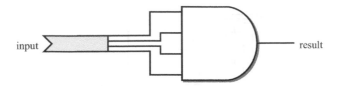

Figure 5.3 Reducing and operator.

There are seven boolean operators in VHDL. The boolean operators have self-explanatory names but nevertheless, this is what they are and what they do:

not	inversion – `true` when the argument is `false`
and	and function – `true` when both arguments are `true`
nand	not-and function – inverse of result from `and`
or	or function – `true` if either argument is true
nor	not-or function – inverse of result from `or`
xor	exclusive-or function – `true` if one argument is `true`
xnor	exclusive-nor function - inverse of result from `xor`

The not operator is a unary operator, that is it only has one argument. The remainder are binary operators in that they have two arguments. In VHDL-2008, all the binary boolean operators have a unary form that is the reduction operator.

The boolean operators allow boolean equations to be described. For example:

```
sum <= a xor b xor c;
```

The example is a boolean equation that forms the exclusive-or of three inputs a, b and c.

5.3.1 Synthesis Interpretation

The interpretation of boolean operators for synthesis is straightforward. Synthesisers work on a circuit representation made up of boolean equations. This internal circuit representation is usually, but not necessarily, in sum-of-products form. For synthesis, all boolean equations are transformed directly into this internal representation.

The following equivalences are used to convert each of the logical operators into sum-of-products form:

```
not a = not a
a and b = a and b
a nand b = not a or not b
a or b = a or b
a nor b = not a and not b
a xor b = (not a and b) or (a and not b)
a xnor b = (a and b) or (not a and not b)
```

However, logic minimisation during synthesis will restructure the circuit anyway, so these equivalences will generally not be seen in the final synthesised circuit.

5.4   Comparison Operators

There are six comparison operators, all of the same precedence. The six comparisons are:

=	equality
/=	inequality
>=	greater-than-or-equal
>	greater-than
<=	less-than-or-equal
<	less-than

All types have equality and inequality operators; most types have all six operators.

The most obvious use of comparison operators is for testing numeric types. However, they can also be applied to any other types, including arrays.

The form of a comparison is:

```
if a < b then
```

The result of the comparison is type `boolean`, which was dealt with in Section 4.5 `Boolean` is a logical type and so relationships can be combined using the boolean operators.

For example:

```
if a = 0 and b = 1 then
```

The example shows how two comparisons have been combined with an `and` function. The result of this test will only be true if both conditions are true, that is if a is 0 and b is 1.

The relational operators have a higher precedence than logical operators. This means that the comparisons are calculated first and then combined with the logical operators. No parentheses are needed to enforce this ordering, but parentheses would be needed to override it.

The example above is equivalent to:

```
if (a = 0) and (b = 1) then
```

5.4.1   Synthesis Interpretation

The circuits produced from the comparison operators are different for integer types (including enumeration types that are synthesised as small unsigned integers) and for array types. It is important to realise that array comparison is different from numeric comparison. For example, when using a `bit_vector` to represent an integer, the comparison operators will not give a numeric ordering. It is considered bad practice to use bit_vector to represent integer values anyway, since the array types described in Chapter 6 should be used for this purpose.

5.4.2   Integer and Enumeration Types

There are two basic circuits used to perform comparisons. One circuit performs the equality tests ("=", "/="), the other performs the ordering tests ("<", "<=", "> ", ">=").

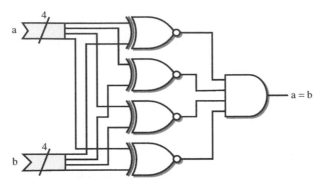

Figure 5.4 Four-bit equality.

Equality is performed as a bit-by-bit comparison. Each bit is compared using an `xnor` function and the results of the individual bit comparisons are then anded together. For example, Figure 5.4 shows a 4-bit comparison.

This circuit is the same regardless of the type being compared. One-bit types such as `bit` for example are the simplest form of this circuit, using just a single `xnor` function.

Inequality uses exactly the same circuit with the output inverted.

The circuit used for the four ordering operators is based on a subtracter. The sign bit of the result of a subtraction can be used to test whether the result is negative. This fact is used to perform the less-than operation. If the first operand is subtracted from the second and the result is negative, then the second operand must be less than the first.

The circuit used for comparison is shown in Figure 5.5, which shows a 4-bit less-than operation. Of course, the synthesiser will optimise the circuit to remove the logic for generating the unused subtracter outputs.

The exact circuit used for the subtracter depends on the type being compared. For integer types it will be an integer subtracter, for fixed-point types a fixed-point subtracter and so on.

The remaining three comparisons are all performed by the less-than operation, with various permutations of exchanging the inputs and inverting the output.

The equivalences are:

```
a > b = b < a
a >= b = not (a < b)
a <= b = not (b < a)
```

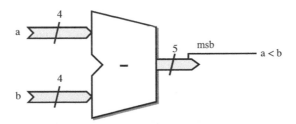

Figure 5.5 Four-bit less-than circuit.

Synthesis can be simplified further when comparing with zero. Bear in mind that the sign bit alone is a test for a number being less than zero (if it is 2's-complement that is). It follows that the inverted sign bit is a test for greater-than-or-equal to zero. Both of these tests require no comparator at all.

Contrast this with the test for a signal being greater than zero. Using the equivalences above, the test for signal a being greater than zero is:

```
0 < a
```

The sign bit is no help here, so this is implemented as a subtracter. This is clearly a much larger circuit than that used for the test for a being less than zero. The same subtraction circuit with inverted output is used for testing less-than-or-equal to zero.

In summary, the test for less-than zero (< 0) and for greater-than-or-equal to zero ($>= 0$) are very efficient, whereas the tests for greater-than zero (> 0) and for less-than-or-equal to zero ($<= 0$) are much less efficient.

5.4.3   Array Types

The equality operator ("=") has two interpretations when comparing arrays. If the two arrays being compared are of different lengths, then they cannot be equal and the equality operator is implemented by the false value – that is, by logic 0. If the two arrays are of equal length, then the equality of the arrays is the individual equalities of the elements all anded together. The circuit is illustrated in Figure 5.6.

If the array element type is a one-bit type such as bit or boolean, then the equality operator for the element is just an xnor function and so the array equality becomes the same as the integer equality described earlier. In other words, array *equality* of an array of one-bit values will give a numeric comparison provided the two arrays are the same length.

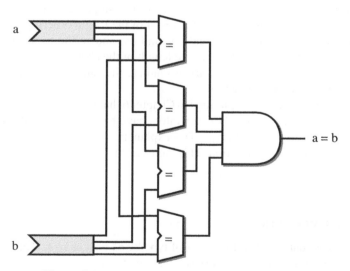

Figure 5.6   Array equality for arrays of equal length.

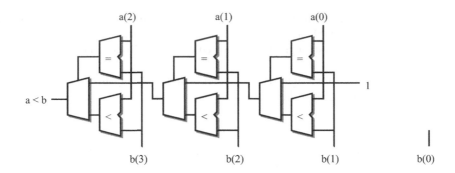

Figure 5.7   Array less-than operator.

Inequality (" / =") uses the same circuit but with the output inverted.

The real difference between integer comparison and array comparison is with the ordering operators ("<", "> ", "<=" and " >=").

The algorithm for the less-than operator ("<") will be used as an example. The elements of the arrays are compared from left to right regardless of their ranges. Each bit is tested for equality until unequal elements are found. If a pair of elements are not equal, then the array with the smaller element at that position is regarded as less than the other array. If the end of either array is reached before any unequal elements are found, then the shorter array is regarded as less than the longer array.

The array less-than circuit is illustrated in Figure 5.7. It has been drawn with an unconventional circuit flow from right to left to reinforce the fact that the elements are compared from left to right. In other words, the leftmost elements are the highest priority and so are shown nearest to the circuit output. If the leftmost elements are unequal, then that determines the output. If they are equal, then the result from the previous bit is rippled through. This circuit is iterated for each bit in the arrays, stopping at the end of the shorter array. The input of the rightmost circuit is the result of the length ordering. In this case, the first operand (a) is shorter than the second operand (b) and so the result of the less-than comparison is true if all the elements up to the length of the shorter operand are equal. The true value is fed in at the end of the comparison chain as the value 1. The trailing element of b does not participate in the comparison.

Note that, for the numeric types described in Chapter 6 that implement integer, fixed-point and floating-point values as arrays, these operators have been replaced with ones that give the correct comparison circuit for the type. This explanation only applies to other array types that do not have a numeric interpretation. This includes bit_vector and any user-defined array types.

5.5   Shifting Operators

Shifting operators are only built-in for arrays of type boolean or bit. This means that the only standard type that has shift operators is bit_vector.

The description here describes the built-in shift operations for bit_vector.

In fact, these operators have also been added to `std_logic_vector` and the numeric packages but with a slightly different interpretation – see Chapter 6 for details.

The shifting operators are:

sll	shift-left logical
srl	shift-right logical
sla	shift-left arithmetic
sra	shift-right arithmetic
rol	rotate-left
ror	rotate-right

The general form of a shift expression is:

```
z <= a sll 1;
```

The result z must be the same array type and the same length as the left-hand operand a, whilst the right operand is an `integer` value representing the shift distance.

The *logical* shifts simply shift the operand, discarding bits that are shifted off one end and filling the other end with the left-most value of the element type. This means that arrays of type `bit` will be filled with the `'0'` value.

Assuming that a is a `bit_vector` with the value `"00001111"`, then the value of z after the one-bit left shift will be `"00011110"`.

The right shift works in a symmetrical way:

```
z <= a srl 1;
```

In this case, with the same value of a, then the value of z after the one-bit right shift will be `"00000111"`.

The *arithmetic* shifts are quite unusual and of very little use on most array types. They are overloaded for array types that do have a numeric representation so that they do perform arithmetic shifts. Again, see Chapter 6 for details of the numeric array types.

The left arithmetic shift extends the rightmost bit just as if it was a sign bit. This rightmost bit is shifted in at the right-hand end as the array is shifted to the left. For example, if a is a `bit_vector` and has the value `"00001111"`, then a left shift of one bit will give the value `"00011111"`. In right shifts it is the leftmost bit that is duplicated and shifted in at the left-hand end.

The *rotate* operators take elements off one end of the array and shift them in at the other end. In left shifts, elements are shifted off the left end and in at the right, whilst in right shifts, elements are shifted off the right end and in at the left.

For example, `"11110000"` rotated left by one bit is `"11100001"` and shifted right by one bit is `"01111000"`. This has no numeric equivalent – it is purely a logical operation.

5.5.1 Synthesis Interpretation

The shift operators have two different synthesis interpretations depending on whether the shift distance is a constant value or a signal or variable.

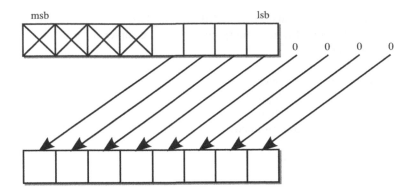

Figure 5.8   Shift-left logical (sll) by 4 bits.

5.5.1.1 Constant Shift Distance

For a constant shift distance, the shift operators are implemented as a fixed rearrangement of the bits of the bus. There is no logic circuitry generated.

The shift-left logical (sll) operator is implemented as a left offset of the bits of the bus, with the rightmost bits connected to zero to represent the shifting in of zeros. The circuit is illustrated by Figure 5.8 which shows a shift-left of 4 bits.

The shift-right logical (srl) operator is symmetrical to this, as would be expected for a logical operation.

The shift-left arithmetic (sla) is also implemented as a left offset of the bits of the bus. The rightmost bit is replicated by simply fanning out this bit to all the bits left open by the offset. The resulting circuit for a shift of 4 bits is shown in Figure 5.9.

The shift-right arithmetic (sra) operator is symmetrical to this.

The rotate operators cause a crossover in the bits of the bus, because the bits shifted off one end of the bus are shifted back in at the other end. The resulting circuit for a rotate-left (rol) of 1 bit is shown in Figure 5.10.

Once again, the rotate-right (ror) operator is symmetrical to this.

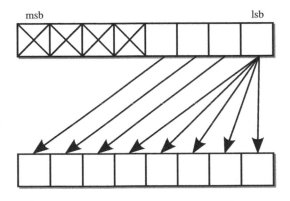

Figure 5.9   Shift-left arithmetic (sla) by 4 bits.

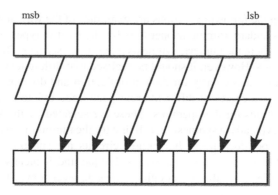

Figure 5.10   Rotate-left (rol) by 1 bit.

5.5.1.2 Variable Shift Distance

If the shift distance is a variable quantity, such as a signal, variable or complex expression, then the shift is implemented as a barrel shifter logic block. This is a predefined circuit provided by the synthesis vendor for performing variable shifts.

5.6   Arithmetic Operators

The arithmetic operators are:

+	addition
−	subtraction
+	plus sign
−	minus sign
*	multiplication
/	division
mod	modulo arithmetic
rem	remainder after division
**	exponentiation
abs	absolute value

5.6.1   Synthesis Interpretation

The circuits that logic synthesis will use for the arithmetic operators will vary according to the type, so for example integer types will have integer operators, fixed-point types will have fixed-point operators and so on. The exact circuits can also vary from synthesiser to synthesiser and from technology to technology, although the calculated results will always be the same between technologies. In general, the circuit used for an arithmetic operator will be the minimum area circuit that implements that function entirely in combinational logic. The operator must be implemented combinationally because registers must be specified explicitly for logic synthesis.

The circuits given in the following sections are examples of the way in which a synthesiser might implement the standard arithmetic operators for the built-in types. This subject will be revisited in Chapter 6 where the numeric synthesis types are described. They are all minimum or near-minimum circuits. However, it should be realised that different synthesisers will give slightly different circuits, particularly for the multiplication and division operators. Furthermore, synthesisers allow a choice of different area/speed trade-offs to be made by providing a range of implementations for the operators. These are selected by the synthesiser to meet timing requirements. Typically, a synthesiser will first try the minimum area circuit given here, then if the circuit is too slow, faster but bigger circuits are substituted until the timing requirements are met. This is a very powerful technique and it means that designers will probably never have to design arithmetic circuits at the gate level or even the boolean equation level, even for high-speed applications.

Part of the learning process of using synthesis is to allow the synthesiser to make this choice of implementation rather than trying to control every detail of the design yourself. So you don't need to know which implementation is used as long as the timing requirements are met.

5.6.2   Plus Sign

The plus sign has no effect whatsoever on the value of a signal and so has no circuit to implement it; it can be thought of as simply a feedthrough.

5.6.3   Minus Sign

The minus sign is implemented as a 2's-complement negation. 2's-complement negation is performed by subtracting the input from zero.

5.6.4   Abs Operator

The abs operator is simply a combination of the two sign operators with a multiplexer controlled by the sign bit, to choose between whichever of the operand and its negation is non-negative. The circuit for the abs operator is shown in Figure 5.11.

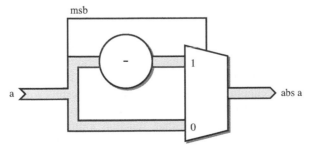

Figure 5.11   Abs operator.

5.6.5   Add Operator

The add operator is implemented in its minimum form as a ripple-carry adder. However, there are many other faster implementations such as carry lookahead adders. Each technology will provide a range of adder circuits for the synthesiser to use with different speed/area tradeoffs. It is the synthesiser's job to select the appropriate implementation for each add operator.

If the operands of the addition are of different lengths, then the shorter operand will be extended to the same length as the longer one. Unsigned numbers are zero-extended, whereas signed numbers are sign-extended. This means that, in practice, signed and unsigned addition can require slightly different circuits.

It is questionable whether it is worth optimising the number of adders in a circuit since the hardware area and delay cost of multiplexing several different inputs onto a single adder may outweigh the saving in adder circuitry.

5.6.6   Subtract Operator

The subtracter is a very similar circuit to the adder. The technology library will provide a similar range of subtracters as adders. Also, the synthesiser can map the subtraction onto an adder/subtracter circuit.

For example, consider the following code fragment:

```
if do_add = '1' then
  z <= a + b;
else
  z <= a - b;
end if;
```

Since the inputs of the addition and the subtraction are the same, and the conditional makes them mutually exclusive (i.e. only one can be required at any one time), the synthesiser may be able to map this onto a single adder/subtracter circuit rather than separate circuits.

5.6.7   Multiplication Operator

Multiplication has a different implementation depending on whether the operands are signed or unsigned. However, for both cases, the minimum-area combinational implementation of multiplication is a essentially a matrix of full-adders. The point is that the area of a multiplier is proportional to the square of the word length, whilst the delay is proportional to the word length.

Multipliers can take arguments of different length and there is no extending of the shorter argument. Instead, this is used as an optimisation. For example, given a 32-bit input and a 16-bit input, a 32×16 multiplier can be generated, rather than normalising the inputs and generating a 32×32 bit multiplier that has twice the area.

A multiplier is a large circuit and has many possible implementations, far more than an adder. Most technology libraries will provide a wide range of designs with different speed/area tradeoffs. It is usually not necessary to try to re-implement multiplication as a multi-cycle operation since combinational multipliers are sufficiently efficient for most designs.

Nevertheless, one of the key optimisations that you can do as a designer is to reduce the number of multipliers by multiplexing arguments onto a single multiplier block.

5.6.8   Division Operator

For many years after the advent of synthesis tools, division was considered an unsynthesisable operator. This is because, in RTL synthesis, all operators must be implemented as combinational logic. The exception was when the right-hand operand (the divisor) was a constant power of two, which allowed the division to be replaced by a shift operation.

However, both ASIC and FPGA technologies have advanced and division is now considered a synthesisable operator. It is typically implemented as a series of subtract and shift operations, bearing in mind that they are all combinational.

This is a larger and slower circuit than multiplication and it may make more sense to use a circuit that implements division over several clock cycles. Such circuits may be provided by the technology vendor or can be licensed from core design services.

Nevertheless, division can be synthesised and will result in a combinational division circuit. It is still the case that if the right-hand operand is a constant power of two, the division will be replaced by a shift operation.

5.6.9   Modulo Operator

The mod operator performs modulo arithmetic. It maps a number onto a restricted range specified by its second operand. It is closely related to division, and requires a similar circuit.

Like division, if the right-hand argument is a constant power of two, the modulo can be mapped onto a simpler operation, in this case a masking operation.

To give an example, the following is a modulo-4 addition:

```
(a + b) mod 4
```

The result of this expression will be mapped onto the range 0–3.

Figure 5.12 shows the mapping from an integer value (shown on the x-axis) onto its modulo 4 (shown on the y-axis).

The implementation is very simple. The modulus of a number is found by simply discarding the m.s.b.s and keeping the number of l.s.b.s required to represent the range of the modulo arithmetic. This is a masking operation in which the m.s.b.s are removed, leaving just the l.s.b.s intact. For example, modulo 4 requires 2 bits to be kept and the rest to be masked. Figure 5.13 shows the modulo 4 conversion of a 4-bit bus.

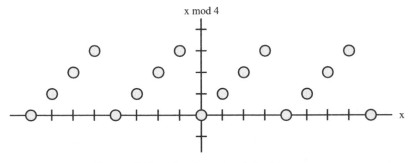

Figure 5.12   Mapping of modulo-4 operator.

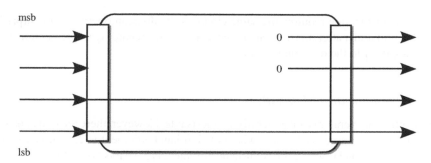

Figure 5.13   Unsigned and signed modulo-4.

Examination of the 2's-complement representation of negative numbers shows that in fact, the masking operation used for unsigned modulo arithmetic also works correctly for signed modulo arithmetic.

When the right-hand argument is variable, the operator will be mapped onto a combinational divider circuit. Again there will be a set of implementations specific to the technology being synthesised to.

5.6.10   Remainder Operator

The rem operator calculates the remainder after a division. The difference between modulus and remainder only applies to negative numbers: the remainder preserves the sign of the number being divided, whereas the modulus preserved the sign of the divisor, in this case positive.

For unsigned numbers, there is no sign and the remainder is exactly the same as modulus, giving exactly the same circuit.

For signed numbers, the result can be negative. Figure 5.14 shows the mapping from an integer value (shown on the x-axis) onto its remainder for division by 4 (shown on the y-axis).

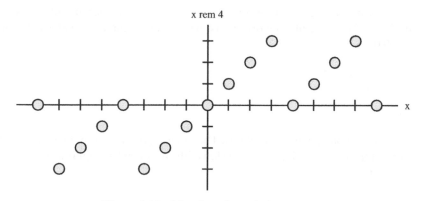

Figure 5.14   Mapping of remainder operator.

The remainder function requires the same circuit is the division operator, not surprisingly since it is the remainder after division. So the same arguments regarding implementation of the remainder apply as to the division operator.

5.6.11   Exponentiation Operator

The exponentiation operator is not generally synthesisable. However, there are two constrained forms where it can be used. The first is where the left hand argument is 2, so it is possible to calculate 2^n. The second is where the right-hand argument is 2, so it is possible to calculate the square, x^2.

The calculation of 2^n is transformed into a shift operation, since this is equivalent to shifting the value 1 by n bits.

The calculation of the square, x^2, is done by mapping the exponent onto the multiplication operator:

```
x**2 = x * x
```

The appropriate multiplier is then used to form the square. Note that, even though the square can never be negative, the result will be a signed number if the argument is a signed number, since that is the behaviour of the multiplication operator.

5.7   Concatenation Operator

The concatenation operators allow an array to be built up out of smaller arrays and elements. There are in fact four concatenation operators defined for any one-dimensional array type. All return the array type; the difference between the operators is in the argument types that they take. There are four operators because there are four possible permutations of the array type and its element type. All these permutations are provided.

For example, type `bit_vector` has an element of type `bit`. Therefore, there are operators to concatenate two `bit_vector` operands, a `bit` with a `bit_vector`, a `bit_vector` with a `bit` and finally an operator to concatenate two `bit` elements, all producing a `bit_vector`.

In hardware terms, concatenation of arrays is equivalent to the merging of buses to form a larger bus. There is no circuitry involved, just wires. For example, consider the following fragment of VHDL:

```
signal a, b : bit_vector (3 downto 0);
signal z : bit_vector (7 downto 0);
...
z <= a & b;
```

This merges two 4-bit buses into an 8-bit bus. The ordering of the bits in the result is a left to right ordering, with the leftmost bit of a on the left of the result and the rightmost bit of b on the right of the result.

6

Synthesis Types

The basic set of types (Chapter 4) and operators (Chapter 5) provided with VHDL are limited when using the language for synthesis. For example, the integer types are limited to 32 bits and they are incapable of bitwise operations such as indexing and logic operations. The basic logic type `bit` is incapable of modelling metalogical values, particularly the `'Z'` value needed for modelling tristate buses.

For this reason a standard set of logical and numeric types have been added to the language that are particularly suited to RTL synthesis. They allow numbers of any size to be created (known as arbitrary-precision numeric types) and used to perform arithmetic operations. These types also provide a full set of logical, comparison, masking and other operations – they are intended to be used as universal types.

Initially, when work started on the synthesis types in the early 1990s, only basic logic types, arrays of those types for buses and arbitrary-precision numeric (integer) types were provided. However, the VHDL-2008 standard has extended this to arbitrary-precision fixed-point and floating-point types. Unfortunately, the new fixed- and floating-point types are provided in packages written in VHDL-2008 so are not compatible with most synthesis tools at the time of writing. Fortunately, a compatibility version of the packages are available, written in VHDL-1993 for use with tools that do not support the new standard yet.

These types together are referred to as the IEEE synthesis types or just the synthesis types. This chapter will explain how to use all of these types using a VHDL-1993-based synthesiser. Designs written this way will be compatible with the VHDL-2008 versions of the packages when they become available.

6.1 Synthesis Type System

The result of the standardisation effort is a collection of packages that between them implement all the synthesis types:

Package `std_logic_1164`
> A basic 9-value logic type with metalogical values suitable for modelling one-bit datapaths in both gate-level and RTL design.
> Also arrays of this type for modelling multi-bit datapaths and buses.

VHDL for Logic Synthesis, Third Edition. Andrew Rushton.
© 2011 John Wiley & Sons, Ltd. Published 2011 by John Wiley & Sons, Ltd.

Package `numeric_std`
> An arbitrary-precision numeric package providing `signed` (2's complement) and `unsigned` (magnitude) types.

Package `numeric_bit`
> An arbitrary-precision numeric package providing similar functionality to `numeric_std` but based on type `bit`. This package is rarely used and won't be discussed any further.

Package `fixed_float_types`
> Support package defining types used in the fixed-point and floating-point packages. Described where relevant in the sections on those packages.

Package `fixed_generic_pkg` (VHDL-2008 only)
> A generic package providing arbitrary-precision fixed-point arithmetic in signed (2's complement) and unsigned (magnitude) types. Generic parameters allow the default behaviour for overflow and underflow to be parameterised.

Package `fixed_pkg`
> An instantiation of `fixed_generic_pkg` with the generic parameters set to the most commonly used default values. Provides the `sfixed` and `ufixed` fixed-point types. May be replaced by a VHDL-1993 compatibility version of the package with the default behaviour hard-coded.

Package `float_generic_pkg` (VHDL-2008 only)
> A generic package providing arbitrary-precision floating-point types. Generic parameters allow the default behaviour to be parameterised.

Package `float_pkg`
> An instantiation of `float_generic_pkg` with the generic parameters set to the most commonly used default values. Provides the arbitrary-precision floating-point type `float` as well as subtypes `float32`, `float64` and `float128`. May be replaced by a VHDL-1993 compatibility version of the package with the default behaviour hard-coded.

This chapter will focus on the packages that are commonly used, available for use in a wide range of synthesis tools and that provide a coherent set of types. Thus, `numeric_bit` is not covered and only the VHDL-1993 compatibility versions of `fixed_pkg` and `float_pkg` will be described. The VHDL-2008 generic packages `fixed_generic_pkg` and `float_generic_pkg` will not be described. So, the packages and types that will be covered are:

```
package std_logic_1164
   type std_logic
   type std_logic_vector
package numeric_std
   type signed
   type unsigned
package fixed_pkg
   type sfixed
   type ufixed
package float_pkg
   type float
```

All of the synthesis types use arrays of `std_logic` to represent numbers. By using arrays, there is effectively no limit to the length of a number represented by these types. Arithmetic can be performed to any precision. Furthermore, bitwise logical and shift operators are provided.

Table 6.1   The synthesis type system

Type	Purpose
std_logic	one-bit paths such as clock and control lines
std_logic_vector	multi-bit paths with no numeric interpretation
signed	multi-bit paths with 2's-complement integer notation
unsigned	multi-bit paths with unsigned integer notation
sfixed	multi-bit paths with signed fixed-point notation
ufixed	multi-bit paths with unsigned fixed-point notation
float	multi-bit paths with floating-point notation
integer	used only for indexing of the multi-bit array types

Finally, buses can be combined using concatenation and split using slices or indexing of the array.

Note: the VHDL-1993 compatibility versions of fixed_pkg and float_pkg use std_ulogic instead of std_logic, which makes it impossible to have tristate buses of these types. In the VHDL-2008 versions the relationship between std_logic and std_ulogic has changed such that it will be possible to have fixed-point and floating-point tristate buses. For now though, tristate buses (Section 12.1) should be implemented using std_logic_vector.

These packages together provide a coherent *type system* of eight types listed in Table 6.1.

The synthesis type system provides practically all of the types needed for RTL modelling of any design. Furthermore, the types are compatible with each other and the packages provide type conversions between them, an important functionality that is lost if you use any other combination of packages.

One problem with the synthesis packages is that they do not necessarily follow the normal conventions for shift operations and for overflow. Nor are they entirely consistent between themselves in the conventions that they do use. These inconsistencies are potential pitfalls and will be described in the relevant sections of this chapter.

6.2   Making the Packages Visible

The synthesis packages are part of a wider range of packages provided by the VASG (VHDL Analysis and Standardisation Group) and all collected into a library with the name ieee.

However, because some of these packages use VHDL-2008 features that, at the time of writing, are not supported by most synthesis systems, compatibility versions have been produced by the package authors to fill the gap. If the standard packages are not available, the compatibility versions of the packages can be used instead, which is done by compiling them into a library called ieee_proposed. This alternative library is used partly because it is considered bad practice to compile non-standard versions of packages into library ieee and partly because some tools forbid write access to library ieee anyway. The name shows the original purpose of the library – to contain packages that are under development and proposed for standardisation. The following two sub-sections explain two scenarios: the first explains how to use the synthesis packages in their official version; the second explains the more likely scenario of how to use the compatibility packages.

Note that packages `std_logic_1164` and `numeric_std` are well established and will always be provided in library `ieee`. It is only the newer `fixed_pkg` and `float_pkg` that might be missing.

6.2.1 Scenario 1: Vendor-Provided VHDL-2008 Packages

If the simulator and synthesiser you are using support all the synthesis packages, they will be found in library `ieee`. To use them, there needs to be a `library` clause to make the library visible and a `use` clause to make the contents of the appropriate package visible. Furthermore, when you use a package such as `numeric_std`, the packages that it in turn uses are not inherited, so it is necessary to explicitly `use` them as well.

Therefore, to use the package `numeric_std` to provide numeric types only, the following declarations will be needed before the entity or architecture that is to use the packages:

```
library ieee;
use ieee.std_logic_1164.all;
use ieee.numeric_std.all;
```

Similarly, to use `fixed_pkg` you must also use the `std_logic_1164, numeric_std` and `fixed_float_types` packages:

```
library ieee;
use ieee.std_logic_1164.all;
use ieee.numeric_std.all;
use ieee.fixed_float_types.all;
use ieee.fixed_pkg.all;
```

Finally, to add the floating-point types to this needs all of these use clauses plus `float_pkg`:

```
library ieee;
use ieee.std_logic_1164.all;
use ieee.numeric_std.all;
use ieee.fixed_float_types.all;
use ieee.fixed_pkg.all;
use ieee.float_pkg.all;
```

So this is the complete set of context clauses needed to use the whole synthesis type system.

6.2.2 Scenario 2: Using VHDL-1993 Compatibility Packages

At the time of writing, most VHDL tools do not provide the fixed-point and floating-point packages. In that case, and only if the tool is missing these packages, you can still use them by downloading the VHDL-1993 compatibility versions from the EDA Industry Working Groups Website [EDA, 2009]. There are slightly different versions for different tools. The packages should then be compiled into a new library called `ieee_proposed`. This library is a temporary location for compatibility versions that will eventually be replaced by VHDL-2008 versions in library `ieee`.

To use the compatibility versions in a design you will need a modified form of context clause before each design unit:

```
library ieee;
use ieee.std_logic_1164.all;
use ieee.numeric_std.all;
library ieee_proposed;
use ieee_proposed.std_logic_1164_additions.all;
use ieee_proposed.numeric_std_additions.all;
use ieee_proposed.fixed_float_types.all;
use ieee_proposed.fixed_pkg.all;
use ieee_proposed.float_pkg.all;
```

Note: the 'additions' packages add some of the VHDL-2008 features to the VHDL-1993 versions of the `std_logic_1164` and `numeric_std` packages. The features added are selecting boolean operators and reduction boolean functions, and a few other functions and operators to make the synthesis packages more consistent.

The rest of the chapter will use the standard packages from library `ieee`, with the understanding that, if you are using the compatibility packages, you need to modify the package and use clauses to use library `ieee_proposed` for the new packages only.

6.2.3 VHDL-2008 Context Declarations

In VHDL-2008 there is a new design unit called a *context declaration* that can be used as a short-cut to include all of these packages in one go. If you have context declarations available, you can gather all these context clauses into one context declaration.

The VHDL-2008 standard pre-defines a context for using the `numeric_std` package:

```
context ieee_std_context is
  library ieee;
  use ieee.std_logic_1164.all;
  use ieee.numeric_std.all;
end;
```

This context is also to be found in library `ieee`.

Then, in the design unit where you want to use the numeric types, you just use a single context clause:

```
library ieee;
context ieee.ieee_std_context;
entity ...
```

Unfortunately, the VHDL-2008 standard does not specify a predefined fixed-point or floating-point context.

A context declaration can be written that contains all the packages needed for the numeric-type system and compiled into the current work library. In this example it will be called `synthesis_types` since it provides all of the synthesis types.

The declaration for `synthesis_types` in the case where you have built-in versions of the numeric packages is:

```
context synthesis_types is
  library ieee;
  use ieee.std_logic_1164.all;
```

```
    use ieee.numeric_std.all;
    use ieee.fixed_float_types.all;
    use ieee.fixed_pkg.all;
    use ieee.float_pkg.all;
  end;
```

The context declaration is a design unit – you write it in a separate design file (for example, `synthesis_types.vhdl`) and compile it into your work library (or any library in your design).

If you are using the compatibility packages, this needs a slightly different set of context clauses:

```
  context synthesis_types is
    library ieee;
    use ieee.std_logic_1164.all;
    use ieee.numeric_std.all;
    library ieee_proposed;
    use ieee_proposed.std_logic_1164_additions.all;
    use ieee_proposed.numeric_std_additions.all;
    use ieee_proposed.fixed_float_types.all;
    use ieee_proposed.fixed_pkg.all;
    use ieee_proposed.float_pkg.all;
  end;
```

Then, in the design unit where you want to use the synthesis types, you just use a single context clause:

```
  context work.synthesis_types;
  entity ...
```

If you are using the compatibility packages and an update to your development tools means the VHDL-2008 versions subsequently become available in library ieee, you only need to edit the context declaration and recompile it. The rest of your code will remain unchanged.

Note: The downside of context declarations is that they may not be available yet in both your simulator and synthesiser, in which case you simply cannot use them. At the time of writing context declarations were not available in any of the tools used to test the examples in this book. So check carefully before trying to use them.

6.3   Logic Types – Std_Logic_1164

The basic one-bit type that underlies all of the synthesis types is `std_ulogic`. This is a nine-value logic type. It was originally designed for gate-level simulation and supports such features as resistive pullups and pulldowns. It has been adopted as the standard type for synthesis models. There is also a subtype of `std_ulogic` called `std_logic` (note the missing 'u') which has the same logic behaviour but is *resolved* so that it can also be used for tristate signals. The subtype `std_logic` is almost universally used for all operations and datapaths, not just tristate operations, and the unresolved basetype `std_ulogic` is rarely used directly.

Note: there is an alternative school of thought that `std_ulogic` and arrays of it should be used throughout a design and `std_logic` should only be used for tristate buses. The arguments between these two schools of thought seem to have been going on forever and have become

Table 6.2 Std_Logic_1164 types

Type	Class	Synthesisable
std_ulogic	multi-value logic type	yes
std_logic	resolved subtype of std_ulogic	yes
std_ulogic_vector	array of std_ulogic	yes
std_logic_vector	array of std_logic	yes

extremely tiresome. There isn't a right answer, since both schools have pros and cons, but the std_logic convention described in this book is the most common convention followed by most designers using the synthesis packages, is therefore the tried and tested approach least likely to cause any problems with any VHDL tools, and is therefore recommended.

The std_logic type was not originally part of the VHDL language, but was an IEEE standard extension to the language under standard number 1164 [IEEE-1164, 1993]. It exists in a library called ieee, in a package called std_logic_1164, and it is listed in Appendix A.3. The 2008 update to the VHDL standard incorporates the std_logic_1164 package into the main VHDL standard, confusingly still with the old standard number 1164 in the name.

The full set of types in std_logic_1164 is listed in Table 6.2.

The std_logic_1164 package has been extended in VHDL-2008. The extensions are available for use with earlier versions and are found in the compatibility package std_logic_1164_additions.

6.3.1 std_logic – One-Bit Logical Type

The basic logical type defined in std_logic_1164 is std_ulogic. This is defined as a character enumeration type with nine values. The literals are character literals that are case-sensitive and are all uppercase for this type:

```
type std_ulogic is ('U','X','0','1','Z','W','L','H','-');
```

The meanings of the nine values are explained in Table 6.3.

The use of multi-valued logic types is full of potential pitfalls. The majority of them fall into the category of using metalogical values as if they were real values, for example,

Table 6.3 The meanings of the std_logic values

Value	Meaning	Synthesisable
'U'	Uninitialised	no
'X'	Forcing Unknown	no
'0'	Forcing 0	yes
'1'	Forcing 1	yes
'Z'	High Impedance	yes
'W'	Weak Unknown	no
'L'	Weak 0	no
'H'	Weak 1	no
'-'	Don't care	no

assigning a metalogical value such as the weak driving value 'L' or an unknown 'X' to a
signal. Synthesis will either treat these as errors (the safest interpretation) or map them onto
one of the two real values. This quite arbitrary mapping may result in a subtle change in the
behaviour of the circuit.

Note that the don't *care* value in std_logic is '-'. The value 'X' means don't *know*, in
other words the signal has a value but that value cannot be determined. This is not the same as
don't care. In a sense, the '-' character relates to the definition of the required circuit
behaviour, whereas 'X' relates to the observed behaviour.

For example, if there *was* don't care handling in the language, then the following test for
equality using std_logic would always be true:

```
if s = '-' then...
```

In fact, the language interprets this, not as a don't care match with any value, but as a test for an
exact match with the literal '-'. In a synthesised circuit, only the real values '0' and '1'
exist, so the test always evaluates to false.

It is not the user's job to specify don't cares, since most don't care information can be
automatically generated by the synthesis tool. It is very difficult to fully simulate logic
containing don't care information, because all the potential permutations of real values and
don't cares must be covered. If the system is not fully simulated in this way, it may contain
subtle errors that will result in the synthesiser changing the behaviour of the circuit because of
incorrectly specified don't care information. This kind of error is notoriously difficult to find
and correct. In general, therefore, it is recommended that don't care behaviour is *not* specified
in the RTL model.

The column in Table 6.3 labelled 'Synthesisable' indicates which values should be used in
a synthesisable design. There are only three values marked as synthesisable. This is because the
safest rule in using std_logic for synthesis is to use it as if it was type bit. That is, not to
refer to the metalogical values at all. Assignments and comparisons of literal values should only
refer to the '0' and '1' values. The sole exception is the high-impedance value 'Z' that
should only ever be used in specifying tristate drivers (see Section 12.1).

Tristate buses require the use of a multi-valued logic type that can model a high-impedance value
and that can also handle multiple drivers driving the same bus. In VHDL terms, this means that the
type must be *resolved*, that is, capable of being the target of more than one signal assignment.

```
subtype std_logic is resolved std_ulogic;
```

Type std_ulogic models high-impedance values as the value 'Z', but only the subtype
std_logic is resolved in a way that models tristate drivers. This is why std_logic is
chosen as the universal logic type.

6.3.2   *std_logic_vector – Multi-Bit Logical Type*

The std_logic_1164 package also provides unconstrained arrays of both std_ulogic and
std_logic:

```
type std_ulogic_vector is
  array (natural range <>) of std_ulogic;
```

```
type std_logic_vector is
  array (natural range <>) of std_logic;
```

So, a `std_ulogic_vector` can be used to model multi-bit datapaths, whereas `std_logic_vector` can be used to model both multi-bit datapaths and tristate buses.

As mentioned earlier, there are two schools of thought regarding whether to predominantly use `std_ulogic` or `std_logic`. The `std_ulogic` school of thought is that `std_ulogic` and `std_ulogic_vector` should be used most of the time and `std_logic` and `std_logic_vector` only used to model tristate buses. The `std_logic` school of thought is that only two of these types should be used – `std_logic` and `std_logic_vector` – for all paths.

The `std_logic` school is the dominant one and is recommended practice. Following this convention means that only `std_logic` and `std_logic_vector` are used. This is the convention used throughout this book.

6.3.3   Operators

The package defines a small set of operators for the `std_logic` and `std_logic_vector` logical types.

The set of operators available for these types is:

comparison:	`=, /=, <, < =, > , >=`
boolean:	`not, and, or, nand, nor, xor, xnor`
shift:	`srl, sll, rol, ror`
concatenation:	`&`

6.3.4   Comparison Operators

The logical types have the complete set of comparison operators:

comparison: `=, /=, <, < =, > , >=`

Bearing in mind that `std_logic` is interpreted in synthesis as a one-bit logical type, the only comparison operators that have any use are the = and /= operators. Performing ordering operations does not make sense and these operators are only present because the language defines them automatically for an enumeration type.

Similarly, the array type `std_logic_vector` has the full set of comparison operators, but only the = and /= operators have useful meaning. Bear in mind the problems with the ordering operators on array types that was discussed in Section 5.4. However, the synthesisable numeric types have ordering operators that do give correct results. These will be discussed in the relevant sections for each type later in this chapter.

6.3.5   Boolean Operators

The logical types have the full set of basic boolean operators:

boolean: `not, and, or, nand, nor, xor, xnor`

The boolean operators on types `std_logic` and `std_logic_vector` work in exactly the same way as for type `bit` as described in Section 5.3.

Basic boolean operators perform bitwise logic on each element of two parameters of the same size to produce a result of the same size. This set of operators is provided in all versions of the `std_logic_1164` package.

Selecting boolean operators combine a single-bit input with each element of an array to produce an array of the same size. They are provided in the VHDL-2008 version of package `std_logic_1164` and are not part of the original package. However, they are provided in the `std_logic_1164_additions` VHDL-1993 compatibility package.

Reducing boolean operators combine all the elements of an array to produce a single-bit output. They are provided as operators in the VHDL-2008 version of the package, but as reduction *functions* (i.e. `and_reduce` etc.) in the `std_logic_1164_additions` VHDL-1993 compatibility package.

These bitwise logical operators are very convenient for performing masking operations. For example, to write a conditional signal assignment that uses a test on whether any one of the four least significant bits of an 8-bit signal (c) are set, the following comparison can be used:

```
z <= a when (c and "00001111") /= "00000000" else b;
```

The mask in this case is represented by a string value, which must be the same length as the signal being masked because this is a requirement of the logical operators. The logical result of the `and` operation, also an 8-bit result, is then compared with a string of zeros, which also needs to be the same length as the masked result, because the comparison operators for `std_logic_vector` only work on same-length operands.

6.3.6 Shift Operators

The original `std_logic_vector` type had no shift operators.

However, the VHDL-2008 version has a partial set of shift operators:

    shift: srl, sll, rol, ror

These have the following meanings:

    sll shift-left logical
    srl shift-right logical
    rol rotate-left
    ror rotate-right

These are provided in the `std_logic_1164_additions` VHDL-1993 compatibility package for use with earlier versions.

When available, these operators are defined to have the same functionality as the built-in operators for bit, described in Section 5.5.

The logical shifts simply shift the operand, discarding bits that are shifted off one end and filling the other end with `'0'`.

The rotate operators take elements off one end of the array and shift them in at the other end. In left shifts, elements are shifted off the left end and in at the right, whilst in right shifts, elements are shifted off the right end and in at the left.

6.4 Numeric Types – Numeric_Std

Package `numeric_std` provides arbitrary-precision numeric (integer) types with a full set of arithmetic, comparison, logical and shifting operators.

The `numeric_std` package has been extended in VHDL-2008. The extensions are available for use with earlier versions and are found in the compatibility package `numeric_std_additions`.

6.4.1 Types Provided

The package `numeric_std` defines two types, both of which are unconstrained arrays of the element type, `std_logic`. The types are called `unsigned` and `signed`:

```
type signed is array (natural range <>) of std_logic;
type unsigned is array (natural range <>) of std_logic;
```

Note that the index type is `natural`, which means that you cannot use negative array indices. So a `signed(7 downto 0)` is legal, but `signed(0 downto -7)` is not.

Type `signed` represents signed numbers using 2's-complement notation, whilst type `unsigned` represents unsigned magnitude notation. The arithmetic operators provided with the package of course implement these notations.

The range of a number is defined by the number of bits used in its array representation. For example, here are two examples of 8-bit signals defined using the two types.

```
signal a : signed(7 downto 0);
signal b : unsigned(7 downto 0);
```

Both declarations create 8-bit buses. However, the first declaration creates a bus that will be interpreted as a signed number with the range −128 to 127, whereas the second, although exactly the same size, will be interpreted as an unsigned number with the range 0 to 255.

This is the convention used throughout the numeric packages – the type determines how a bus is interpreted and therefore determines which operators will be used. A signed type will get signed arithmetic operators, whereas an unsigned type will get unsigned arithmetic operators.

The use of two completely separate types means that there is never any confusion over the interpretation of a particular bit pattern. The types are also completely distinct from `std_logic_vector`, which does not have a numeric interpretation, so there is no possible confusion there either. To make use of the extended operator set provided by the package, signals or variables must be of type `unsigned` or `signed`.

In order to use the packages effectively, it is important to understand how the bit-array representation is interpreted by the package.

There is only one rule:

    1. Arrays are read from left to right, with the most-significant bit on the left. This will always be true, regardless of the range or direction of the array indices.

In addition to this rule, there are also some conventions that help to make VHDL models clear and easy to understand. Correct results will be obtained if these conventions are not followed, but nevertheless, you are strongly recommended to follow them:

    2. Use a descending range. This gives the most-significant bit (the left-most bit) the highest index.

    3. The least-significant bit should be numbered 0.

So, for example:

```
signal a : signed(7 downto 0);
```

This follows the conventions above, which means that the left-most bit is the most-significant bit and has index 7. The right-most bit is the least-significant bit and has index 0.

All the examples in this book follow these rules and conventions.

6.4.2   Resize Functions

The package defines two functions, both called `resize`, which are used to truncate and extend values of type `signed` and `unsigned`. These can be used, for example, to assign an 8-bit number to a 16-bit number.

To give an example of how to use a `resize` function, here is such an example of an 8-bit `signed` signal being assigned to a 16-bit signal:

```
library ieee;
use ieee.std_logic_1164.all;
use ieee.numeric_std.all;
entity signed_resize_demo is
  port (a : in signed(7 downto 0);
        z : out signed(15 downto 0));
end;
architecture behaviour of signed_resize_demo is
begin
  z <= resize(a, 16);
end;
```

The `signed resize` function takes two arguments. The first is the signal or variable to resize (in this case the signal a) and the second is the size to resize it to. This must be a constant value if it is to be synthesisable, since a synthesiser must know what size the result is in order to create a bus with the right number of bits. It does this by examining the second argument's value.

In this example, the signal is being resized to a larger size. The function does this by sign-extending the argument, as shown in Figure 6.1.

The other `resize` function, which operates on `unsigned` numbers, is almost identical to the `signed resize` function and is used in exactly the same way:

```
library ieee;
use ieee.std_logic_1164.all;
use ieee.numeric_std.all;
```

```
entity unsigned_resize_demo is
  port (a : in unsigned(7 downto 0);
        z : out unsigned(15 downto 0));
end;
architecture behaviour of unsigned_resize_demo is
begin
  z <= resize(a, 16);
end;
```

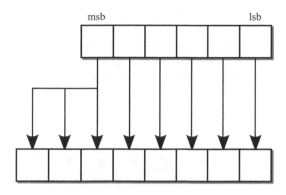

Figure 6.1   Signed resize to a larger size.

The main difference is that it resizes unsigned values to a larger size by zero-extension, as illustrated in Figure 6.2.

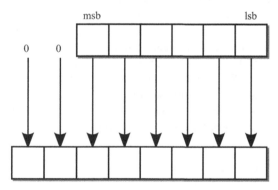

Figure 6.2   Unsigned resize to a larger size.

The resize functions also work the other way round, to make a smaller value by truncation of a large one. Thus, the opposite of the above signed example is:

```
library ieee;
use ieee.std_logic_1164.all;
use ieee.numeric_std.all;
entity signed_resize_demo2 is
  port (a : in signed(15 downto 0);
        z : out signed(7 downto 0));
end;
```

```
architecture behaviour of signed_resize_demo2 is
begin
  z <= resize(a, 8);
end;
```

This example reduces a 16-bit signal to an 8-bit result and assigns this to the output port. Reduction is carried out by the resize function by truncating the most-significant bits of the argument, but keeping the sign bit. In other words, in this case, bits 14 downto 7 would be discarded. Figure 6.3 illustrates the action of signed resize to a smaller size.

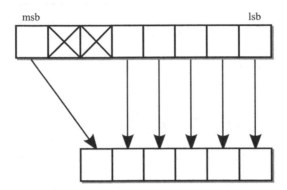

Figure 6.3 Signed resize to a smaller size.

Note: this is unusual behaviour for truncation. Usually signed truncation is done by just discarding the most-significant bits, allowing the result to change sign if necessary. In other words the normal convention is for signed numbers to wrap round if they overflow as a result of the truncation, so that the largest negative number is adjacent to the largest positive number. For some reason, the numeric_std package does not follow this convention for resizing the signed type. Beware! Conventional resizing can be achieved by slicing the source, as will be illustrated later in this section.

The final combination is the resize of a large unsigned number to a smaller size, carried out by the unsigned version of the resize function:

```
library ieee;
use ieee.std_logic_1164.all;
use ieee.numeric_std.all;
entity unsigned_resize_demo2 is
  port (a : in unsigned(15 downto 0);
        z : out unsigned(7 downto 0));
end;
architecture behaviour of unsigned_resize_demo2 is
begin
  z <= resize(a, 8);
end;
```

The unsigned version of resize simply discards the most-significant bits, since there is no sign bit. This behaviour is illustrated by Figure 6.4.

This is the normal behaviour for truncation. It is only the signed resize that has unexpected behaviour.

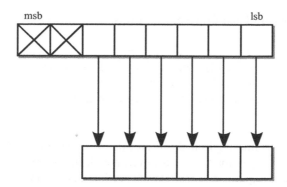

Figure 6.4   Unsigned resize to a smaller size.

Of course, it is not necessary to use these functions to achieve these results, but they are a considerable convenience. Probably the only operation where it makes sense to bypass the `resize` functions is where conventional truncation is required for a `signed resize` to a smaller size. In other words, where a `signed` number is to be simply truncated by the removal of its most-significant bits. This can be achieved most simply by using the slice operation which is available for use on any array type:

```
library ieee;
use ieee.std_logic_1164.all;
use ieee.numeric_std.all;
entity signed_slice_demo is
  port (a : in signed(15 downto 0);
        z : out signed(7 downto 0));
end;
architecture behaviour of signed_slice_demo is
begin
  z <= a(7 downto 0);
end;
```

A common convention is to use the length attribute of the target to specify the size:

```
architecture behaviour of unsigned_resize_demo2 is
begin
  z <= resize(a, z'length);
end;
```

This guarantees that the result of the resize is the right size and adjusts automatically if the circuit is redesigned to use a different datapath width.

6.4.3   Operators

The package defines a set of operators for both types. Furthermore, there are guidelines on how to interpret and use the types to get the correct functionality from the packages.

The set of operators available for type `signed` is:

comparison:	=, /=, <, <=, >, >=
boolean:	not, and, or, nand, nor, xor, xnor
shift:	srl, sll, rol, ror
arithmetic:	sign -, abs, +, -, *, /, mod, rem
concatenation:	&

In fact, the only operators missing from this list are the sign operator "+" that would have no effect, and the exponent operator "**". However, not all the operators in this list are fully synthesisable as will be explained in the relevant section that follows.

Type `unsigned` has a slightly more restricted set of operators, in that the sign operator "-" and the `abs` operator are not present, since they are meaningless for unsigned arithmetic.

6.4.4 Comparison Operators

The numeric types have the complete set of comparison operators:

comparison: =, /=, <, <=, >, >=

The comparison operators for `signed` and `unsigned` have been defined to give correct numeric comparisons. Furthermore, they have been defined to give correct comparisons even with arguments of different lengths. This makes the comparison operators extremely easy and natural to use.

As an example of the usage of comparison operators, consider the following entity that returns the maximum of two `signed` values.

```
library ieee;
use ieee.std_logic_1164.all;
use ieee.numeric_std.all;
entity max is
   port (a, b : in signed(7 downto 0);
         z : out signed(7 downto 0));
end;
architecture behaviour of max is
begin
   z <= a when a > b else b;
end;
```

This example shows how the greater-than operator can be used to compare two parameters of type `signed`. All the other comparison operators can be used in the same way.

6.4.5 Boolean Operators

The numeric types have the full set of boolean operators:

boolean: not, and, or, nand, nor, xor, xnor

The boolean operators on types `signed` and `unsigned` work in exactly the same way as on type `std_logic_vector`.

Basic boolean operators perform bitwise logic on each element of two parameters of the same size to produce a result of the same size. This set of operators is provided in all versions of the `std_logic_1164` package.

Selecting boolean operators combine a single-bit input with each element of an array to produce an array of the same size. They are provided in the VHDL-2008 version of package `numeric_std` and are not part of the original package. However, they are provided in the `numeric_std_additions` VHDL-1993 compatibility package.

Reducing boolean operators combine all the elements of an array to produce a single-bit output. They are provided in the VHDL-2008 version of the package, but reduction *functions* (i.e. `and_reduce`, etc.) are provided in the `numeric_std_additions` VHDL-1993 compatibility package.

6.4.6 Shift Operators

The original `std_logic_vector` type has a partial set of shift operators:

>   shift: `srl, sll, rol, ror`

The VHDL-2008 version extends these to the full set:

>   shift: `sra, sla, srl, sll, rol, ror`

The arithmetic shifts for earlier versions are provided in the `numeric_std_additions` compatibility package.

The meaning of the operators as defined for package `numeric_std` is explained in Table 6.4.

The operators all take two arguments, the first being the value, signal or variable to shift and the second being an integer specifying how far to shift the value. If the shift distance is a constant value the synthesis interpretation is a hardwired rearrangement of the bits of the bus with no hardware overhead. If the shift distance is non-constant, it is implemented by a barrel shifter circuit.

The result of the shift is the same size as the left argument, that is the value being shifted.

The logical shifts simply shift the operand, discarding bits that are shifted off one end and filling the other end with `'0'`.

Table 6.4 Shift operators

Operator	Name	Description
sll	Shift Left Logical	discard left bits, zero fill on the right
srl	Shift Right Logical	discard right bits, zero fill on the left
sla	Shift Left Arithmetic	discard left bits, zero fill on the right
sra	Shift Right Arithmetic	discard right bits: signed: sign extend on the left unsigned: zero extend on the left
rol	Rotate Left	remove bits from left, reinsert on right
ror	Rotate Right	remove bits from right, reinsert on left

The arithmetic shifts are the same as the logical shifts when shifting left or shifting an unsigned value right, but sign-extend the result when shifting a signed value right.

The rotate operators take elements off one end of the array and shift them in at the other end. In left shifts, elements are shifted off the left end and in at the right, whilst in right shifts, elements are shifted off the right end and in at the left.

6.4.7 Arithmetic Operators

The numeric types have a restricted set of arithmetic operators:

arithmetic: `sign -, abs, + , -, *, /, mod, rem`

There are no + sign or ** operators. Type unsigned does not have the sign "-" or the abs operators.

Type signed has been given a set of operators that performs 2's-complement integer arithmetic, whilst type unsigned has been given a set of operators that performs magnitude arithmetic.

The operators have also been defined so that, unlike integer arithmetic and unlike the signed resize function, the numbers wrap-round on overflow. For example, adding one to the most positive signed number gives the most negative signed number. Similarly, adding one to the most positive unsigned number gives zero.

The unary (sign) "-" operator forms the 2's-complement negation of a number. It gives a result that is the same size as the argument, so that it is possible to negate a number and assign it back to itself, or to another signal of the same size. For example, to negate a signal:

```
library ieee;
use ieee.std_logic_1164.all;
use ieee.numeric_std.all;
entity negation_demo is
  port (a : in signed(7 downto 0);
        z : out signed(7 downto 0));
end;
architecture behaviour of negation_demo is
begin
  z <= -a;
end;
```

Of course, the 2's-complement integer range is asymmetrical, so the negation of the most negative value will overflow and therefore wrap-round. This means in practice that the negation of the most negative number is itself! This strange behaviour could have been avoided in the package design by returning a result that was one bit bigger, but this would have sacrificed the characteristic of these operators that they preserve the size and type of the datapath. If the extra bit is needed, then the way to achieve this is to combine the negation operator with the resize function:

```
library ieee;
use ieee.std_logic_1164.all;
use ieee.numeric_std.all;
entity negation_resize_demo is
  port (a : in signed(7 downto 0);
        z : out signed(8 downto 0));
```

```
end;
architecture behaviour of negation_resize_demo is
begin
  z <= -resize(a, 9);
end;
```

Note that the resize must happen *before* the negation.

The abs operator is only available for the signed type. It forms the absolute value by negating the argument if it is negative. The result will be the same size as the argument, which means that it has the same characteristics as the negation operator in that the negation of the most negative value overflows and therefore wraps round onto the most negative number. This gives the rather strange side effect that the result of the abs operator can be negative in this one special case. Once again the solution if this is to be avoided is to resize to a size at least one bit larger than the operand before taking the absolute value.

Alternatively, the result can be type converted to an unsigned value of the same size, in which case the range is sufficient to store all the values:

```
library ieee;
use ieee.std_logic_1164.all;
use ieee.numeric_std.all;
entity abs_demo is
  port (a : in signed(7 downto 0);
        z : out unsigned(7 downto 0));
end;
architecture behaviour of abs_demo is
begin
  z <= unsigned(abs a);
end;
```

Surprisingly, this does give the correct result, even for the most negative value. For example, the bit pattern "10000000", representing -128, will be mapped onto -128 by the abs operator, but the type conversion will then map it onto the unsigned value "10000000", representing $+128$.

The add and subtract ("+" and "-") operators also preserve the length of the datapath; if two 8-bit numbers, for example, are added together, then the result will also be an 8-bit wide number. This means that addition and subtraction wrap-round on overflow. The following example shows an 8-bit addition:

```
library ieee;
use ieee.std_logic_1164.all;
use ieee.numeric_std.all;
entity add_demo is
  port (a, b : in signed(7 downto 0);
        z : out signed(7 downto 0));
end;
architecture behaviour of add_demo is
begin
  z <= a + b;
end;
```

If it is not desirable for the addition to wrap-round, in other words, if the result is required to be one bit larger than the arguments, then the arguments should be resized before they are added:

```
library ieee;
use ieee.std_logic_1164.all;
use ieee.numeric_std.all;
entity add_resize_demo is
  port (a, b : in signed(7 downto 0);
        z : out signed(8 downto 0));
end;
architecture behaviour of add_resize_demo is
begin
  z <= resize(a,9) + resize(b,9);
end;
```

In fact, these operators have been defined so that they can be used on arguments that are different sizes. In this case, the shorter argument is extended to the size of the larger argument before the add, giving a result size that is the same as the larger of the arguments. For example, if an 8-bit number is added to a 16-bit number, the result is a 16-bit number.

The multiplication operator ("* ") is defined for both signed and unsigned types such that they can take arguments of different lengths. For example, an 8-bit number can be multiplied by a 16-bit number. The result size is calculated by adding the sizes of the arguments, so in this example, the result would be 24 bits long. This makes the multiplication operator the exception to the general rule that the numeric operators preserve the datapath width. It also means that multiplication never overflows and therefore never wraps round the result, so it should never be necessary to resize arguments before multiplication to avoid losing overflow bits. However, it often means that the result will need to be either truncated by slicing or by resizing.

The following example shows how two 8-bit signed numbers can be multiplied to give an 8-bit result with no overflow handling and without preserving the sign of the result. It uses a slice rather than the resize function to avoid the unusual behaviour when truncating. If the sign was required to be preserved, the slice would be replaced by a resize operation.

```
library ieee;
use ieee.std_logic_1164.all;
use ieee.numeric_std.all;
entity multiply_demo is
  port (a, b : in signed(7 downto 0);
        z : out signed(7 downto 0));
end;
architecture behaviour of multiply_demo is
  signal product : signed(15 downto 0);
begin
  product <= a * b;
  z <= product(7 downto 0);
end;
```

In fact, this can be expressed in a single assignment to signal z.

```
z <= (a * b)(7 downto 0);
```

This simplified form relies on the multiplication returning an array with a conventional descending range ending at zero. In theory, an operator can return any range and using such a slice is making a dangerous assumption. With other, less well-defined packages it would not be safe to do this. However, all the arithmetic operators in numeric_std have been defined so that they must return a normalised range.

This means that it is safe to slice (or index) the return value of any of the arithmetic operators.

The division operators (`"/"`, `mod` and `rem`) are all synthesisable with the same constraints discussed for integer types in Section 5.6. The convention adopted by the numeric packages is that the result size is the same as the first argument for division, but the same as the second argument for modulo and remainder. This makes no difference if the arguments are the same size, for example, dividing a 16-bit value by a 16-bit value gives a 16-bit result for all three operators. However, dividing a 32-bit value by a 16-bit value will give a 32-bit result for division, but a 16-bit result for modulo and remainder.

6.5   Fixed-Point Types – Fixed_Pkg

Package `fixed_pkg` provides arbitrary-precision fixed-point arithmetic. It is one of the new packages provided in VHDL-2008 and didn't exist before that. Nevertheless, it has a VHDL-1993 compatibility version for use on VHDL systems that do not provide it.

A common mistake for designers coming from a more software-orientated background is to assume that floating-point is the only solution to representing wide-range and fractional numbers. However, floating-point arithmetic requires a lot of circuitry and a lot of calculation time. It can be argued that floating-point operations should only be performed by a specialist floating-point unit (FPU) which is integrated into an ASIC or FPGA alongside the programmable logic.

Fixed-point arithmetic is an intermediate form of arithmetic between integer and floating-point arithmetic. It generates circuits that are typically the same size or slightly larger than integer arithmetic with the same delay or only slightly slower. Fixed-point arithmetic allows fractions to be represented without incurring the size and slowness of floating-point arithmetic that is typically 3–4 times the area and 2–3 times the delay of fixed-point.

Fixed-point arithmetic is commonly used in Digital Signal Processing (DSP) applications but can be applied to many areas of hardware design. A key skill in hardware design, especially in DSP, is in converting a specification (which may be a software model) expressed in floating-point into a fixed-point implementation in order to reduce the chip size and increase the performance, without introducing too much calculation error (i.e. noise) due to the limited dynamic range of fixed-point types.

Traditionally, designers have implemented fixed-point arithmetic using integer arithmetic and manually kept track of the position of the binary point for each datapath. This can be difficult to get right and is a potential source of design errors. The fixed-point package allows the position of the binary point to be specified in the design and visualised in simulation output.

Fixed-point arithmetic has a similar implementation to integer arithmetic, with a little extra circuitry to support rounding and overflow modes. It is possible to choose these modes such that there is no overhead. The fixed-point package is largely implemented using `numeric_std` and therefore benefits from the optimisations that synthesis tools already provide for that package.

6.5.1   Types Provided

The package `fixed_pkg` defines two main types, both of which are unconstrained arrays of the element type, `std_logic`. The types are called `sfixed` and `ufixed`:

```
type sfixed is array (integer range <>) of std_logic;
type ufixed is array (integer range <>) of std_logic;
```

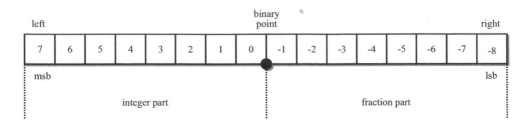

Figure 6.5 Fixed-point storage format.

Type `sfixed` represents signed fixed-point numbers using 2's-complement notation, whilst type `ufixed` represents unsigned fixed-point numbers using a simple magnitude notation.

Note that the range of the type uses integer, not natural as with `signed` or `unsigned`. This means that the range can include negative indices.

Note: In the full VHDL-2008 version, these are resolved arrays that can implement tristate buses. However, in the compatibility packages only, these types are based on `std_ulogic`, not `std_logic` so they cannot implement tristate buses. This is a strange design decision because it makes the two versions of the package slightly incompatible. It is doubly strange because package `numeric_std` is based on arrays of `std_logic` and so it would make sense for the fixed-point packages to use the same convention. It is triply strange because the most common convention amongst designers is to use arrays of `std_logic` throughout a design.

The fixed-point notation requires signals and variables to be declared with a descending range and places the binary point after element 0 of the array, with bits to the left with natural indices representing the integer part of the number and bits to the right with negative indices representing the fraction part. Figure 6.5 illustrates this for a fixed-point number with 8-bit integer part and an 8-bit fraction part.

The range of a fixed-point number is defined by the number of bits used in its array representation and by the offset of the binary point. For example, here are two examples of 16-bit signals with equal-sized integer and fraction parts defined using the two types.

```
signal a : sfixed(7 downto -8);
signal b : ufixed(7 downto -8);
```

Both declarations create 16-bit buses. Both have 8-bit integer parts (7 downto 0) and 8-bit fractional parts (-1 downto -8). However, the first declaration creates a bus that will be interpreted as a signed number with the range -128.000 to 127.996 (i.e. 127 and 255/256ths), whereas the second, although exactly the same size, will be interpreted as an unsigned number with the range 0.000 to 255.996 (i.e. 255 and 255/256ths).

The resolution of a fixed-point number is the difference between two adjacent numbers. For this example, with an 8-bit fractional part, the resolution is 1/256 (i.e. 2^{-8}).

The use of two completely separate types for signed and unsigned fixed-point values means that there is never any confusion over the interpretation of a particular bit pattern. The types are also completely distinct from `std_logic_vector` and the `numeric_std` types, so there is no possible confusion there either.

In order to use the packages effectively, it is important to understand how the bit-array representation is interpreted by the package.

There are several rules:

1. Arrays are read from left to right, with the most-significant bit on the left.
2. You must use a descending range. This gives the most-significant bit (the left-most bit) the highest index.
3. The integer part must have natural indices (i.e. descending to 0).
4. The fractional part must have negative indices (i.e. descending from -1).

These rules are followed in all the examples.

Note in particular that it is illegal to declare a fixed-point number with an ascending range and such use will result in an error being raised during simulation. This is different from `signed` and `unsigned` where the use of a descending range was merely a recommended convention and ascending ranges could be used in principle. With the fixed-point types it is a rule that ranges must be descending.

A convenient shorthand for describing a fixed-point type is to refer to it by the number of bits before and after the binary point. In this shorthand, the examples above are referred to as '8.8 bit' signed and unsigned fixed-point types. That is, they have 8 bits before the point and 8 bits after it. Note that the notation gives the number of bits, not the bounds. So an 8.8-bit number has the bounds `7 downto -8`. The upper bound is always one less than the number of bits in the integer part.

Note: there is another notation for fixed-point numbers that doesn't count the sign bit in the integer part of signed numbers, so would describe the signed example above as 7.8 bit, while still calling the unsigned example 8.8 bit. This convention is confusing and unnecessarily obfuscates the issue. In this book, the notation that includes the sign bit (i.e. 8.8-bit signed) is used throughout.

6.5.2 Overflow and Underflow Modes

There are two ways in which a fixed-point calculation can introduce inaccuracies in the result. The calculation can overflow if the result is too large to be represented by the result type. Or the calculation can underflow if there aren't enough fraction bits to represent the result accurately.

6.5.2.1 Overflow

A fixed-point type can represent a limited range of values. For example, an 8.8-bit signed value can only represent numbers in the range -128 to 127.996 (binary `10000000.00000000` to `01111111.11111111`). Overflow occurs if the result of the calculation is outside this range.

Overflow can happen at either end of the number range, the term isn't just used for overflow at the positive end. So if the 8.8-bit signed value fell below -128, that would be an overflow. Unsigned numbers overflow if they go below zero.

The fixed-point package provides options such that, whenever overflow is possible, you can specify which one of two outcomes should happen:

Wrap Mode
: With this option the value wraps round, so for example incrementing the largest number wraps round to the smallest. Wrap mode is the natural mode for computer arithmetic and has no additional hardware overhead.

Saturation Mode

With this option the value is held at the saturation value, so for example if a signed 8.8-bit number overflows in a positive direction it will be held at the positive saturation value that is the largest value of the type, that is `01111111.11111111` but if it overflows in a negative direction it will be held at the negative saturation value `10000000.00000000` regardless of the actual result of the calculation. Saturation mode requires a test for overflow and a multiplexer to switch in the saturation value, so this does have an additional hardware overhead.

Saturation mode is the default overflow mode.

6.5.2.2 Underflow

Underflow can happen when the fractional part does not have enough resolution to represent the result. For example, if you successively shift a value to the right to divide by two, bits will be lost from the least significant end of the value creating results that are not exactly half the input value.

Whenever underflow is possible, you can specify which one of two possible outcomes happens:

Truncate Mode

With this option the value simply loses the least-significant bits. This is known as Truncate Mode because the value is simply cut off at the lower end. Truncation mode has no hardware overhead associated with it since bits are simply not connected to the output.

Rounding Mode

With this option the fractional part is rounded to the nearest value that can be represented by the number's resolution. This takes into account all of the bits in the fractional part of the value in deciding whether to round up or down. Rounding mode requires a comparison and a rounding operation that is similar to an addition, so this does add extra hardware overhead.

Rounding mode is the default underflow mode.

There is an extra detail to the use of rounding mode in division. Because division (i.e. operator `"/"`, `mod` or `rem`) is an iterative operation, errors due to underflow can accumulate. That is, division is conceptually a series of shift and subtract operations and each subtraction can add an underflow error, so all the subtracts together can add a large underflow error. In the division operators, or any other operators implemented using division, you can specify a number of guard bits to add to the fractional part during the operation so that underflow errors are made smaller. For example, adding 3 guard bits will make the errors on average 1/8 the size. At the end of the operation the guard bits are removed by rounding according to the underflow mode.

6.5.3   Resize Functions

The resize functions are used to convert from one size of fixed-point number to another, performing overflow and underflow calculations in the process.

The resize functions can also be seen as the overflow and underflow operators.

The fixed-point package allows for two different strategies for working with overflow and underflow. Either you can perform basic operations on values large enough to avoid errors, then reduce the size of the result at the end of the calculation using a resize function, or you can resize after every operation. These two strategies will be described in this section.

There are four variants of the resize function altogether, two for `sfixed` and two more for `ufixed`.

The two variants for each type represent two different ways of specifying the size of the result. One variant takes the upper and lower bounds of the array range as integer values. For example, to represent a 16.16-bit number requires an array range of `(15 downto -16)` and so the bounds are the values 15 and -16. These must be constant values to be synthesisable, since a synthesiser must know what size the result is in order to create a bus with the right number of bits.

The other variant takes a signal or variable as the argument and extracts the bounds from that. This works for any signal or variable in VHDL-2008 but does not work for an `out` port in older versions of VHDL since the port cannot be passed as an `in` parameter to a function. In that instance, the first variant should be used, since the `left` and `right` attributes of an `out` port can be read.

The two variants can be illustrated by examples. The first example shows how the bounds are specified as integer values:

```
library ieee;
use ieee.std_logic_1164.all;
use ieee.numeric_std.all;
use ieee.fixed_float_types.all;
use ieee.fixed_pkg.all;
entity sfixed_resize_demo is
  port (a : in sfixed(7 downto -8);
        z : out sfixed(15 downto -16));
end;
architecture behaviour of sfixed_resize_demo is
begin
  z <= resize(a, 15, -16);
end;
```

This could also have been written using attributes:

```
  z <= resize(a, z'left, z'right);
```

It would seem that an even simpler way of expressing this is to use the second variant of the resize function that takes a signal or variable and extracts the bounds itself:

```
  z <= resize(a, z);
```

However, this can't be used in this case because it reads an `out` port. If z had been an internal signal, then this version would have worked.

Note: one of the main pitfalls in using the fixed-point package is that the assignment does not check the bounds, just the size, of the target. So, assigning an 8.8-bit number to a 12.4-bit number is perfectly legal, and effectively shifts the result four bits left, introducing a design error that neither the simulator nor the synthesiser can detect. The above forms of resize where the sizes are read directly from the target ensures that the source and target of the assignment always match and so is the recommended form.

The resize functions also implement overflow and underflow modes. Note that values can only overflow when reducing the number of bits in the integer part and can only underflow when reducing the number of bits in the fractional part.

The default modes are Saturation Mode for overflow and Rounding Mode for underflow. These default modes can be overridden by specifying additional parameters to the resize functions. For example, to get a minimum area implementation, use Wrap Mode and Truncate Mode:

```
z <= resize(a, z, fixed_wrap, fixed_truncate);
```

The first additional argument is the overflow style to use and can be either `fixed_saturate` or `fixed_wrap`. The second additional argument is the underflow style and can be either `fixed_round` or `fixed_truncate`. These values are defined in the `fixed_float_types` package:

```
package fixed_float_types is
   type fixed_overflow_style_type is (fixed_saturate, fixed_wrap);
   type fixed_round_style_type is (fixed_round, fixed_truncate);
   ...
end;
```

When resizing to a larger integer part, signed numbers are sign-extended, whereas unsigned numbers are zero extended. When enlarging the fractional part, it is always zero-filled. Neither of these extension operations change the value.

Note: unlike with `numeric_std` (Section 6.4), there are no quirks to the `resize` functions that would cause them to be bypassed and replaced with a slice.

6.5.4   Operators

The fixed-point package defines a comprehensive set of operators for both types. Furthermore, there are guidelines on how to interpret and use the types to get the correct functionality from the packages.

The set of operators available for type `sfixed` is:

comparison:	=, /=, <, <=, >, >=
boolean:	not, and, or, nand, nor, xor, xnor
arithmetic:	sign -, abs, +, -, *, /, mod, rem

The operators missing from this list are the sign operator "+", which would have no effect, the concatenation operator " & ", which would be hard to define given how important it is to be exact with bounds, the exponent operator "**", and all the shift operators.

Type `ufixed` has a slightly more restricted set of operators, in that the sign operator "-" and the abs operator are not present, since they are meaningless for unsigned arithmetic.

6.5.5   Comparison Operators

The fixed-point types have the complete set of comparison operators:

comparison:	=, /=, <, <=, >, >=

The comparison operators for `sfixed` and `ufixed` have been defined to give correct numeric comparisons for values with different-sized integer parts or fractional parts. This makes the comparison operators extremely easy and natural to use.

As an example of the usage of comparison operators, consider the following entity that returns the maximum of two `sfixed` values.

```
library ieee;
use ieee.std_logic_1164.all;
use ieee.numeric_std.all;
use ieee.fixed_float_types.all;
use ieee.fixed_pkg.all;
entity max is
  port (a : in sfixed(7 downto -8);
        b : in sfixed(7 downto -8);
        z : out sfixed(7 downto -8));
end;
architecture behaviour of max is
begin
  z <= a when a > b else b;
end;
```

This example shows how the greater-than operator can be used to compare two parameters of type `sfixed`. All the other comparison operators can be used in the same way.

6.5.6   Boolean Operators

The fixed-point types have the full set of boolean operators:

boolean: `not, and, or, nand, nor, xor, xnor`

All three kinds of boolean operator are provided.

Basic boolean operators perform bitwise logic on each element of two parameters of the same size to produce a result of the same size.

Selecting boolean operators combine a single-bit input with each element of an array to produce an array of the same size.

Reducing boolean operators combine all the elements of an array to produce a single-bit output. They are provided as operators in the VHDL-2008 version of the package, but reduction *functions* (i.e. `and_reduce`, etc.) are provided in the VHDL-1993 compatibility version of the package.

In other words, in the compatibility package, the following expression is the reduction `and` of a fixed-point value:

```
result <= and_reduce(input);
```

The same expression in the official VHDL-2008 version is:

```
result <= and input;
```

The basic bitwise logical operators are very convenient for performing masking operations. For example, to write a conditional signal assignment that uses a test on whether any one of the fraction bits of a signed 4.4-bit sign4-bit signal (`c`) are set, the following comparison can be used:

```
library ieee;
use ieee.std_logic_1164.all;
use ieee.numeric_std.all;
use ieee.fixed_float_types.all;
use ieee.fixed_pkg.all;
entity mask_demo is
  port (a, b, c : in sfixed(3 downto -4);
             z : out sfixed(3 downto -4));
end;
architecture behaviour of mask_demo is
  constant mask : sfixed(3 downto -4) := "00001111";
  constant zero : sfixed(3 downto -4) := "00000000";
begin
  z <= a when (c and mask) /= zero else b;
end;
```

If the masked result is not zero, then one or more of the four fraction bits in signal c must have been set. In other words, this tests if the value of c is not an integer value.

This example also demonstrates a work-around for a common problem in using string literals with the synthesis types. If this had been written without the constant, for example:

```
z <= a when (c and "00001111") /= "00000000" else b;
```

This would not work. The reason is that the analyser cannot deduce a range for the string literals from their context, so defaults to giving them the ascending range (integer'low to integer'low + 7). This clearly breaks the rules for specifying fixed-point ranges and will cause an error to be raised.

The simplest way to get a string literal to take a specific range is to declare a local constant with the right range and value and then use it instead of the string literal. This is a general rule for all fixed-point literals.

The logical result of the and operation, also a 4.4-bit result, is then compared with zero, specified using another 4.4-bit signed constant for the same reason.

However, this can be expressed as a comparison with integer value 0.

```
z <= a when (c and mask) /= 0 else b;
```

The fixed-point package allows comparisons with integer values like this in which case it converts the integer into the same fixed-point type as the other argument (in this case a 4.4-bit value) and then performs the comparison. See Section 6.9 for more on mixing types like this.

6.5.7 Shift Operators

The fixed-point packages provide a complete set of shift operators:

    shift: sll, srl, sla, sra, rol, ror

The operators all take two arguments, the first being the value, signal or variable to shift and the second being an integer specifying how far to shift the value. The shift distance can be a constant, in which case the implementation is just a rearrangement of the wires, or it can be a variable or signal in which case the implementation will use a barrel shifter.

These operations are implemented using the `numeric_std` package and have the same functionality. The result range is exactly the same as the first argument, that is the value being shifted. So shifting a 9.5-bit argument gives a 9.5-bit result.

A typical use of the shift functions is shown in the following example which shows a left shift of four bits being performed on an unsigned 8.8-bit signal:

```
library ieee;
use ieee.std_logic_1164.all;
use ieee.numeric_std.all;
use ieee.fixed_float_types.all;
use ieee.fixed_pkg.all;
entity shift_demo is
  port (a : in ufixed(7 downto -8);
        z : out ufixed(7 downto -8));
end;
architecture behaviour of shift_demo is
begin
  z <= a sll 4;
end;
```

The meaning of the operators is explained in Table 6.5.

For more explanation, see Section 6.4 where the shift operators for `numeric_std` are explained.

6.5.8 Arithmetic Operators

The fixed-point types have a near-complete set of arithmetic operators:

arithmetic: `sign -, abs, + , -, *,/, mod, rem`

Type `sfixed` has been given a set of operators that performs 2's-complement fixed-point arithmetic, whilst type `ufixed` has been given a set of operators that performs unsigned fixed-point arithmetic. Type `ufixed` does not have the sign "-" or the `abs` operators.

Most of these operators have been defined such that they neither overflow nor underflow. Instead they generate results large enough to contain all the possible values of the arguments.

Table 6.5 Shift and rotate operators for Fixed_Pkg

Operator	Name	Description
sll	Shift Left Logical	discard left bits, zero fill on the right
srl	Shift Right Logical	discard right bits, zero fill on the left
sla	Shift Left Arithmetic	discard left bits, zero fill on the right
sra	Shift Right Arithmetic	discard right bits: sfixed: sign extend on the left ufixed: zero extend on the left
rol	Rotate Left	remove bits from left, reinsert on right
ror	Rotate Right	remove bits from right, reinsert on left

Table 6.6   Calculating result sizes for arithmetic operators

Operator	Integer part (I_O)	Fractional part (F_O)
sign "-", abs	$I_L + 1$	F_L
"+" and "-"	$\max(I_L, I_R) + 1$	$\max(F_L, F_R)$
"*"	$I_L + I_R$	$F_L + F_R$
"/" sfixed	$I_L + F_R + 1$	$F_L + I_R - 1$
ufixed	$I_L + F_R$	$F_L + I_R$
mod sfixed	$\min(I_L, I_R)$	$\min(F_L, F_R)$
ufixed	I_R	$\min(F_L, F_R)$
rem	$\min(I_L, I_R)$	$\min(F_L, F_R)$

This is different from the convention adopted by numeric_std that was defined with size-preserving operators that therefore did overflow.

The idea is that you generate oversize results and then call resize to constrain the datapath whenever suitable for the circuit being designed. To get a constant datapath width, you need to resize after every operation.

The exception to this is the division operators /, mod and rem where it isn't possible to define how many bits might be needed to represent all possible values of the result. So in this case a different formula is used to decide how big to make the result. Consequently, both overflow and underflow are possible with these operators.

Each operator has a different formula for the size of the result. This means you have to be careful in defining intermediate signals of the right range to accept the result of an arithmetic operation. Table 6.6 shows the calculations for the result sizes of the arithmetic operators. They are based on a left argument of size $I_L.F_L$ and an optional right argument of size $I_R.F_R$ to produce an output of size $I_O.F_O$.

In most cases, the left and right arguments will be the same size, so Table 6.7 gives a simplified version of this table with identical left and right operand sizes I.F:

The sign "-" and the abs operator are only provided for the signed type sfixed. The sign "-" operator forms the 2's-complement negation of a number, whilst the abs operator forms the negation only if the value is negative. They both generate a result with an integer part that is one bit bigger than the input but with the same size fractional part. So, for example, the negation of an 8.8-bit number is a 9.8-bit number. This is because the 2's complement range is asymmetrical – the most negative value has a slightly larger magnitude than the most positive – so negating the most negative value in the range would cause overflow without the extra bit.

Table 6.7   Result sizes for arithmetic operators with identical input sizes

Operator	Integer part (I_O)	Fractional part (F_O)
sign "-", abs	$I + 1$	F
"+" and "-"	$I + 1$	F
"*"	$2I$	$2F$
"/" sfixed	$I + F + 1$	$I + F - 1$
ufixed	$I + F$	$I + F$
mod	I	F
rem	I	F

Unlike `signed` (Section 6.4), `sfixed` can never give the strange effect of negating a negative value to give a negative result, since the result is always large enough to contain the actual value of the negation. However, this strange effect will be seen if you then resize the result smaller with Wrap Mode. To illustrate this, consider the most negative 4.4-bit number 1000.0000 (−8). Negating this gives the 9.8-bit number 01000.0000 (+8). Resizing back to 4.4 bits in Wrap Mode gives 1000.0000 (−8) again. Note also that, if this example had been resized in Saturation Mode instead, the overflow would have been detected, and the result would have been the positive saturation value 0111.1111 (+7.94).

With the addition and subtraction operators, the situation is a little more complicated. First, these operators have two arguments and the operators are defined so that these arguments can have different ranges. However, the underlying arithmetic operations really require equal-sized arguments. So, within the operator, the arguments are normalised to be the same range. This is done by extending both arguments to a range that is big enough to contain both values. So for example, given an 8.4-bit number and a 6.9-bit number, the two arguments are resized to 8.9 bits. The two normalised values are then added (or subtracted) to generate the result. One bit is added to the integer part of the result to contain the largest sum that can be generated thereby avoiding overflow. Thus, the result in this case will be a 9.9-bit number. This size calculation is the same for both `sfixed` and `ufixed` types.

The multiplication operator is also defined such that its two arguments can be of any range. No normalisation is needed because multiplication is easily carried out on different sized operands. However, the result must be much bigger than the arguments to hold all the possible values. The result size is the sum of the sizes of the inputs; to be more specific, the integer part is the sum of the sizes of the arguments' integer parts and likewise for the fractional parts. So multiplying a 5.6-bit number with a 7.3-bit number gives a 12.9-bit result.

For the division operator, the size formulae are more complicated than the other operations and need careful calculation using Tables 6.5 and 6.6 as a guide. These sizes are calculated so that overflow is impossible. However, underflow is still possible and in that case the default underflow rule is used, Rounding Mode with 3 guard bits. To override these default values, use the divide function instead:

```
z <= divide(a, b, fixed_truncate, 0);
```

This will generate the same size output as the `"/"` operator but with underflow set to use Truncate Mode and the calculation performed with no guard bits. There are similar remainder and modulo functions to provide the same flexibility to the `rem` and `mod` operators. Finally, there is a reciprocal function that simplifies this common use of division.

When you want a constant datapath width, the convention when using the fixed-point package is to resize after every operation. So these width calculations can be eliminated by enclosing the operation in the resize function call:

```
library ieee;
use ieee.std_logic_1164.all;
use ieee.numeric_std.all;
use ieee.fixed_float_types.all;
use ieee.fixed_pkg.all;
entity sum_demo is
  port (a, b, c, d : in sfixed(7 downto -8);
        z : out sfixed(7 downto -8));
```

```
  end;
architecture behaviour of sum_demo is
   signal aplusb : sfixed(z'range);
   signal cplusd : sfixed(z'range);
begin
  aplusb <= resize(a + b, aplusb);
  cplusd <= resize(c + d, cplusd);
  z <= resize(aplusb + cplusd, z'left, z'right);
end;
```

Note how the last resize uses the bounds of z, rather than passing z as a parameter as was done in the other two examples. This is because z is an out port and cannot be passed as an in parameter of a function. However, its attributes can be read and passed as in parameters.

A more extreme version of this nesting is:

```
architecture behaviour of sum_demo is
begin
  z <= resize(
         resize(a + b, z'left, z'right)
         +
         resize(c + d, z'left, z'right),
         z'left, z'right);
end;
```

As this example shows, nesting the calls to resize makes it difficult to keep track of the parameters, but it does eliminate all intermediate signals.

This example uses Saturate Mode for its overflow behaviour because that is the default behaviour of the fixed-point package. In this case there cannot be underflow because the fraction part does not get bigger and so is not shortened by the resize.

This can be rewritten to perform the more efficient Wrap Mode for overflow:

```
architecture behaviour of sum_demo is
   constant aplusb : sfixed(z'range);
   constant cplusd : sfixed(z'range);
begin
  aplusb <= resize(a + b, aplusb, fixed_wrap);
  cplusd <= resize(c + d, cplusd, fixed_wrap);
  z <= resize(aplusb + cplusd, z, fixed_wrap);
end;
```

An alternative approach is to allow the calculation to expand and then to resize in Wrap Mode at the end:

```
architecture behaviour of sum_demo is
begin
  z <= resize((a + b) + (c + d), z'left, z'right, fixed_wrap);
end;
```

So, the result of the first level of add adds two 8.8-bit values to give a 9.8-bit result. The second level increases this to 10.8. Then the resize converts it back to an 8.8-bit result, in this case using Wrap Mode.

The parentheses allow the order of calculation to be controlled and creates a balanced tree with a maximum datapath length of two adders. Without them, the formula a + b + c + d is equivalent to ((a + b) + c) + d, which gives an unbalanced tree with a maximum datapath length of three adders instead of two.

6.5.9   Utility Functions

The fixed-point package contains a few utility functions that don't fit into any of the above classifications.

6.5.9.1 is_negative

The is_negative function is used to test whether the value is less than zero:

```
function is_negative (arg : sfixed) return boolean;
```

This can be used in conditionals:

```
if is_negative(x) then ...
```

This function is only provided for sfixed.

6.5.9.2 add_carry

The add_carry procedure is used to perform an add-with-carry generating an output of the same size as the inputs with a carry output. There are two versions: an unsigned and a signed version.

```
procedure add_carry
   (L, R : in ufixed; c_in : in std_ulogic;
    result : out ufixed; c_out : out std_ulogic);
procedure add_carry
   (L, R : in sfixed; c_in : in std_ulogic;
    result : out sfixed; c_out : out std_ulogic);
```

The usual usage is for L and R to be the same size, in which case result must also be the same size. If the inputs are different sizes, then the maximum integer part and the maximum fractional part are used to calculate the result range. An incrementer can be created by setting one of the inputs to zero and then controlling the increment via the carry input.

Since these are procedures with variable outputs, they can only be used in a process and the result and c_out outputs connected to variables. See Section 11.5 for more on the use of procedures with out parameters.

6.5.9.3 scalb

The scalb function scales the argument by a power of two. This is different from the arithmetic shifts because no bits are lost, instead the bit pattern is preserved but the range is changed, meaning the binary point is shifted by the specified distance. A positive distance represents a shift left and a negative distance a shift right.

There are two versions, one for `sfixed` and one for `ufixed`.

```
function scalb (y : ufixed; n : integer) return ufixed;
function scalb (y : sfixed; n : integer) return sfixed;
```

For synthesis, the shift distance must be a constant so that a result bus of the right range can be created. For example, an 8.8-bit input shifted 4 bits will be converted into a 12.4-bit result but with the same bit-pattern. Shifting by −4 bits converts the input into a 4.12-bit result instead.

6.5.9.4 Maximum and Minimum

The maximum and minimum functions do what their name suggests and return the minimum and maximum values of their inputs.

```
function minimum (l, r : ufixed) return ufixed;
function minimum (l, r : sfixed) return sfixed;
function maximum (l, r : ufixed) return ufixed;
function maximum (l, r : sfixed) return sfixed;
```

The most common usage is to make the two inputs the same range, in which case the result will be the same range.

However, this becomes more complicated if the inputs have different ranges. In that case, the result is resized to be big enough to accommodate either value without underflow or overflow. So the result is the maximum of the input sizes. For example, the result of an 8.4-bit input and a 4.8-bit input will be an 8.8-bit result.

6.5.9.5 Saturate

The saturate function generates the upper saturation value for a given size of array. The upper saturation value is the largest value in the range. This value is used in Saturation Mode but it can also be used directly, for example, to set a variable to the saturation value to indicate that a calculation has failed.

Like the resize functions, there are four permutations of the saturate function. Two operate on signed and two operate on unsigned. For each type, there is one function that takes the left and right bounds as arguments and a second form that takes a sample signal or variable to define the size.

So, for example, to set a signal to the saturation value:

```
z <= saturate(z);
```

The function call generates the saturation value for the range of z, then the assignment assigns it to the signal. An alternative form using attributes would be:

```
z <= saturate(z'left, z'right);
```

To generate the lower saturation value (the smallest value in the range), use the saturation function but then invert it using the `not` operator:

```
z <= not saturate(z);
```

This transformation applies to both `ufixed` and `sfixed` versions of the `saturate` function.

This relationship between the upper and lower saturation values is not obvious and it would have been preferable for the package to supply functions for both rather than relying on this transformation of one onto the other.

6.6 Floating-Point Types – Float_Pkg

Package `float_pkg` provides arbitrary-precision floating-point arithmetic.

Bear in mind when planning to use this package that floating-point operations are very complex and RTL synthesis maps operators onto purely combinational logic. This makes synthesis of floating-point operations expensive in both circuit area and delay. For example, the Case Study in Chapter 15 is an example of a circuit that is approximately 3–4 times bigger and has twice the delay using floating-point arithmetic as the same design using fixed-point arithmetic.

Note: for most designs it makes more sense to either map the problem onto fixed-point types and then synthesise using `fixed_pkg` described in the last section, or to use a device with a built-in Floating-Point Unit (FPU) that most FPGA and ASIC vendors provide as part of their service. An alternative is to use a synthesisable FPU core obtained from a core vendor or from Open Cores [Open Cores, 2010]. With both of the latter two solutions, the task of the RTL designer is to schedule bus operations to feed arguments to the FPU and receive the result from it.

Despite this, it is possible to synthesise floating-point operations using the floating-point package.

6.6.1 Types Provided

The package provides just one floating-point type, `float`:

```
type float is array (integer range <>) of std_logic;
```

Note that the range of the type uses integer, not natural as with `signed` or `unsigned`. This means that the range can include negative indices.

Note: in the compatibility packages only, this type is based on `std_ulogic`, not `std_logic`. In the full VHDL-2008 version, this is an array of `std_logic`. This is a strange decision because it makes the two versions of the package slightly incompatible. One consequence of this choice of `std_ulogic` as the element type is that you cannot use the compatibility version of these types to create tristate buses, because tristate buses can only be made from `std_logic` or arrays of `std_logic`.

The `float` type implements variable-width floating-point arithmetic as defined by IEEE standard number 754 [IEEE-754, 2008]. Most computer FPUs conform to IEEE-754.

It also provides subtypes for the most commonly used sizes:

```
subtype float32 is float(8 downto -23);
subtype float64 is float(11 downto -52);
subtype float128 is float(15 downto -112);
```

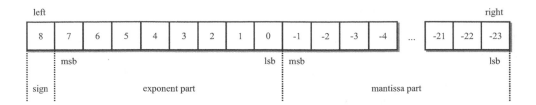

Figure 6.6   Floating-point storage format.

These are the standard floating-point types defined by the IEEE for 32, 64 and 128-bit words.

The floating-point representation is a sign, exponent and mantissa representation. These three fields are stored in the word as shown in Figure 6.6.

In order to use the packages effectively, it is important to understand how the bit-array representation is interpreted by the package.

There are several rules:

1. Arrays are read from left to right, with the most-significant bit on the left.
2. You must use a descending range. This gives the most-significant bit the highest index.
3. The sign bit is always the leftmost bit.
4. The exponent part is the range from the sign bit downto 0.
5. The mantissa is the range from −1 downto the rightmost bit.

These rules are followed in all the examples.

Any combination of exponent and mantissa lengths can be used to define any kind of floating-point type. However, there is a lower bound of three for both exponent and mantissa.

A common notation for floating-point types is of the form 8:23, which means 8-bit exponent part, 23-bit mantissa. The sign bit is not represented in this notation, so the full length of the word is actually one bit longer than the sum of the two parts. So this example represents a 32-bit floating-point number, indeed it represents the IEEE standard 32-bit floating-point number represented by subtype `float32`. This two-value colon-separated notation will be used in this book.

6.6.2   Interpreting Floating-Point Numbers

Given a floating-point number with a sign S, a mantissa M and an exponent E, the value V is:

$$V = S.M.2^E$$

The sign S is binary encoded as 0 for +1 and 1 for −1.

The mantissa M is usually *normalised* so that it is between 1.0 and 2.0. This means that normalised floating-point numbers always start with '1.' in binary, that is the value 1 followed by the binary point, so the representation does not store the first digit (nor the binary point) that is made implicit. Only the remainder after the binary point is actually stored. When calculating the true value of the representation, this implicit first digit must be reinstated to create M.

The exponent E is not a 2's-complement representation as you might expect, but an unsigned representation with an implied offset. So the value of the exponent is calculated by interpreting its bit-pattern as an unsigned representation, then subtracting the offset. The offset is one less than half the unsigned range of the exponent. This means that, for example, with an 8-bit exponent, the offset is 127 ($2^8/2-1$). This offset notation means that the exponent values 1 to 126 represent negative exponents -126 to -1, the value 127 (i.e. the offset value) represents zero, and the values 128 to 254 represent natural exponents 1 to 127.

The exponent can also be seen as a shift offset, so the true value is the mantissa shifted by the number of bits represented by the value of the exponent. This is a left shift for a positive exponent and a right shift for a negative one.

Zero is represented by a special value with an all-zero mantissa and an all-zero exponent part. An oddity of the floating-point representation is that there are two zero values because of the sign bit, so there can be a positive zero and a negative zero value.

6.6.3   Overflow, Underflow and Error Modes

The exponent values of all-zeros and all-ones have special meanings related to the underflow and overflow modes and so are not part of the usable range of the exponent for normalised numbers. The exception is the zero value itself, which has a zero exponent and a zero mantissa.

Underflow and overflow are quite complicated concepts with floating-point numbers.

Underflow occurs when the exponent is at its most negative value and the mantissa becomes too small to represent in normalised form.

To extend the range that can be represented, when underflow occurs to a normalised number, the exponent is set to the all-zeros value to indicate that the number must now be interpreted as a *denormalised* number and the mantissa is then allowed to drop below 1.0. This means the leading digit is now an implied '0.'

When the denormalised number sequence underflows, the value becomes zero.

The package `float_pkg` provides the option of switching off denormalised numbers. So in this simplified form, when a normalised number underflows, the result becomes zero. This can greatly simplify the logic required to implement operations at a cost to the dynamic range. The default is for denormalised numbers to be switched on.

Overflow occurs when the exponent is at its most positive value and the mantissa becomes too large to represent. The result of overflow is infinity. The mantissa of infinity is all-zeros and the exponent all-ones. There are two infinities, positive and negative, which are represented by different sign bits.

The floating-point algorithms can sometimes result in an error, for example dividing by zero. Error modes are indicated by special bit-patterns known as NaNs, where NaN stands for 'Not a Number'. These are also represented by the all-ones exponent, but with a mantissa that is not the all-zeros value.

The package `fixed_pkg` provides the option of switching off checking for NaNs and infinity, thus saving some error-checking logic.

6.6.4   Rounding Modes

Most floating-point operations are performed using an internal, extended form which is then rounded to a smaller result format at the end of the calculation. This is done to minimise rounding errors during the calculation.

The rounding mode is defined by an enumeration in package `fixed_float_types`:

```
type round_type is
   (round_nearest, round_inf, round_neginf, round_zero);
```

There are four rounding modes:

Round Nearest – `round_nearest` (default)
   Round up or down to the nearest representable value.
Round Towards Positive Infinity – `round_inf`
   Round upwards to the next higher representable value. This is a round to a larger mantissa for
   a positive value, to a smaller mantissa for a negative value.
Round Towards Negative Infinity – `round_neginf`
   Round downwards to the next lower representable value. This is a round to a smaller
   mantissa for a positive value, to a larger mantissa for a negative value.
Round Towards Zero – `round_zero` (no hardware cost)
   Round to next smaller representable value regardless of the sign.

The default mode is Round Nearest, but all operations can be individually controlled to use
other modes. Note that for the sign-magnitude representation used in floating-point numbers,
round towards zero is equivalent to truncation, which has zero hardware cost.

Also, the internal extended form used in calculations is created by adding guard bits. The
default is to add 3 guard bits, but the number of guard bits can be controlled for each operation.
Setting the number of guard bits to 0 switches off rounding.

6.6.5   Mode Options

Many of the floating-point operations in `float_pkg` are available in two forms: an operator
itself (e.g. `"*"`) and a function (e.g. `multiply`). The operator form is the same as the function
form with all options set to their default value. The function form is the general case and allows
all the options controlling modes to be specified per-operation. Similarly, the function form
has default values for all the mode parameters, so that if the mode parameters are not specified,
you get the same behaviour as the operator form.

The four options listed in Table 6.8 are available for most operations.

The default combination gives full IEEE-standard floating-point behaviour. However, if the
full standard is not needed some or all of these can be switched off. The smallest, fastest
combination is:

```
round_style => round_zero
guard => 0
check_error => false
denormalize => false
```

6.6.6   Functions and Operators

Most floating-point operations calculate a result that is the size of the largest mantissa and the
largest exponent part of the arguments. If the arguments are the same range, then the result is the

Table 6.8 Options controlling floating-point modes

Name	Type	Default	Description
round_style	round_type	round_nearest	select rounding mode
guard	natural	3	number of guard bits
check_error	boolean	true	check for overflow errors
denormalize	boolean	true	enable denormalised numbers

same range as the arguments. This means that the normal case for floating-point operations is to generate fixed-width datapaths. To generate any other kind, resizing should be used before the operation so that the operation is performed at the higher precision. For example, you can extend the mantissa to reduce rounding errors during a complex calculation, perform a sequence of operations with rounding disabled at this extended format, then reduce the result again at the very end. This will result in a calculation error smaller than adding guard bits would, because guard bits only minimise the error within one operation rather than across a sequence of operations.

6.6.7   Classification Functions

Because there are different classes of value that can be stored in a floating-point number, there is a classification function that enables the class to be checked and used in conditionals.

The classification function is:

```
function classfp (x : float; check_error : boolean := true)
   return valid_fpstate;
```

The function takes two arguments, the number to be checked and an optional switch to determine whether to check for error conditions, that is NaNs and infinity.

The return type is an enumeration type with the values described in Table 6.9.

Table 6.9 Results of classification function classfp

Value	Meaning
nan	Signalling NaN
quiet_nan	Quiet NaN
neg_inf	Negative Infinity
neg_normal	Negative Normalised Number
neg_denormal	Negative Denormalised Number
neg_zero	Negative Zero
pos_zero	Positive Zero
pos_denormal	Positive Denormalised Number
pos_normal	Positive Normalised Number
pos_inf	Positive Infinity
isx	at least one input is unknown

There are two types of NaN, one is meant to raise a signal or interrupt and the other (quiet) NaN is meant to be just returned as a value. Whether this distinction is meaningful in a particular design depends entirely on the application domain and is beyond the scope of the floating-point package.

If error checking is disabled by setting the `check_error` argument to `false`, the function cannot detect either of the NaN values or either of the infinity values.

The `isx` value has no meaning in synthesis since X values cannot exist in a real circuit.

There are some short-cut functions to test for certain common cases:

```
function isnan (x : float) return boolean;
function finite (x : float) return boolean;
```

The `isnan` function tests whether the value is one of the two NaN values `nan` or `quiet_nan`. If a value is a NaN, then it is unordered and cannot be meaningfully compared or used in a calculation.

The `finite` function doesn't test for finite values, it actually tests for the value not being infinite, that is not being either `neg_inf` or `pos_inf`. Unfortunately, this means that a NaN would be classed as finite, which is a strange conclusion, so this test should always be used in conjunction with the `isnan` function. Alternatively, use the `classfp` function and a case statement.

There is one final test function – `unordered` – which takes two arguments and can be used to determine whether two values can be meaningfully compared or used in a calculation:

```
function unordered (x, y : float) return boolean;
```

This can be used to decide whether to perform an operation or bypass it. It takes two arguments and returns true if the argument values are unordered in respect of one another, in other words if either of them is a NaN. It is more efficient to test two inputs for unordered values before a complex calculation than to test for them at every stage of the calculation. If the inputs of a circuit have been pre-tested with the `unordered` function, the `error_check` parameter of all operations within the calculation can be set to `false`.

6.6.8   Operators

The set of operators available for type `float` is:

comparison: `=, /=, <, <=, >, >=`
boolean: `not, and, or, nand, nor, xor, xnor`
arithmetic: `sign-, abs, +, -, *, /, mod, rem`

As stated above, most operators are available in two forms, the operator itself and the function form with mode options.

There are no shift operators because this doesn't make sense given the specific meaning of the different fields in the floating-point representation.

6.6.9   Comparison Operators

The floating-point type has the complete set of comparison operators:

comparison: $=$, $/=$, $<$, $<=$, $>$, $>=$

It also provides the function forms of these:

```
eq, ne, lt, le, gt, ge
```

Each function provides extra mode parameters:

```
function eq (l, r : float;
             check_error : boolean := true;
             denormalize : boolean := true)
   return boolean;
```

It seems strange that the comparison operators need to have a function form, since they just perform comparisons. However, the mode options need to be user-configurable for the case where the two operands are of different sizes, in which case they are resized to the same size before the comparison takes place. The function form allows denormalised numbers and error checking to be switched on or off during this resizing. The default is for both to be switched on and this is the behaviour of the operators.

Clearly, if the parameters of comparisons are always the same size, there is no need to use the function form and the operators should be used throughout.

The comparison operators rely on the concept of *ordered* values. The ordering depends first on the class of the number, then on its value. The lowest class is negative infinity, then the negative normalised numbers, then the negative denormalised numbers. Then there is zero, in which case positive zero and negative zero are equal. This is followed by the positive denormalised numbers, the positive normalised numbers and finally positive infinity. When comparing, if the classes are different, that is sufficient to determine the ordering (with some special handling of the zero values). The sign and mantissas of the arguments are only compared if they are the same class and that class is either a normalised or a denormalised number.

A NaN is not ordered, so a NaN cannot be equal, less-than or greater-than anything, even itself. All the comparison operators return `false` if either argument is a NaN.

As an example of the usage of comparison operators, consider the following entity that returns the maximum of two `float32` values.

```
library ieee;
use ieee.std_logic_1164.all;
use ieee.numeric_std.all;
use ieee.fixed_float_types.all;
use ieee.fixed_pkg.all;
use ieee.float_pkg.all;
entity max is
  port (a : in float32;
        b : in float32;
        z : out float32);
end;
```

```
architecture behaviour of max is
begin
  z <= a when a > b else b;
end;
```

The function form of the comparison operators have a parameter profile like the following:

```
function gt (l, r : float;
                check_error : boolean := true;
                denormalize : boolean := true)
   return boolean;
```

This is the functional form of the greater than operator ("`>`"). Only two of the mode options are provided in this case, not the full set of four. Those provided are the `check_error` and `denormalise` options, both of them switched on (`true`) by default. So, the example above can be rewritten using the functions:

```
architecture behaviour of max is
begin
  z <= a when gt(a,b) else b;
end;
```

This has identical behaviour to the operator form. The example can be rewritten with the mode options switched off:

```
architecture behaviour of max is
begin
  z <= a when gt(a,b,false,false) else b;
end;
```

This gives a more efficient form without error checking or support for denormalised numbers.

6.6.10   Boolean Operators

All three kinds of boolean operator are provided: basic, selecting and reducing as described in Section 5.3. They work in exactly the same way as described in that section:

    boolean: not, and, or, nand, nor, xor, xnor

Basic boolean operators perform bitwise logic on each element of two parameters of the same size to produce a result of the same size.

  Selecting boolean operators combine a single-bit input with each element of an array to produce an array of the same size.

  Reducing boolean operators combine all the elements of an array to produce a single-bit output. They are provided as operators in the VHDL-2008 version of the package, but reduction *functions* (i.e. and_reduce etc.) are provided in the VHDL-1993 compatibility version of the package.

  In other words, in the compatibility package, the following expression is the reduction and of a fixed-point value:

```
result <= and_reduce(input);
```

The same expression in the official VHDL-2008 version is:

```
result <= and input;
```

The VHDL-2008 version is therefore not backwards-compatible with the VHDL-1993 version.

6.6.11   Arithmetic Operators

The floating-point type has the following set of arithmetic operators:

   arithmetic: `sign -, abs, + , -, * , / , mod, rem`

It also has the following set of function versions of these operators:

   `add, subtract, multiply, divide, modulo, remainder, reciprocal`

Each operator is equivalent to the function form with the default set of mode parameters. For example, the add function has the following profile:

```
function add (l, r : float;
                round_style : round_type := round_nearest;
                guard       : natural    := 3;
                check_error : boolean    := true;
                denormalize : boolean    := true)
    return float;
```

Mode options can be controlled by using positional notation to list all the options:

```
x <= add(a,b,round_zero,0,false,false);
```

This gives the minimum-area implementation. The parameters can alternatively be given by named association, in which case only those parameters being changed from the default need to be specified:

```
x <= add(a,b, guard => 0);
```

This switches off rounding, so the rounding mode becomes irrelevant, but leaves error checking and support for denormalised numbers switched on.

There is no size calculation needed for floating-point operations as there was for fixed-point since the convention is to generate a result the same size as the arguments. The exception is where the arguments are different sizes, in which case the result is normalised to a size sufficient to contain either argument. So, for example, adding an 8:23 value to a 10:21 value will result in a 10:23 value.

One difference with the floating-point representation is that it is a sign-magnitude rather than 2's-complement representation. It therefore has a symmetrical range, with the same number of negative and positive values.

The negation (sign -) and absolute (abs) operators only change the sign bit. Since the floating-point representation is symmetrical, this cannot cause overflow and so there are no strange side effects as there are with type signed.

Addition and subtraction are performed by aligning the binary points of the two values by shifting the mantissa of the number with the smaller exponent until both values have the same exponent. Once aligned, the mantissas can be added or subtracted. Finally, the result is normalised to the range 1.0–2.0 if possible.

Multiplication is performed by multiplying the mantissas and adding the exponent parts. The result is then normalised, if possible

Division is similar in that the result is generated by dividing the mantissas and subtracting the exponents. The division of the mantissas uses a similar algorithm to the division of fixed-point numbers. The divide function or "/" operator returns the quotient and the remainder function or rem operator returns the remainder of this division. The result is then normalised, if possible.

The reciprocal function performs 1/X but in a more efficient way than the divide function.

All of these operations may result in overflow or underflow in which case the overflow and underflow options determine the outcome.

Since the convention for the floating-point package is to generate constant datapath width, there is little need for the resize functions:

```
library ieee;
use ieee.std_logic_1164.all;
use ieee.numeric_std.all;
use ieee.fixed_float_types.all;
use ieee.fixed_pkg.all;
use ieee.float_pkg.all;
entity sum_demo is
  port (a, b, c, d : in float32;
        z : out float32);
end;
architecture behaviour of sum_demo is
  signal aplusb : float32;
  signal cplusd : float32;
begin
  aplusb <= a + b;
  cplusd <= c + d;
  z <= aplusb + cplusd;
end;
```

Or, more simply:

```
architecture behaviour of sum_demo is
begin
  z <= (a + b) + (c + d);
end;
```

6.6.12   Resize Functions

The resize functions are used to convert from one size of floating-point number to another, performing overflow and underflow calculations in the process.

There are two variants of the resize function altogether. The two variants represent two different ways of specifying the size of the result. One variant takes the exponent size and mantissa size as integer values. For example, to represent a 8:23 bit number the size values 8 and 23 are used. Unfortunately, the use of sizes instead of bounds is inconsistent with the fixed-point package where the resize functions take the array bounds as their arguments.

The size arguments must be constant values to be synthesisable, since a synthesiser must know what size the result is in order to create a bus with the right number of bits.

The other variant takes a signal or variable as the argument and extracts the sizes from that. This works for any signal or variable in VHDL-2008 but does not work for an `out` port in older versions of VHDL since the port cannot be passed as an `in` parameter to a function. In that instance, the first variant should be used.

The two variants can be illustrated by examples. The first example shows how the bounds are specified as integer values:

```
library ieee;
use ieee.std_logic_1164.all;
use ieee.numeric_std.all;
use ieee.fixed_float_types.all;
use ieee.fixed_pkg.all;
use ieee.float_pkg.all;
entity float_resize_demo is
  port (a : in float32;
        z : out float64);
end;
architecture behaviour of float_resize_demo is
begin
  z <= resize(a, 11, 52);
end;
```

Note that the bounds are specified as sizes, so the exponent size is 11 and the mantissa size is 52. Note that the mantissa size is positive, even though the lower array bound will be negative.

This could also have been written using attributes:

```
z <= resize(a, z'left, -z'right);
```

Note how the lower (right) bound is turned into a size by negation. The upper (left) bound is already the same as the exponent size because of the sign bit, so there is no need to add one to get the size from the upper bound as you do with the other synthesisable types.

It would seem that an even simpler way of expressing this is to use the second variant of the resize function that takes a signal or variable and extracts the bounds itself:

```
z <= resize(a, z);
```

However, this can't be used in this case because it reads an `out` port. If z had been an internal signal, then this version would have been usable in this way.

Note: one of the main pitfalls in using the floating-point package is that the assignment does not check the bounds, just the size, of the target. So, assigning an 8:23 bit number to a 10:21 bit number is perfectly legal because they have the same number of bits, and effectively shifts the result two bits left, causing some of the mantissa to be misinterpreted as part of the exponent.

This introduces a design error that neither the simulator nor the synthesiser can detect. The above forms of resize where the sizes are read directly from the target ensures that the source and target of the assignment always match and so is the recommended form.

The resize functions also implement overflow and underflow modes. They take up to four extra parameters that define the options for rounding mode, error-checking and support for denormalised numbers whilst performing the resize.

The full declaration of the resize that takes two bounds is:

```
function resize (arg : float;
    exponent_width : natural     := 8;
    fraction_width : natural     := 23;
    round_style    : round_type := round_nearest;
    check_error    : boolean    := true;
    denormalize_in : boolean    := true;
    denormalize    : boolean    := true)
  return float;
```

The size parameters have default values, so that if you don't specify a size, the result will be a float32.

The default rounding style is Round to Nearest, so this can be overridden to any of the four rounding modes. The three remaining options control the behaviour for handling of infinity and NaNs and support for denormalised numbers. Setting check_error to false means that there will be no special handling of infinity or NaNs in the conversion. The denormalize_in parameter informs the function whether the input value could possibly have denormalised numbers (i.e. was generated with denormalised numbers enabled). The denormalize parameter determines whether the output of the resize will have denormalised numbers enabled.

There are three alternatives to the resize function for resizing to one of the standard (32-bit, 64-bit and 128-bit) sizes:

```
function to_float32(arg    : float;
                    round_style    : round_type := round_nearest;
                    check_error    : boolean    := true;
                    denormalize_in : boolean    := true;
                    denormalize    : boolean    := true)
  return float32;

function to_float64(arg : float;
                    round_style    : round_type := round_nearest;
                    check_error    : boolean    := true;
                    denormalize_in : boolean    := true;
                    denormalize    : boolean    := true)
  return float64;

function to_float128(arg : float;
                    round_style    : round_type := round_nearest;
                    check_error    : boolean    := true;
                    denormalize_in : boolean    := true;
                    denormalize    : boolean    := true)
  return float128;
```

These are simply calls to resize with the appropriate sizes inserted, so a float32 is an 8:23 bit float, a float64 is 11:52 and a float128 is 15:112.

If the design is expressed entirely in terms of standard sized floating-point types, these functions mean that it is never necessary to remember the standard ranges. For example:

```
library ieee;
use ieee.std_logic_1164.all;
use ieee.numeric_std.all;
use ieee.fixed_float_types.all;
use ieee.fixed_pkg.all;
use ieee.float_pkg.all;
entity float_resize_demo is
  port (a : in float32;
        z : out float64);
end;
architecture behaviour of float_resize_demo is
begin
  z <= to_float64(a);
end;
```

6.6.13   Utility Functions

The package contains many utility functions, although not all are really useful for synthesis. The following are those that are sometimes useful.

6.6.13.1   Constant Values

There are several functions that return constant values, adjusted to a particular range of floating-point number:

```
function zerofp(exponent_width : natural := 8;
                fraction_width : natural := 23)
   return float;

function neg_zerofp(exponent_width : natural := 8;
                    fraction_width : natural := 23)
   return float;

function nanfp(exponent_width : natural := 8;
               fraction_width : natural := 23)
   return float;

function qnanfp(exponent_width : natural := 8;
                fraction_width : natural := 23)
   return float;

function pos_inffp(exponent_width : natural := 8;
                   fraction_width : natural := 23)
   return float;
```

```
function neg_inffp(exponent_width : natural := 8;
                   fraction_width : natural := 23)
   return float;
```

So, `zerofp` returns the zero value of the designated size, with `neg_zerofp` returning negative zero. Similarly, the `nanfp` function gives a signalling NaN, the `qnanfp` a quiet NaN, `pos_inffp` gives positive infinity (effectively the upper saturation value) and `neg_inffp` gives negative infinity (effectively the lower saturation value).

With all of these functions, the value can be generated for any size of floating-point number as defined by the size arguments. If no size is given, a `float32`-sized result is generated.

All of these functions have a second form that takes a sample variable or signal and extracts the sizes from that, in a similar way to the resize functions.

6.6.13.2 is_negative

The `is_negative` function is used to test whether the value is less than zero:

```
function is_negative (arg : float) return boolean;
```

This can be used in conditionals:

```
if is_negative(x) then ...
```

The test only checks the sign bit, it does not take into account the class of the value.

6.6.13.3 scalb

The `scalb` function scales the argument by a power of two, returning the scaled value. This is performed by adding the power of two to the exponent, then handling overflow or underflow if necessary.

```
function scalb(y : unresolved_float;
               n : integer;
               round_style : round_type := round_nearest;
               check_error : boolean    := true;
               denormalize : boolean    := true)
   return float;
```

There are two versions, one that takes an integer second argument, the other a `signed` second argument.

The extra options control the rounding mode, error checking and support for denormalised numbers as for the resize function.

6.6.13.4 logb

The `logb` function gets the exponent value:

```
function logb (x : float) return integer;
function logb (x : float) return signed;
```

The exponent is corrected from the internal representation to a `signed` number by subtracting the offset.

6.6.13.5 Maximum and Minimum

The maximum and minimum functions do what their name suggests and return the minimum and maximum values of their inputs.

```
function minimum (l, r : float) return float;
function maximum (l, r : float) return float;
```

The most common usage is to make the two inputs the same range, in which case the result will be the same range.

However, this becomes more complicated if the inputs have different ranges. In that case, the result is resized to be big enough to accommodate either value without underflow or overflow. So the result is the maximum of the input sizes. For example, the result of an 8:23 bit input and a 10:21 bit input will be an 10:23 bit result.

6.6.13.6 nextafter

The `nextafter` function is the increment/decrement operation for floating-point numbers. It returns the next value that can be represented by the precision of the argument. The full form of the `nextafter` function is:

```
function nextafter(x : float;
                   y : float;
                   check_error : boolean := true;
                   denormalize : boolean := true)
   return float;
```

The first argument is the value to be incremented or decremented. The second argument is a reference – the function generates the next value towards the reference. This is often a constant value as generated by one of the constant value functions – for example, the next value towards positive infinity:

```
library ieee;
use ieee.std_logic_1164.all;
use ieee.numeric_std.all;
use ieee.fixed_float_types.all;
use ieee.fixed_pkg.all;
use ieee.float_pkg.all;
entity float_increment_demo is
 port (a : in float32;
       z : out float32);
end;
architecture behaviour of float_increment_demo is
begin
  z <= nextafter(a,pos_inffp);
end;
```

The result is the same size as the first argument. The `check_error` and `normalise` arguments allow control of these overflow options.

6.7   Type Conversions

As explained at the start of the chapter, the synthesis packages provide a type system of eight types:

```
ieee.std_logic_1164.std_logic
ieee.std_logic_1164.std_logic_vector
ieee.numeric_std.signed
ieee.numeric_std.unsigned
ieee.fixed_pkg.sfixed
ieee.fixed_pkg.ufixed
ieee.float_pkg.float
std.standard.integer
```

It is good practice to choose types for each datapath so as to use the type most appropriate for the job. Thus, conversions between types should be minimised. Nevertheless, it is necessary sometimes to convert between types. To this end, the synthesis packages contain a set of type-conversion functions between the different types in use.

Type conversions can be classed as bit-preserving or value-preserving.

bit-preserving
: The conversion simply copies the value bit by bit without trying to interpret the bits as a numeric value. This may change the value in the process.

value-preserving
: The conversion interprets the value according to its type and generates an equivalent value in the target type.

Sometimes a bit-preserving conversion also preserves the value, in which case the conversion is classed as value-preserving.

The convention is for type-conversion functions to be named to_*type*, where *type* is the name of the type being converted to. This convention is mostly followed by the synthesis packages, but with some exceptions.

However, this isn't the whole story. There are also built-in type conversions that extend the range of possibilities. Built-in type conversions (see Section 11.4) use the name of the target type as the name of the conversion function. These are sometimes used as bit-preserving conversions where no conversion function is provided.

6.7.1   Bit-Preserving Conversions

A bit-preserving conversion simply copies the value bit by bit without trying to interpret the bits as a numeric value. Consequently, the bit-preserving conversions have no hardware overhead, they are implemented as just wires.

The type conversions between the synthesis types and `std_logic_vector` are all bit-preserving, since there is no numeric interpretation of the type. In fact, `std_logic_vector` is the source or target type for all bit-preserving conversions provided by the synthesis packages.

The following sub-sections explain how to perform common bit-preserving conversions.

6.7.1.1 Conversion of Numeric Types and std_logic_vector

For the numeric types `signed` and `unsigned`, type conversions to and from `std_logic_vector` are implicitly provided by the built-in type conversions between similar array types. This means that, for example, to convert from `std_logic_vector` to `unsigned`, the name of the type, `unsigned`, is used as the type-conversion function.

```
library ieee;
use ieee.std_logic_1164.all;
use ieee.numeric_std.all;
entity type_conversion_demo is
  port (slv_in       : in  std_logic_vector(7 downto 0);
        unsigned_in  : in  unsigned(7 downto 0);
        signed_in    : in  signed(7 downto 0);
        slv_out1     : out std_logic_vector(7 downto 0);
        slv_out2     : out std_logic_vector(7 downto 0);
        unsigned_out : out unsigned(7 downto 0);
        signed_out   : out signed(7 downto 0));
end;
architecture behaviour of type_conversion_demo is
begin
  -- convert to std_logic_vector
  slv_out1 <= std_logic_vector(signed_in);
  slv_out2 <= std_logic_vector(unsigned_in);
  -- convert from std_logic_vector
  unsigned_out <= unsigned(slv_in);
  signed_out <= signed(slv_in);
end;
```

It is also possible to perform a bit-pattern preserving conversion between `signed` and `unsigned` using the type name as the conversion function. This is because they are also similar array types. For example:

```
library ieee;
use ieee.std_logic_1164.all;
use ieee.numeric_std.all;
entity to_unsigned_demo is
  port (a : in signed(6 downto 0);
        z : out unsigned(6 downto 0));
end;
architecture behaviour of to_unsigned_demo is
begin
  z <= unsigned(a);
end;
```

The opposite conversion is similar to this.

6.7.1.2 Conversion of Type Integer and std_logic_vector

For the built-in type integer, it is not possible to directly convert to the array type std_logic_vector in a bit-preserving way because that functionality is not provided by the std_logic_1164 package. Instead, convert the integer to signed or unsigned using a value-preserving conversion from numeric_std, then convert the signed or unsigned value to std_logic_vector using the built-in bit-preserving conversions for that type as described above.

Similarly for converting the other way, convert the std_logic_vector to signed or unsigned using the built-in conversion, then use the value-preserving conversion to integer provided by numeric_std.

```
library ieee;
use ieee.std_logic_1164.all;
use ieee.numeric_std.all;
entity type_conversion_demo is
  port (slv_in      : in  std_logic_vector(7 downto 0);
        integer_in  : in  integer range -128 to 127;
        slv_out     : out std_logic_vector(7 downto 0);
        integer_out : out integer range -128 to 127);
end;
architecture behaviour of type_conversion_demo is
begin
  -- convert to std_logic_vector
  slv_out <= std_logic_vector(to_signed(integer_in, 8));
  -- convert from std_logic_vector
  integer_out <= to_integer(signed(slv_in));
end;
```

Note: any conversion from integer to an array type will need extra information about the size of the array result to generate. In this case the size has been hard-coded. However, the convention is to use the length attribute on the target to specify the size of the result:

```
slv_out <= std_logic_vector(to_signed(integer_in, slv_out'length));
```

6.7.1.3 Conversion Between Fixed/Float Types and std_logic_vector

For the fixed-point types sfixed and ufixed and the floating-point type float the built-in type conversions do not work because these types can have negative ranges, whereas the std_logic_vector type cannot. For this reason, type-conversion functions have been provided to perform the bit-preserving conversions.

Table 6.10 lists the bit-preserving type-conversion functions provided by fixed_pkg.

The column labelled 'size?' indicates whether the function needs extra information to determine the size of the result. If the column says 'no', then the function can deduce the size from the input. This is true of all the conversions *to* std_logic_vector. If it says 'bounds' it needs two arguments giving the upper and lower bounds. Finally, if it says 'sample' then a sample variable or signal is used to provide the bounds. An entry with more than one option means there are several versions of the function, one for each option. For example, an entry of 'bounds/ sample' indicates two variants of the function, one that takes bounds and one that takes a sample.

Table 6.10 Bit-preserving type conversions in fixed_pkg

Function	From type	To type	Size?
to_ufixed	std_logic_vector	ufixed	bounds/sample
to_sfixed	std_logic_vector	sfixed	bounds/sample
to_slv	ufixed	std_logic_vector	no
to_slv	sfixed	std_logic_vector	no

These type conversions work by offsetting the range but preserving the bit-pattern. When converting to std_logic_vector, these types are offset until they have a lower bound of 0. When converting from std_logic_vector, a negative range is created by offsetting by a distance specified by bounds parameters.

For example:

```
library ieee;
use ieee.std_logic_1164.all;
use ieee.numeric_std.all;
use ieee.fixed_float_types.all;
use ieee.fixed_pkg.all;
entity type_conversion_demo is
  port (slv_in        : in  std_logic_vector(15 downto 0);
        ufixed_in     : in  ufixed(7 downto -8);
        sfixed_in     : in  sfixed(7 downto -8);
        slv_out1      : out std_logic_vector(15 downto 0);
        slv_out2      : out std_logic_vector(15 downto 0);
        ufixed_out    : out ufixed(7 downto -8);
        sfixed_out    : out sfixed(7 downto -8));
end;
architecture behaviour of type_conversion_demo is
begin
  -- convert to std_logic_vector
  slv_out1 <= to_slv(sfixed_in);
  slv_out2 <= to_slv(ufixed_in);
  -- convert from std_logic_vector
  ufixed_out <=
    to_ufixed(slv_in, ufixed_out'left, ufixed_out'right);
  sfixed_out <=
    to_sfixed(slv_in, sfixed_out'left, sfixed_out'right);
end;
```

Note how the conversions to_ufixed and to_sfixed take extra arguments that provide the bounds for the result just as with the resize functions. The bounds arguments must be constant for synthesis, since the synthesiser must be able to deduce the size of bus to create by examining the value of these arguments.

A similar set of conversions exist for float types, except that the extra arguments are sizes, not bounds.

Table 6.11 lists the bit-preserving conversions provided by float_pkg.

The column labelled 'size?' indicates whether the function needs extra information to determine the size of the result. If the column says 'no', then the function can deduce the size from the input. This is true of all the conversions *to* std_logic_vector. If the size column

Table 6.11 Bit-preserving type conversions in float_pkg

Function	From type	To type	Size?
to_float	std_logic_vector	float	sizes/sample
to_slv	float	std_logic_vector	no

says 'sizes', then two sizes are needed. Finally, if it says 'sample' then a sample variable or signal is used to provide the bounds. An entry with more than one option means there are several versions of the function, one for each option. For example, an entry of 'sizes/sample' indicates two variants of the function, one that takes sizes and one that takes a sample.

For example:

```
library ieee;
use ieee.std_logic_1164.all;
use ieee.numeric_std.all;
use ieee.fixed_float_types.all;
use ieee.fixed_pkg.all;
use ieee.float_pkg.all;
entity type_conversion_demo is
  port (slv_in    : in  std_logic_vector(31 downto 0);
        float_in  : in  float32;
        slv_out   : out std_logic_vector(31 downto 0);
        float_out : out float32);
end;
architecture behaviour of type_conversion_demo is
begin
  -- convert to std_logic_vector
  slv_out <= to_slv(float_in);
  -- convert from std_logic_vector
  float_out <=
    to_float(slv_in, float_out'left, -float_out'right);
end;
```

Note how the output bounds are converted into exponent and mantissa sizes.

6.7.2 Value-Preserving Conversions

A value-preserving conversion interprets the value according to its type and generates an equivalent value in the target type, a process that may require wrapping, saturating, rounding, truncation, zero filling or sign extension. For example, converting from an sfixed to a signed value is a value-preserving conversion where the fixed-point value is truncated or rounded to the nearest integer value during the conversion.

The value-preserving conversions are provided by the three synthesis packages. They are built-up in layers such that each layer provides conversions only to simpler types in lower layers. The simplest type is built-in integer. Then, numeric_std provides conversions between its types and integer. The next layer is provided by fixed_pkg that provides conversions between its types and the lower types, namely the numeric types and integer. Finally, the top layer is float_pkg that provides conversions between its types and all the other layers.

Table 6.12 Type conversion functions in numeric_std

Function	From type	To type	Size?
to_integer	signed	integer	no
to_integer	unsigned	integer	no
to_signed	integer	signed	size
to_unsigned	natural	unsigned	size

6.7.2.1 Conversions Provided by numeric_std

Table 6.12 lists the value-preserving type conversions provided by numeric_std.

The column labelled 'size?' indicates whether the function needs extra information to determine the size of the result. If the column says 'no', then the function can deduce the size from the input. If the size column says 'size', then a single size argument is needed.

Numeric_std only provides conversions between each numeric type and integer using these type-conversion functions. It does not provide a conversion between unsigned and signed.

To convert between subtype integer and signed, the type-conversion functions are used.

For example:

```
library ieee;
use ieee.std_logic_1164.all;
use ieee.numeric_std.all;
entity signed_conversions is
  port (integer_in  : in integer range -128 to 127;
        signed_in   : in signed(7 downto 0);
        integer_out : out integer range -128 to 127;
        signed_out  : out signed(7 downto 0));
end;
architecture behaviour of signed_conversions is
begin
  integer_out <= to_integer(signed_in);
  signed_out <= to_signed(integer_in, 8);
end;
```

The to_integer function can overflow if the input value has more than 32 significant bits. Overflow of an integer raises an error during simulation and is undefined in synthesis. It is best to avoid this scenario by only trying to convert signed values of no more than 32 bits, if necessary resizing first.

The to_signed can overflow the numeric type if the target is not big enough to take the full range of the input. Package numeric_std doesn't give overflow options but implements wrap mode on overflow.

The size argument of to_signed must be a constant argument for synthesis, since the synthesiser must be able to deduce the size of bus to create by examining the value of this argument. It can be a literal value as in the example, or a common convention is to use the target signal size:

```
signed_out <= to_signed(integer_in, signed_out'length);
```

The value-preserving conversion from `unsigned` to `signed` is made complicated by the fact that the `numeric_std` package does not provide the relevant function. So, conversion is a two-step operation of zero-extending using the `resize` function and then doing a bit-preserving conversion using the type name:

```
library ieee;
use ieee.std_logic_1164.all;
use ieee.numeric_std.all;
entity to_signed_demo is
  port (a : in unsigned(6 downto 0);
        z : out signed(7 downto 0));
end;
architecture behaviour of to_signed_demo is
begin
  z <= signed(resize(a,z'length));
end;
```

6.7.2.2 Conversions Provided by fixed_pkg

Table 6.13 lists the value-preserving type conversions provided by `fixed_pkg`.

The column labelled 'size?' indicates whether the function needs extra information to determine the size of the result. If the column says 'no', then the function can deduce the size from the input. If the size column says 'size', then a single size argument is needed. If it says 'bounds' it needs two arguments giving the upper and lower bounds. Finally, if it says 'sample' then a sample variable or signal is used to provide the bounds. An entry with more than one

Table 6.13 Type conversion functions in fixed_pkg

Function	From type	To type	Size?
to_ufixed	unsigned	ufixed	bounds/sample/no
to_ufixed	real[a]	ufixed	bounds/sample
to_ufixed	natural	ufixed	bounds/sample
to_sfixed	signed	sfixed	bounds/sample/no
to_sfixed	real[a]	sfixed	bounds/sample
to_sfixed	integer	sfixed	bounds/sample
to_sfixed	ufixed	sfixed	no
to_unsigned	ufixed	unsigned	size/sample
to_signed	sfixed	signed	size/sample
to_real	ufixed	real[a]	no
to_real	sfixed	real[a]	no
to_integer	ufixed	natural	no
to_integer	sfixed	integer	no

[a]*Note:* the conversions to and from real are not yet supported by most synthesis systems. These conversions have been included in the table because they may be supported, at least for floating-point literals, in the near future.

option means there are several versions of the function, one for each option. For example, an entry of 'bounds/sample' indicates two variants of the function, one that takes bounds and one that takes a sample.

To convert between subtype `integer` and `sfixed`, functions from `fixed_pkg` are used much like the `numeric_std` versions. However, for fixed-point types the function can take rounding options:

```
function to_integer (x : sfixed;
   overflow_style : fixed_overflow_style_type := fixed_saturate;
   round_style : fixed_round_style_type := fixed_round)
   return integer;
```

For example:

```
library ieee;
use ieee.std_logic_1164.all;
use ieee.numeric_std.all;
use ieee.fixed_float_types.all;
use ieee.fixed_pkg.all;
entity signed_conversions is
   port (int_in  : in integer range -128 to 127;
         sf_in   : in sfixed(7 downto -8);
         int_out : out integer range -128 to 127;
         sf_out  : out sfixed(7 downto -8));
end;
architecture behaviour of signed_conversions is
begin
   int_out <=
     to_integer(sf_in, fixed_wrap, fixed_truncate);
   sf_out <=
     to_sfixed(int_in, sf_out'left, sf_out'right, fixed_wrap);
end;
```

The `to_integer` function removes the fractional part using one of the underflow modes to decide how to convert the result. In Truncate Mode, the fractional part is simply discarded, whereas in Rounding Mode the integer part is rounded to the nearest integer value. Also, since type integer is a 32-bit type, the conversion could overflow. In this case the overflow behaviour is determined by the overflow mode, just as with the resize functions. In Wrap Mode the number is wrapped round, whereas in Saturation Mode it stops at the saturation value.

The `to_sfixed` function converts the integer input to a fixed-point value with the same integer part and zero-filled fractional part. It takes extra arguments that give the bounds of the value to create. As usual, you can either use a sample signal to specify this, or two values to specify the bounds as in the example. If the target is smaller than the input type, overflow can take place. In this case the overflow behaviour is determined by the overflow mode.

The bounds arguments must be constant for synthesis, since the synthesiser must be able to deduce the size of bus to create by examining the value of this argument.

The example overrides the default overflow and underflow modes to give the minimum-area conversion using Wrap Mode and Truncate Mode, respectively. The default modes of Saturate Mode and Rounding Mode can be used by omitting the extra argument:

```
int_out <= to_integer(sf_in);
sf_out <= to_sfixed(int_in, sf_out'left, sf_out'right);
```

There are similar conversions between `signed` and `sfixed`. The main difference is that the conversion to `signed` needs a size parameter or a sample signal to determine the size of the result.

There is a conversion function provided for the conversion from `ufixed` to `sfixed` so this is implemented by a single function call to `to_sfixed`.

```
library ieee;
use ieee.std_logic_1164.all;
use ieee.numeric_std.all;
use ieee.fixed_float_types.all;
use ieee.fixed_pkg.all;
entity to_sfixed_demo is
  port (a : in ufixed(7 downto -8);
        z : out sfixed(8 downto -8));
end;
architecture behaviour of to_sfixed_demo is
begin
  z <= to_sfixed(a);
end;
```

No size argument is needed, but the conversion function increases the size of the value by 1 bit to contain the whole range of the result so you must size the target of the assignment correctly.

There is no reverse conversion because there is no value-preserving interpretation of a negative number in an unsigned representation. A bit-preserving conversion can still be used if necessary.

6.7.2.3 Conversions Provided by float_pkg

Table 6.14 lists the value-preserving type-conversion functions provided by `float_pkg`.

The column labelled 'size?' indicates whether the function needs extra information to determine the size of the result. If the column says 'no', then the function can deduce the size from the input. If the size column says 'size', then a single size argument is needed. If it says 'sizes' it needs two arguments giving the exponent and mantissa sizes. Finally, if it says 'sample' then a sample variable or signal is used to provide the bounds.

To convert between subtype `integer` and `float`, functions from `float_pkg` are used much like the `fixed_pkg` versions. However, for floating-point types the function can take error-checking and rounding options:

```
function to_integer (arg : float;
    round_style : round_type := round_nearest;
    check_error : boolean := true)
  return integer;

function to_float (arg : integer;
    exponent_width : natural := 8;
    fraction_width : natural := 23;
    round_style : round_type := round_nearest)
  return float;
```

Table 6.14 Type conversion functions in float_pkg

Function	From type	To type	Size?
to_float	sfixed	float	sizes/sample
to_float	ufixed	float	sizes/sample
to_float	signed	float	sizes/sample
to_float	unsigned	float	sizes/sample
to_float	real[a]	float	sizes/sample
to_float	integer	float	sizes/sample
to_sfixed	float	sfixed	sizes/sample
to_ufixed	float	ufixed	sizes/sample
to_signed	float	signed	size/sample
to_unsigned	float	unsigned	size/sample
to_real	float	real[a]	no
to_integer	float	integer	no

[a]*Note:* the conversions to and from real are not yet supported by most synthesis systems. These conversions have been included in the table because they may be supported, at least for floating-point literals, in the near future.

For example:

```
library ieee;
use ieee.std_logic_1164.all;
use ieee.numeric_std.all;
use ieee.fixed_float_types.all;
use ieee.fixed_pkg.all;
use ieee.float_pkg.all;
entity float_conversions is
  port (int_in    : in integer range -128 to 127;
        float_in  : in float32;
        int_out   : out integer range -128 to 127;
        float_out : out float32);
end;
architecture behaviour of float_conversions is
begin
  int_out <=
    to_integer(float_in, round_zero, false);
  float_out <=
    to_float(int_in, float_out'left, -float_out'right, round_zero);
end;
```

The to_integer function removes the fractional part using one of the four rounding modes to decide how to convert the result.

The to_float function converts the integer input to a floating-point value with the same integer value and then normalises it. It takes extra arguments that give the bounds of the value to create. As usual, you can either use a sample signal to specify this, or two values to specify the sizes as in the example. Note how the array bounds are converted into field sizes. If the target is smaller than the input type, overflow can take place. In this case the overflow behaviour is determined by the rounding mode.

The bounds arguments must be constant for synthesis, since the synthesiser must be able to deduce the size of bus to create by examining the value of this argument.

The example overrides the default error-checking and rounding modes to give the minimum-area conversion. The default mode of Round to Nearest can be used by omitting the extra argument:

```
int_out <= to_integer(float_in);
float_out <= to_float(int_in, float_out'left, -float_out'right);
```

There are similar conversions between sfixed or ufixed and float. The main difference is that the conversion to a fixed-point type needs two bounds parameters or a sample signal to determine the size of the result.

```
function to_sfixed (arg : float;
  left_index     : integer;
  right_index    : integer;
  overflow_style : fixed_overflow_style_type := fixed_saturate;
  round_style    : fixed_round_style_type    := fixed_round;
  check_error    : boolean                    := true;
  denormalize    : boolean                    := true)
  return sfixed;
```

Since conversion from floating-point to fixed-point can underflow or underflow, the fixed-point modes need to be specified as well as the optional checking for errors and support for denormalised numbers for the float being converted, giving four mode parameters altogether.

Finally, there are conversions between unsigned or signed and float. The conversion to a numeric type needs one size parameter or a sample signal to determine the size of the result.

```
function to_signed (arg : float;
  size        : natural;
  round_style : round_type := round_nearest;
  check_error : boolean    := true)
  return signed;
```

This is very similar to the to_integer conversion.

6.8   Constant Values

Constant values of synthesis types can be represented by string values or by bit-string values.

The interpretation of the numeric types means that string values are written exactly as you would expect them to be written – a bit pattern with the most-significant bit on the left. For example, to initialise an 8-bit signal, the following assignment would be used:

```
library ieee;
use ieee.std_logic_1164.all;
use ieee.numeric_std.all;
entity zero is
  port (z : out signed (7 downto 0));
end;
architecture behaviour of zero is
begin
  z <= "00000000";
end;
```

When assigning to a signal, or variable, of a synthesis type, the string value must be exactly the right length. However, in this case, an alternative would have been to use an aggregate with an `others` choice:

```
z <= (others => `0');
```

When adding a `signed` or `unsigned` signal to a constant value, there is no need to match the size of the arguments, because the add operator has been designed to take arguments of different lengths. In this case, an `unsigned` signal is incremented by simply adding the string value `"1"` to the input.

```
library ieee;
use ieee.std_logic_1164.all;
use ieee.numeric_std.all;
entity unsigned_increment is
  port (a : in unsigned (7 downto 0);
        z : out unsigned (7 downto 0));
end;
architecture behaviour of unsigned_increment is
begin
  z <= a + "1";
end;
```

Beware when using type `signed`, because then the string value `"1"` has the value -1. Remember that type `signed` is a signed type and that therefore the sign bit must be included in the string value. For the value `"1"`, the single bit is the sign bit and, since the bit is `'1'`, then the value must be negative. In fact a 1-bit signed number has the range -1 to 0. To get the value 1, use the 2-bit value `"01"`.

The easiest way to avoid such potential pitfalls is to use integer values directly. The mixing of numeric types with type `integer` in arithmetic and comparisons is covered by the next section.

When using `sfixed`, `ufixed` and `float`, literal values can only be used where the range can be deduced from the context. This is because of the strict rules about fixed-point ranges and their interpretation in placing the binary point.

The simple assignment is one case where the range can be deduced, so the following is a legal assignment to an 8.8-bit `sfixed`:

```
library ieee;
use ieee.std_logic_1164.all;
use ieee.numeric_std.all;
use ieee.fixed_float_types.all;
use ieee.fixed_pkg.all;
entity zero is
  port (z : out sfixed (7 downto -8));
end;
architecture behaviour of zero is
begin
  z <= "0000000000000000";
end;
```

However, the bit-string literal is a more convenient choice because it allows the use of an underscore to represent the binary point for a bit more readability:

```
z <= B"00000000_00000000";
```

However, literal values cannot be used in expressions, either as part of an assignment or in a conditional:

```
library ieee;
use ieee.std_logic_1164.all;
use ieee.numeric_std.all;
use ieee.fixed_float_types.all;
use ieee.fixed_pkg.all;
entity illegal_increment is
  port (a : in sfixed (7 downto -8);
        z : out sfixed (8 downto -8));
end;
architecture behaviour of illegal_increment is
begin
  z <= a + B"01_0"; -- error
end;
```

The reason is that the range of this literal cannot be deduced from the context. It is an argument of the $+$ operator and this can take any range as its input, so the rules of VHDL say that it must then take the range (integer'left to integer'left+2), which breaks the rules for fixed-point ranges and is certainly not what was intended.

Bear in mind that the fixed-point packages are providing an interpretation of a sequence of bits that is not part of the language and there is no built-in support for fixed-point representations in the language. It is therefore necessary to work round the limitations that this imposes.

The solution to this problem is either to add the integer value 1 or to declare a constant value and use that in the addition:

```
architecture behaviour of legal_increment is
  constant one : sfixed(1 downto -1) := B"01_0";
begin
  z <= a + one;
end;
```

The use of integer literals is usually preferred provided the values being added are integral values, but if fractions are involved, constants with string or bit-string literal values must be used since there is no other way of representing fractions.

6.9   Mixing Types in Expressions

The operators in the synthesis packages have also been defined so that it is possible to mix the array types with type integer. This is mainly of use when adding or comparing with constant integer values, since it avoids the need to convert the integer value into its string value. This is easier to understand since integer values are expressed in decimal, whereas string values are binary. Having said that, sometimes a binary representation is clearer (for example in masking operations) or is the only way (for example in representing fractions) and in those cases string values should be used.

As an example of using `integer` values, the previous example of an 8-bit incrementer could have been written using the `integer` value 1 in the addition:

```
architecture behaviour of increment is
begin
  z <= a + 1;
end;
```

The second advantage of this form of the addition is that there is no possibility of falling into the trap of omitting the leading sign bit on `signed` numbers.

It is also possible to use `integer` values in comparisons. For example, a test for a signal being negative could be written using a test against the `integer` value 0:

```
library ieee;
use ieee.std_logic_1164.all;
use ieee.numeric_std.all;
entity compare is
  port (a : in signed (7 downto 0);
        negative : out std_logic);
end;
architecture behaviour of compare is
begin
  negative <= '1' when a < 0 else '0';
end;
```

In this example, the comparison is being made between a signal of type `signed` and the `integer` value 0.

Note: The fixed-point packages *appear* to allow real values to be used instead of integer values in mixed-type expressions, so it would appear that the following would be legal:

```
z <= a + 0.1;
```

The problem is that 0.1 is a `real` value, and `real` is a floating-point type not a fixed-point type. So this usage would require the synthesis tool to convert the floating-point value into the fixed-point value 0.1 in such a way as to incur no hardware cost. It may be that some synthesisers will provide this functionality at some time in the future for constant values only, but it isn't generally available at the time of writing and probably won't ever be guaranteed to be available. So this usage is not recommended for synthesis.

6.10   Top-Level Interface

Synthesis tools generate a gate-level netlist in VHDL that has the same basic interface as the RTL model's top-level entity but reduced down to `std_logic` and `std_logic_vector` types. All the multi-bit synthesis types get mapped to `std_logic_vector`, as does `integer`.

Some designers choose to design with a top-level entity using only these two types so that the interface does not change between the RTL model and the gate-level model, thus allowing the same test-bench to be used for both.

So, to use this convention, design the system using the most appropriate types throughout as shown throughout this book, then for each component that needs a test bench, add an extra top-level entity and architecture with simple ports.

This top-level model will have an entity with the same port names as the original design, but each reduced to these two simple types while maintaining the same number of bits. The architecture of the top-level model then contains a single component instance and a set of bit-preserving type conversions on the port map of the component.

Consider the `unsigned_increment` design from the last section. It has the following interface:

```
library ieee;
use ieee.std_logic_1164.all;
use ieee.numeric_std.all;
entity unsigned_increment is
  port (a : in unsigned(7 downto 0);
        z : out unsigned(7 downto 0));
end;
```

A top-level circuit can be added as a wrapper around this:

```
library ieee;
use ieee.std_logic_1164.all;
entity unsigned_increment_top is
  port (A : in std_logic_vector(7 downto 0);
        Z : out std_logic_vector(7 downto 0));
end;
```

Note that I have made the ports uppercase to make the explanation clearer. The architecture is:

```
use ieee.numeric_std.all;
architecture behaviour of unsigned_increment_top is
begin
  d1: entity work.unsigned_increment(behaviour)
    port map (a => unsigned(A),
              std_logic_vector(z) => Z);
end;
```

Notice how the type conversions have been added to the port map of the component instance. For `in` mode ports the type conversion is on the right-hand side of the port map, that is converting the top-level entity port, but for `out` mode ports, the type conversion goes on the left-hand side, that is converting the lower-level component ports.

Conceptually, for an input the top-level entity port is type converted before the component and the converted signal connected to the component port, whereas for an output, the component port produces a result that is type converted after leaving the component and then the converted signal connected to the top-level entity port.

An `inout` port (i.e. a tristate bus) uses a combination of the two methods. Consider a tristate driver component with the following partly defined interface:

```
library ieee;
use ieee.std_logic_1164.all;
```

```
use ieee.numeric_std.all;
entity tristate is
  port (...
        z : inout unsigned(7 downto 0);
        ...);
end;
```

Then, the following is a top-level wrapper for this component:

```
library ieee;
use ieee.std_logic_1164.all;
entity tristate_top is
  port (...
        Z : inout std_logic_vector(7 downto 0);
        ...);
end;

use ieee.numeric_std.all;
architecture behaviour of tristate_top is
begin
  d1: entity work.tristate(behaviour)
    port map (...
              std_logic_vector(z) => unsigned(Z),
              ...);
end;
```

So, data travelling into the component gets converted from `std_logic_vector` to `unsigned` by the conversion on the right-hand side, whereas data travelling out of the component gets converted from `unsigned` to `std_logic_vector` by the conversion on the left-hand side.

When converting one of the floating-point or fixed-point types, use the type-conversion functions rather than the built-in type conversions to handle the negative indices correctly. Array ports with negative indices then get mapped onto a natural range.

For example, consider another simple example using the fixed-point types and having the following interface:

```
library ieee;
use ieee.std_logic_1164.all;
use ieee.numeric_std.all;
use ieee.fixed_float_types.all;
use ieee.fixed_pkg.all;
entity increment is
  port (a : in sfixed(7 downto -8);
        z : out sfixed(8 downto -8));
end;
```

The wrapper circuit uses normalised `std_logic_vector` ports and type-conversion functions from `fixed_pkg`:

```
library ieee;
use ieee.std_logic_1164.all;
entity increment_top is
```

```
   port (A : in std_logic_vector(15 downto 0);
         Z : out std_logic_vector(15 downto 0));
end;

use ieee.numeric_std.all;
use ieee.fixed_float_types.all;
use ieee.fixed_pkg.all;
architecture behaviour of increment_top is
begin
  d1: entity work.increment(behaviour)
    port map (a => to_sfixed(A, 7, -8),
              to_slv(z) => Z);
end;
```

Note that the `to_sfixed` conversion takes three arguments, because it also needs the bounds of the value to convert to. In this case there is no signal available with the right range so integer constants have been used.

7

Std_Logic_Arith

The last chapter described the synthesis packages for performing arbitrary-precision arithmetic. Unfortunately, these are not the only numeric packages in use. The reason for this is historical: the numeric packages were added to VHDL several years after synthesis tools using VHDL became available. The need for arbitrary-precision integer arithmetic was obvious, so synthesis vendors filled the gap by providing their own proprietary packages.

The package that came to be adopted by most synthesis vendors and for a while became accepted as a de-facto standard, is `std_logic_arith`. This package originated from Synopsys Incorporated, for use with their synthesis system. This package is now available with most synthesis and simulation systems and is public domain.

The main problem with `std_logic_arith` was that it was originally a proprietary package, the copyright in which rested with Synopsys Incorporated. This led to a standardisation effort to replace the package with a non-proprietary one of comparable functionality. This effort resulted in the IEEE standard packages `numeric_std` and `numeric_bit`. Package `numeric_std` is almost a direct replacement of `std_logic_arith` and has the advantage of being standardised. `Numeric_bit` has the same functionality but uses type `bit` and so is rarely used.

It has taken many years for the new IEEE standard to come into widespread use and package `std_logic_arith` is still in use in some designs. However, its use is deprecated and it is strongly recommended that new designs use `numeric_std` and the other synthesis packages described in Chapter 6. This chapter is only included in order to understand and modify legacy designs that still use it.

7.1 The Std_Logic_Arith Package

Package `std_logic_arith` represents numeric values as arrays of `std_logic`. Operators are provided such that it is possible to perform bitwise logical operations, arithmetic operations and numeric comparisons on the same type. Furthermore, since the type is an array, it is also possible to directly access individual bits, take slices to implement bus rippers, concatenate arrays to merge buses together and all the other bus-like operations that you expect in hardware design. In effect, the types defined in `std_logic_arith` are universal types.

VHDL for Logic Synthesis, Third Edition. Andrew Rushton.
© 2011 John Wiley & Sons, Ltd. Published 2011 by John Wiley & Sons, Ltd.

The choice of logic type for these packages was `std_logic` because it can also be used to represent unknown and high-impedance values so that bidirectional buses and tristate drivers can be modelled.

This package, combined with std_logic_1164, provide a *type system* of five types:

>`std_logic` – for one-bit paths such as clock and control lines
>`std_logic_vector` – for multi-bit paths with no numeric interpretation
>`signed` – for multi-bit paths with 2's-complement integer interpretation
>`unsigned` – for multi-bit paths with unsigned integer notation
>`integer` – for array indexing of the multi-bit types

Note: it is difficult to mix `std_logic_arith` with the fixed-point and floating-point packages because they are implemented using the `numeric_std` types `unsigned` and `signed`. If you want to use the new packages, you should use `numeric_std` for your numeric types.

7.1.1   Making the Package Visible

The package is usually found in library `ieee`. This is a bit cheeky on the part of the originators because the package is not an IEEE standard. However, this is the most sensible place to put the package since it uses package `std_logic_1164`, which is in library `ieee`. Some vendors keep it separate in a different library, or provide two versions of library `ieee`, one with non-standard additions and one that is in strict keeping with the philosophy of IEEE standardisation.

To use the package, there needs to be a `library` clause to make the library visible and a `use` clause to make the contents of the package visible. It will also be necessary to have a `use` clause that makes the contents of the `std_logic_1164` package visible. When you use a package such as `std_logic_arith`, the packages that it in turn uses are not inherited, so it is necessary to explicitly `use` them. Therefore, the following declarations will be needed before the entity or architecture that is to use the packages:

```
library ieee;
use ieee.std_logic_1164.all;
use ieee.std_logic_arith.all;
```

7.2   Contents of Std_Logic_Arith

The package `std_logic_arith` defines two types, both of which are unconstrained arrays of the element type, `std_logic`. The types are called `unsigned` and `signed`:

```
type signed is array (natural range <>) of std_logic;
type unsigned is array (natural range <>) of std_logic;
```

The convention is that type `signed` represents 2's-complement signed numbers, whilst type `unsigned` represents unsigned magnitudes. The arithmetic operators provided with the package of course rely on this convention.

The use of two completely separate types means that there is never any confusion over the interpretation of a particular bit pattern.

For example, here are two examples of 8-bit signals defined using the two types.

```
signal a : signed(7 downto 0);
signal b : unsigned(7 downto 0);
```

Both declarations create 8-bit buses. However, the first declaration creates a bus that will be interpreted as a signed number with the range −128 to 127, whereas the second, although exactly the same size, will be interpreted as an unsigned number with the range 0 to 255.

The types are also completely distinct from `std_logic_vector`, which does not have a numeric interpretation, so there is no possible confusion there either. To make use of the extended operator set provided by the package, signals or variables must be of type `unsigned` or `signed`.

7.2.1 Resize Functions

The package defines a set of functions, called `conv_signed` and `conv_unsigned`, which are used to resize – that is, to truncate or extend values of type `signed` and `unsigned`. These can be used, for example, to assign an 8-bit number to a 16-bit number. There are in fact many functions with these names: the others also perform type conversions. This section will focus on the functions that perform resizing without changing the type. The other functions that combine resizing with type conversion will be described separately in Section 7.3, which describes type conversions.

To give an example of how to use the resize functions, here is such an example of an 8-bit `signed` signal being assigned to a 16-bit signal:

```
library ieee;
use ieee.std_logic_1164.all;
use ieee.std_logic_arith.all;
entity signed_resize_larger is
  port (a : in signed(7 downto 0);
        z : out signed(15 downto 0));
end;
architecture behaviour of signed_resize_larger is
begin
  z <= conv_signed(a, 16);
end;
```

The `conv_signed` function takes two arguments. The first is the signal or variable to resize (in this case the signal a) and the second is the size to resize it to (in this case 16 bits). This must be a constant value if it is to be synthesisable, since a synthesiser must know what size the result is in order to create a bus with the right number of bits.

However, that constant value can be derived from the `length` attribute of the target of the assignment, and this gives a very convenient notation:

```
z <= conv_signed(a, z'length);
```

This is effectively saying: resize to the size of z, whatever that happens to be. It is possible to read the length of z, even though it is an out port. It is only the value of out ports that are unreadable, not their dimensions. The advantage of this notation is that there is no possibility of an accidental mismatch of signal sizes in the assignment if this notation is used.

In this example, the signal is being resized to a larger size. The function does this by sign-extending the argument.

The conv_unsigned function, which operates on unsigned numbers, is almost identical to the conv_signed function and is used in exactly the same way:

```
library ieee;
use ieee.std_logic_1164.all;
use ieee.std_logic_arith.all;
entity unsigned_resize_larger is
  port (a : in unsigned(7 downto 0);
        z : out unsigned(15 downto 0));
end;
architecture behaviour of unsigned_resize_larger is
begin
  z <= conv_unsigned(a, 16);
end;
```

The main difference is that it resizes unsigned values to a larger size by zero extension rather than by sign extension.

These functions also work the other way round, to make a smaller value by truncation of a large one. Thus, the opposite of the above signed example is:

```
library ieee;
use ieee.std_logic_1164.all;
use ieee.std_logic_arith.all;
entity signed_resize_smaller is
  port (a : in signed(15 downto 0);
        z : out signed(7 downto 0));
end;
architecture behaviour of signed_resize_smaller is
begin
  z <= conv_signed(a, 8);
end;
```

This example reduces a 16-bit signal to an 8-bit result and assigns this to the output port. Reduction is carried out by the conv_signed function by truncation: the most-significant bits of the argument are simply stripped off. In other words, in this case, bits 15 downto 8 would be discarded.

Note: this follows the normal convention when truncating. If the value is small enough to fit in the truncated size, then the value is preserved for both positive and negative numbers. However, for large values that don't fit into the truncated size, signed numbers wrap round so that for example, a large negative number can become positive as a result of the truncation. Similarly, a large positive number can become negative. This is of course the expected behaviour in 2's-complement notation.

As an example, consider the truncation of the 16-bit negative value "1000000000000001" (−32767) being truncated to 8-bits to give "00000001" (+1).

The final combination is the use of `conv_unsigned` of a large `unsigned` number to a smaller size:

```
library ieee;
use ieee.std_logic_1164.all;
use ieee.std_logic_arith.all;
entity unsigned_resize_demo2 is
  port (a : in unsigned(15 downto 0);
        z : out unsigned(7 downto 0));
end;
architecture behaviour of unsigned_resize_demo2 is
begin
  z <= conv_unsigned(a, 8);
end;
```

The `conv_unsigned` function also simply discards the most-significant bits as expected.

7.2.2 Operators

The `std_logic_arith` package defines a reasonable set of synthesisable operators for both types. Furthermore, there are guidelines on how to interpret and use the types to get the correct functionality from the packages.

The set of operators available for type `signed` is:

comparison:	`=, /=, <, <=, >, >=`
arithmetic:	`sign -, sign +, abs, +, -, *`
concatenation:	`&`

The operators missing from this list are all the boolean operators, the exponent operator "**" and the division operators "/", `mod` and `rem`. The shift operators are not present, but they are provided instead as functions with non-standard names.

Type `unsigned` has a slightly more restricted set of operators, in that the sign operator "-" and the `abs` operator are not present, since they are meaningless for unsigned arithmetic.

One of the characteristics of package `std_logic_arith` is that there are several copies of each operator provided, each supporting a different permutation of operands. For the comparison and arithmetic operators, all permutations of type `signed`, type `unsigned` and type `integer` are supported. This is a method called *overloading* and it allows the types to be mixed without type conversions. For example, an `unsigned` and an `integer` can be added without first converting one or other of the operands.

Unfortunately, overloading in `std_logic_arith` is excessive and can lead to ambiguities that in turn lead to compilation errors for what appears to be legitimate VHDL. This problem affects all of these operators, but the discussion of the problem and its solution will be addressed in Section 7.4 on constant values, because it is a particularly irritating problem when trying to use string values as constants in expressions.

7.2.3   Comparison Operators

The numeric types have the complete set of comparison operators:

comparison: =, /=, <, <=, >, >=

Section 4.10 issued a warning that the built-in comparison operators for arrays do not work correctly for bit-arrays representing numbers. However, this warning can be ignored for types in `std_logic_arith`, because the comparison operators for `signed` and `unsigned` have been redefined to give correct numeric comparisons. Furthermore, they have been defined to give correct comparisons even with arguments of different lengths. This makes the comparison operators extremely easy and natural to use.

As an example of the usage of comparison operators, consider the following circuit that returns the maximum of two `signed` values.

```
library ieee;
use ieee.std_logic_1164.all;
use ieee.std_logic_arith.all;
entity max is
  port (a, b : in signed(7 downto 0);
        z : out signed(7 downto 0));
end;
architecture behaviour of max is
begin
  z <= a when a > b else b;
end;
```

This example shows how the greater-than operator can be used to compare two parameters of type `signed`. All the other comparison operators can be used in the same way.

7.2.4   Boolean Operators

The most surprising omission from the `std_logic_arith` package is the lack of bitwise logical operators for the two array types.

The way to work round this omission is to use the logical operators provided by the underlying package `std_logic_1164`, which does define the full range of bitwise logical operators.

To combine two signals of a numeric type, first type convert to `std_logic_vector`, then perform the logical operation, then convert the type back again.

The arrays must be the same size for this to work, although their ranges may be different. This rule is required by the `std_logic_vector` rules.

For example, the following trivial entity and architecture form the logical-and of two 8-bit `unsigned` numbers to form an 8-bit result.

```
library ieee;
use ieee.std_logic_1164.all;
use ieee.std_logic_arith.all;
entity and_demo is
  port (a, b : in unsigned(7 downto 0);
        z : out unsigned(7 downto 0));
```

```
    end;
    architecture behaviour of and_demo is
    begin
      z <= unsigned(std_logic_vector(a)
                    and
                    std_logic_vector(b));
    end;
```

Many designers make the mistake of performing the bit-by-bit logical operations themselves by writing a `for` loop to go through all the elements individually. As the example shows, this is unnecessary and using the array logical operators from `std_logic_1164` is much simpler.

7.2.5   Arithmetic Operators

The numeric types have a restricted set of arithmetic operators:

arithmetic: `sign -, sign +, abs, +, -, *`

These operators have been added to the types in such a way that they can be used as numeric types. Type `signed` has been given a set of operators that performs 2's-complement arithmetic, whilst type `unsigned` has been given a set of operators that performs magnitude arithmetic. Type `unsigned` does not have the sign `"-"` or the `abs` operators since they would be useless.

The operators have also been defined so that, unlike integer arithmetic in VHDL, the numbers wrap-round on overflow. For example, adding one to the most positive `signed` number gives the most negative `signed` number. Similarly, adding one to the most positive `unsigned` number gives zero.

The unary (sign) `"-"` operator forms the 2's-complement negation of a number. It gives a result that is the same size as the argument, so that it is possible to negate a signal or variable and assign it back to itself, or to another signal or variable of the same size. For example, to negate a signal `a`:

```
    library ieee;
    use ieee.std_logic_1164.all;
    use ieee.std_logic_arith.all;
    entity negation_demo is
      port (a : in signed(7 downto 0);
            z : out signed(7 downto 0));
    end;
    architecture behaviour of negation_demo is
    begin
      z <= -a;
    end;
```

Of course, the 2's-complement integer range is asymmetrical, so the negation of the most negative value will overflow and therefore wrap-round onto the most negative value. This means in practice that the negation of the most negative number is itself. This strange behaviour is a side effect of the 2's-complement representation and happens in hardware, so this is not necessarily a problem. It could have been avoided in the package design by returning a result that was one bit bigger, but this would have sacrificed the characteristic of these operators that they preserve the size of the datapath. If the extra bit is needed, then the way to achieve this is to combine the negation operator with the `conv_signed` function such that the resize happens *before* the negation:

```
library ieee;
use ieee.std_logic_1164.all;
use ieee.std_logic_arith.all;
entity negation_resize_demo is
  port (a : in signed(7 downto 0);
        z : out signed(8 downto 0));
end;
architecture behaviour of negation_resize_demo is
begin
  z <= -conv_signed(a, 9);
end;
```

The `abs` operator is only available for the `signed` type. It forms the absolute value by negating the argument if it is negative but leaving it unchanged otherwise. The result will be the same size as the argument, which means that it has the same characteristics as the negation operator in that the negation of the most negative value overflows and therefore wraps-round onto the most negative number. This gives the rather strange side effect that the result of the `abs` operator can be negative in this one special case. Once again the solution if this is to be avoided is to resize to a size at least one bit larger than the operand *before* taking the absolute value.

```
library ieee;
use ieee.std_logic_1164.all;
use ieee.std_logic_arith.all;
entity abs_resize_demo is
  port (a : in signed(7 downto 0);
        z : out signed(8 downto 0));
end;
architecture behaviour of abs_resize_demo is
begin
  z <= abs conv_signed(a, 9);
end;
```

Note that `abs` is an operator, not a function, so it doesn't have parentheses around its argument.

The add and subtract ("+" and "-") operators also preserve the length of the datapath; if two 8-bit numbers, for example, are added together, then the result will also be an 8-bit wide number. This means that addition and subtraction wrap-round on overflow. The following example shows an 8-bit addition:

```
library ieee;
use ieee.std_logic_1164.all;
use ieee.std_logic_arith.all;
entity add_demo is
  port (a, b : in signed(7 downto 0);
        z : out signed(7 downto 0));
end;
architecture behaviour of add_demo is
```

```
begin
  z <= a + b;
end;
```

If it is not desirable for the addition to wrap-round, in other words, if the result is required to be one bit larger than the arguments, then the arguments should be resized *before* they are added:

```
library ieee;
use ieee.std_logic_1164.all;
use ieee.std_logic_arith.all;
entity add_resize_demo is
  port (a, b : in signed(7 downto 0);
          z : out signed(8 downto 0));
end;
architecture behaviour of add_resize_demo is
begin
  z <= conv_unsigned(a,9) + conv_unsigned(b,9);
end;
```

In fact, the add and subtract operators have been defined so that they can be used on arguments that are different sizes. In the case where they are different sizes, the result size is the same as the largest of the arguments. For example, if an 8-bit number is added to a 16-bit number, the result is a 16-bit number.

The final synthesisable arithmetic operator is the multiplication operator ("* "). As with the other operators, this is defined for both signed and unsigned types. The multiplication operators have also been defined so that they can take arguments of different lengths. For example, an 8-bit number can be multiplied by a 16-bit number. Unlike the other arithmetic operators, however, the result size is calculated by adding the sizes of the arguments, so in this example, the result would be 24 bits long. This makes the multiplication operator the exception to the general rule that the arithmetic operators preserve the datapath width. It also means that multiplication never overflows and therefore never wraps-round the result, so it should never be necessary to resize arguments before multiplication to avoid losing overflow bits. However, it often means that the result *after* the multiplication will need to be either truncated by slicing or by resizing.

The following example shows how two 8-bit signed numbers can be multiplied to give an 8-bit result by truncating the 16-bit result with conv_signed *after* the multiplication.

```
library ieee;
use ieee.std_logic_1164.all;
use ieee.std_logic_arith.all;
entity multiply_demo is
  port (a, b : in signed(7 downto 0);
          z : out signed(7 downto 0));
end;
architecture behaviour of multiply_demo is
  signal product : signed(15 downto 0);
begin
  product <= a * b;
  z <= conv_unsigned(product, 8);
end;
```

In fact, the whole operation can be reduced to a single line:

```
z <= conv_unsigned(a * b, 8);
```

7.2.6   Shift Functions

Package std_logic_arith was developed soon after the original VHDL-1987 standard
was introduced, so is based on the version of the language with no shift operators built-in. To
overcome this lack the package provides a set of shift *functions* on the signed and
unsigned types. There are two shift operations on each type:

    shl shift-left
    shr shift-right

The shl and shr functions each take two arguments, the first being the value, signal or
variable to shift and the second being an unsigned value specifying how far to shift the value.
Note that, unlike the VHDL shift operators, where the shift distance is given by an integer value,
in package std_logic_arith the shift distance is given by the array type unsigned.
For synthesis, the shift distance must be a constant value because the synthesis interpretation
of a shift is a hardwired rearrangement of the bits of the bus.

The shift operations perform straightforward arithmetic shifts. There is no ambiguity about
the shift operations performed on the unsigned type. Both left-shift and right-shift fill with
'0'. In other words, when shifting to the left, zeros are shifted in at the right and when shifting
right, zeros are shifted in at the left.

A typical use of the shift functions is shown in the following example that shows a left shift
of four bits being performed on an 8-bit signal:

```
library ieee;
use ieee.std_logic_1164.all;
use ieee.std_logic_arith.all;
entity shift_demo is
  port (a : in unsigned(7 downto 0);
        z : out unsigned(7 downto 0));
end;
architecture behaviour of shift_demo is
begin
  z <= shl(a, "100");
end;
```

Note that shl is a function, not an operator, so it needs the parentheses. Note also the use of
a string value to specify the shift distance as an unsigned value, in this case 4.

For the signed type, the shifts are slightly different, so that in a right shift the sign bit is
extended. In other words, in a right shift the sign bit is shifted in at the left. There is no difference
to the left shift, which is exactly the same as the unsigned left shift. Note that this implies that
overflow results in the value being wrapped around, potentially changing the sign of the result.
This is consistent with the design of the other arithmetic operators and is expected behaviour
for the 2's-complement notation.

Table 7.1 Type-conversion functions in std_logic_arith

Function name	Argument type	Result type	Second argument?
conv_integer	integer	integer	yes
conv_integer	signed	integer	no
conv_integer	unsigned	integer	no
conv_integer	std_ulogic	integer	no
conv_unsigned	integer	unsigned	yes
conv_unsigned	signed	unsigned	yes
conv_unsigned	unsigned	unsigned	yes
conv_unsigned	std_ulogic	unsigned	yes
conv_signed	integer	signed	yes
conv_signed	signed	signed	yes
conv_signed	unsigned	signed	yes
conv_signed	std_ulogic	signed	yes
conv_std_logic_vector	integer	std_logic_vector	yes
conv_std_logic_vector	signed	std_logic_vector	yes
conv_std_logic_vector	unsigned	std_logic_vector	yes
conv_std_logic_vector	std_ulogic	std_logic_vector	yes

7.3   Type Conversions

In all, there are up to four different types in use when using package std_logic_arith. There are the numeric types, signed and unsigned, and then there is the built-in type integer and the basic array type std_logic_vector.

It is generally good practice to choose types carefully for each datapath so as to use the type most appropriate for the job. Thus, conversions between types should be minimised. Nevertheless, it is necessary sometimes to convert between types. To this end, the std_logic_arith package contains a comprehensive set of type-conversion functions between the four different types in use.

Type-conversion functions are named conv_*type*, where *type* is the name of the type being converted to. This is slightly different from the normal convention of calling type-conversion functions to_*type*. Table 7.1 shows the set of functions available in the package and the types that they convert to and from.

Note that there are no type conversions *from* std_logic_vector to the other types, although there are type conversions from the single-bit type std_ulogic. Type conversions between std_logic_vector and the signed and unsigned types are implicitly provided by the built-in type conversions between similar array types. These implicit type conversions were used to show how to access the logical operators for std_logic_vector so that they could be used on types signed and unsigned. For example, to convert from std_logic_vector to unsigned, the name of the type, unsigned, is used as the type-conversion function.

```
library ieee;
use ieee.std_logic_1164.all;
use ieee.std_logic_arith.all;
```

```
entity conversion_demo is
  port (value : in std_logic_vector(7 downto 0);
        result : out unsigned(7 downto 0));
end;
architecture behaviour of conversion_demo is
begin
  result <= unsigned(value);
end;
```

The implicit type conversions cannot change the size of the array. The source signal must be the same size as the target, although once again their ranges can be different.

Also note that some of the type conversions convert from and to the same type. These are the resize functions covered in Section 7.4. They allow, for example, an 8-bit signed value to be converted to a 16-bit signed value.

As the third column shows, some of these functions take just one argument, the value to be converted, and yield the same value converted into the target type. This is only true of the functions that convert to an integer type, since there is no size information needed to convert to an integer. However, those functions that convert to an array type take a second argument that gives the size of the array to create. To give an example, the following circuit shows a straightforward conversion from an integer subtype to an unsigned number.

```
library ieee;
use ieee.std_logic_1164.all;
use ieee.std_logic_arith.all;
entity conv_unsigned_demo is
  port (value : in natural range 0 to 255;
        result : out unsigned(7 downto 0));
end;
architecture behaviour of conv_unsigned_demo is
begin
  result <= conv_unsigned(value, 8);
end;
```

The second argument must be a constant value for synthesis, since the synthesiser must be able to deduce the size of bus to create by examining the value of this argument.

7.4 Constant Values

Constant values of signed and unsigned types can be represented by string values.

The interpretation of the numeric types means that string values are written exactly as you would expect them to be written – with the most-significant bit on the left. For example, to initialise an 8-bit signal, the following assignment would be used:

```
z <= "00000000";
```

When assigning to a signal, or variable, of a numeric type, the string value must be exactly the right length. However, in this case, an alternative would have been to use an aggregate with an others choice:

```
z <= (others => '0');
```

Another way of avoiding long strings is to use the `conv_type` functions to extend a short number to the length required. Using the `conv_unsigned` that converts from a single `std_ulogic` gives the simplest solution:

```
z <= conv_unsigned('0', 8);
```

Note that the argument has single quotes because it is a character value of type `std_logic`, not a string value of type `unsigned`.

When adding a signal to a constant value, there is no need to match the size of the arguments, because the add operator has been designed to take arguments of different lengths. In this case, an `unsigned` signal is incremented by simply adding the string value `"1"` to the input. The problem is that the string `"1"` is ambiguous. It could represent type `signed`, `unsigned`, `std_logic_vector` or for that matter `string`. There is no add operator for adding a `string` to a `signed`, but there is an add operator for the other three array types. This illustrates one of the fundamental problems with the `std_logic_arith` package – there are far too many overloaded operators for the arithmetic operators, so there are many such ambiguities possible. It is not permissible for any VHDL system to just choose one interpretation out of those available so an ambiguous expression like this causes a compilation error.

The way to resolve such ambiguities is with type qualification. Type qualification looks very like a type conversion but has a different action. It tells the VHDL analyser what type to expect an expression to be. Where there is an ambiguity in the model – meaning that there are several possible types that an expression could take – a type qualification resolves the ambiguity by informing the analyser which of the possibilities to choose. It has no impact at all on either simulation or synthesis, it is purely an aid to the analyser to resolve language ambiguities.

A type qualification consists of the name of the expected type, a tick (a single quote mark) and then the expression being resolved enclosed in parentheses. For example, to type qualify the string `"1"` as an `unsigned`, use the expression:

```
unsigned'("1")
```

This qualified expression is used in the following example to increment an `unsigned` signal.

```
library ieee;
use ieee.std_logic_1164.all;
use ieee.std_logic_arith.all;
entity increment is
   port (a : in unsigned (7 downto 0);
         z : out unsigned (7 downto 0));
end;
architecture behaviour of increment is
begin
   z <= a + unsigned'("1");
end;
```

Beware when using type `signed`, because then the string value `"1"` has the value -1. To get the value 1, use the 2-bit value `"01"`. One way to avoid the pitfall is to use the character value `'1'` instead of a string value. There is an add operator that allows the addition of a single `std_ulogic` bit to a `signed` and in this case the bit is interpreted as either $+1$ or 0.

This gives the following:

```
architecture behaviour of increment is
begin
  z <= a + '1';
end;
```

Once again notice the use of single quotes. Note also that there is no type qualification. This expression is not ambiguous, because there is only one add operation that takes a character value, and that is the one that adds a `signed` to a `std_ulogic`. Indeed, there is another operator that can add an `unsigned` to a `std_ulogic`, so this method could have been used to avoid the type qualification in the earlier example.

7.5   Mixing Types in Expressions

The operators in the `std_logic_arith` packages have been extensively overloaded so that it is possible to mix the bit-array types with type `integer`. This is mainly of use when adding or comparing with constant integer values, since it avoids the need to convert the integer value into its string value.

For example, the previous example of an 8-bit incrementer could have been written using the integer value 1 in the addition:

```
architecture behaviour of increment is
begin
  z <= a + 1;
end;
```

The second advantage of this form of the addition is that there is no possibility of falling into the trap of omitting the leading sign bit on `signed` numbers. There is also no need for type qualification.

It is also possible to use `integer` values in comparisons. For example, a test for a signal being negative could be written using a test against the `integer` value 0:

```
library ieee;
use ieee.std_logic_1164.all;
use ieee.std_logic_arith.all;
entity compare is
  port (a : in signed (7 downto 0);
        negative : out std_logic);
end;
architecture behaviour of compare is
begin
  negative <= '1' when a < 0 else '0';
end;
```

In this example, the comparison is being made between a signal of type `signed` and the `integer` value 0.

Table 7.2 lists all the permutations of `signed` and `unsigned` types that can be used with all of the provided arithmetic operators. Note in particular the return types. Generally, if

Table 7.2 Permutations of types for all arithmetic operators

Left operand	Right operand	Result
unsigned	unsigned	unsigned
signed	signed	signed
unsigned	signed	signed
signed	unsigned	signed

Table 7.3 Integer permutations for add, subtract

Left operand	Right operand	Result
unsigned	integer	unsigned
integer	unsigned	unsigned
signed	integer	signed
integer	signed	signed

Table 7.4 std_ulogic permutations for add, subtract

Left operand	Right operand	Result
unsigned	std_ulogic	unsigned
std_ulogic	unsigned	unsigned
signed	std_ulogic	signed
std_ulogic	signed	signed

either operand of the operator is of type signed, then the result is type signed. Otherwise the result is of type unsigned. In all cases the result is the size of the larger of the two arguments.

In addition, the permutations listed in Table 7.3 using type integer are supported for add, subtract only. They are not supported for multiplication. Once again, if either operand is type signed, the result is type signed, otherwise it is unsigned. The size of the result is the size of the array argument. The integer argument does not affect the size of the result, nor its type.

Finally, Table 7.4 lists the set of operators that allow the addition or subtraction of type std_ulogic with the numeric types. These are useful when representing simple increment and decrement operations or when incorporating a carry input. Once again the result size is the same as the size of the array operand.

There are some arithmetic operators also provided that return type std_logic_vector. However, these are rarely used, especially if the recommendation is followed that a numeric type is always used to represent a numeric value. In that case, the std_logic_vector forms will never be used, so will not be covered here.

The comparison operators are also extensively overloaded. It is possible to compare type signed with any permutation of signed, unsigned or integer. Similarly it is possible to compare type unsigned with any permutation of signed, unsigned or integer.

8

Sequential VHDL

VHDL covers two programming domains: the concurrent domain and the sequential domain. In other words, VHDL can be used to describe some activities that occur simultaneously and other activities that must occur in a defined order.

The concurrent domain is the VHDL architecture, containing signal assignments, component instances (covered in Chapter 10) and processes. All of these execute simultaneously.

The sequential domain of VHDL exists within a process. It is analogous to the domain of conventional programming languages. Sequential VHDL is difficult to interpret as hardware, since hardware is inherently concurrent. The interpretation is in fact a conversion from the sequential code to a concurrent equivalent. In general, this is a very difficult (indeed, impossible) task, so is only made possible by imposing rules on coding style, and this is exactly how synthesisers work. It is essential that these rules are followed if sequential VHDL is to be synthesisable. Nevertheless, sequential VHDL is very powerful to the user and leads to simpler, clearer models. It is therefore essential to master the use of sequential VHDL to make the most of synthesis. This chapter covers this sequential domain.

8.1 Processes

A process is a series of sequential statements that must be executed in order. At this level, VHDL is like many software programming languages.

The main difference between a process and a software program is that a process runs repeatedly throughout a simulation run, a bit like a continuous loop. It does not have an end.

8.1.1 Anatomy of a Process

A process can appear anywhere in the body of an architecture (after the `begin`). The basic structure of a process is:

```
process sensitivity list
  declaration part
begin
  statement part
end process;
```

VHDL for Logic Synthesis, Third Edition. Andrew Rushton.
© 2011 John Wiley & Sons, Ltd. Published 2011 by John Wiley & Sons, Ltd.

There are three parts to the process that need further explanation: the sensitivity list, the declaration part and the statement part.

The sensitivity list is the list of signals that the process is sensitive to. This is a set of signals – enclosed in parentheses – which the simulator monitors for events (changes of value). Any events on any of the signals in the sensitivity list causes the process to be executed once. That is, all the statements in the statement part will be executed and then the process will stop and wait for further activity. Every time the process is activated it will run in its entirety. The process is in effect an infinite loop: when the end of the process is reached, the process restarts at the top again. At this point the process will pause until another change on the sensitivity list starts execution again.

The sensitivity list is optional. If it is absent, then the process will run continuously. However, the process must contain `wait` statements in this case to pause the process and wait for further activity. The use of `wait` statements will be dealt with later. If the sensitivity list is present, then no `wait` statements can appear in the process.

The declaration part of the process allows the declaration of types, functions, procedures and variables (Section 8.3) that are local to the process. That is, they can only be used within the process.

The statement part of the process contains the sequential statements to be executed each time the process is activated. The set of statements available for use in a process is called sequential VHDL and includes `if` statements, `case` statements, `for` loops and simple signal assignments, but not conditional or selected signal assignments. The statements in the process are run by the simulator from the top of the process to the bottom and then this is repeated, ad infinitum, or at least until the end of the simulation. All processes in a design are run simultaneously. That is, they are run as if they were simultaneous, but the design of the language is such that they can in fact be run one at a time in any order and it will make no difference to the result.

While a process is running, nothing else is happening in the simulator. In particular, no events are being processed and therefore no signals are being updated. All signals act as constant values during the process execution. The process continues to dominate the simulator until a `wait` statement is reached. This causes the process to suspend and wait until the `wait` condition is satisfied. It is during the process suspension that signals are updated with new values. Only when the `wait` condition is satisfied will the process resume. If a process has no `wait` statements, process suspension happens when the end of the process is reached. A sensitivity list is exactly equivalent to a `wait` statement at the end.

Not all of the sequential statements available in VHDL can be synthesised, since not all of them have a hardware equivalent. The following sections will describe in detail only those that can.

8.1.2   Combinational Process

For a process to model combinational logic, it must be sensitive to all the signals that are read in the process. In other words, the process must be re-evaluated every time one of the inputs to the circuit it represents changes. Clearly, this is the expected behaviour of combinational logic.

If a process is not sensitive to all its inputs and is not a registered process (which will be described in Chapter 9), then it is unsynthesisable. There is no hardware equivalent for such a process. This rule should be enforced by the synthesiser. However, it will not be detected by a simulator because in VHDL it is perfectly legal: this is a synthesis rule.

An example of a combinational process, shown in context within an entity/architecture pair is:

```
library ieee;
use ieee.std_logic_1164.all, ieee.numeric_std.all;
entity adder is
  port (a, b : in unsigned(3 downto 0);
        sum : out unsigned(3 downto 0));
end;
architecture behaviour of adder is
begin
  process (a,b)
  begin
    sum <= a + b;
  end process;
end;
```

The inputs of this process are the signals a and b that are of type unsigned, from package numeric_std. Both these signals are in the sensitivity list, making the process combinational. The process simply adds the two values a and b, a trivial example that could have been achieved with a simple concurrent signal assignment, but it illustrates how a process fits in the VHDL architecture.

In simulation, every time a or b changes, this process will be run and the signal sum therefore updated with a new value. This clearly models combinational logic.

8.1.3   Wait Statements

The alternative form of a combinational process has no sensitivity list, but contains a wait statement that is sensitive to all the process inputs:

```
process
  declaration part
begin
  statement part
  wait on sensitivity list;
end process;
```

This is exactly equivalent to the sensitivity list version. The alternative form of the earlier example is:

```
process
begin
  sum <= a + b;
  wait on a, b;
end process;
```

In fact, VHDL allows any number of wait statements in a process with no sensitivity list. However, when used for synthesis of combinational logic, only one wait statement can be present.

8.1.4   Wait Statement Positioning

In the wait statement version of the combinational process, the wait statement appears at the end of the process. This equivalence is due to the action of a VHDL simulator at elaboration (also known as initialisation). At the elaboration phase of the simulation, all the processes in a model are run once until they pause at a wait statement. If the process has a sensitivity list, this means that all the statements in the process are run once to the end of the process.

It is not necessary to place the wait statement at the end of the process. Indeed the most common place for it is at the start. This changes the elaboration behaviour since it prevents any of the statements from being run at elaboration time. However, since elaboration has no hardware equivalent, this will make no difference to the synthesised circuit.

The wait statement version of the previous combinational example is:

```
process
begin
  wait on a, b;
  sum <= a + b;
end process;
```

The difference between simulation and synthesis due to elaboration can be a pitfall, since it may mean that the circuit, which appears to work correctly in simulation due to initial values set up during the elaboration phase, gets stuck in an unknown state when synthesised to a gate-level circuit. This problem is most likely to manifest itself when using registers, so a more detailed discussion will be deferred to Section 9.12.

8.2   Signal Assignments

Signals are the interface between VHDL's concurrent domain and the sequential domain within the process. Signals are a means of communication between processes: in effect a VHDL model is a network of processes intercommunicating via signals. Indeed, this is how many simulators model a VHDL system – all hierarchy is effectively removed to leave a network of processes and signals.

The simulator alternates between updating signal values and then running processes activated by changes on those signals listed in their sensitivity lists. Whilst a process is running, all signals in the system remain unchanged. Thus, signals are effectively constants during process execution and even after a signal assignment, will not take on a new value. What happens is that a signal assignment in a process causes a signal change (called an event) to be queued for that signal, but the simulator will not process the event until process execution stops.

When modelling in RTL, time delays are generally not used, so a simpler view of what happens to signals can be applied. The simple way of looking at signal processing is that signals will be updated at the end of the process, either literally the end in the case of sensitivity list processes, or at a wait statement.

In Chapter 3, three types of signal assignment were introduced: the simple assignment, the conditional assignment and the selected assignment. These are all concurrent signal assignments – that is, they exist in the concurrent domain of VHDL, outside the process. Within a process, only the simple assignment can be used. The behaviour of the other two kinds of signal assignment can be reproduced by the `if` statement and `case` statement, respectively. These will be covered in Sections 8.4 and 8.5, respectively.

8.3   Variables

Variables are used to store intermediate values within a process. They only exist within sequential VHDL and cannot be declared or used directly in an architecture.

8.3.1   Declaration

A variable declaration is very much like a signal declaration:

```
variable a, b, c : std_logic;
```

Variable declarations appear in the declarative part of a process (before the `begin`). A variable can be of any type.

8.3.2   Initial Values

Like signals, all variables have initial values. Variables are given their initial values at elaboration time during simulation (at time zero), and the value can be either a user-defined value or a default value.

Initial values given in the variable declaration look like:

```
variable a : std_logic := '1';
```

This means that, at the start of simulation, variable `a` will take the value `'1'`.

If a variable does not have an explicit initial value given in its declaration, it will still have an initial value in simulation. This value will be the left value of the type. For type `bit`, the left value is `'0'`, so all variables of type `bit` will be initialised with the value `'0'` unless an explicit initial value is used to override it. For `std_logic` the left value is `'U'`, so all variables will be initialised with `'U'` (which means uninitialised) unless an explicit initial value is used to override it.

For synthesis, there is no hardware interpretation of an initial value, so synthesis ignores initial values. This can be a pitfall if the design relies on variables being initialised.

8.3.3   Using Variables

Signals are not updated during process execution. That is, they will not change value while the process is running. This means that signals cannot be used to store intermediate values of calculations during process execution.

This is where variables come in. Variables are a feature of sequential VHDL that is just as would be expected in a conventional programming language. Variables are updated immediately by a variable assignment statement. It is not even possible to specify a time delay.

Generally, variables are used to accumulate results or to store intermediate values in a calculation within a process. The use of a variable to store intermediate values is shown in the following example:

```
library ieee;
use ieee.std_logic_1164.all;
use ieee.numeric_std.all;
entity add_tree is
  port (a, b, c, d : in signed(7 downto 0);
        result : out signed(7 downto 0));
end entity;
architecture behaviour of add_tree is
begin
  process (a, b, c, d)
    variable sum : signed(7 downto 0);
  begin
    sum := a;
    sum := sum + b;
    sum := sum + c;
    sum := sum + d;
    result <= sum;
  end process;
end;
```

Notice that the variable assignments use the symbol ":=", whereas the signal assignments use the symbol "<=".

This example is a classic multi-input adder built from two-input adders. In this case, the result is built up by adding each of the inputs to the intermediate signal, then finally the output of the circuit is assigned to the output port using a signal assignment, because ports are signals.

8.4 If Statements

The if statement is the sequential equivalent of the conditional signal assignment, although they are not exactly equivalent. The if statement will be familiar to anyone who has used any conventional programming language. It allows the execution of a process to take one of a number of branches depending on a condition.

An if statement can have one or more branches, controlled by one or more conditions. This is illustrated by the following process, containing a three-branch if statement:

```
process (a, b)
begin
  if a = b then
    result <= "00";
  elsif a < b then
    result <= "11";
  else
    result <= "01";
  end if;
end process;
```

The first branch of the if statement tests for the condition that a and b are equal. If this is true, then the contents of the first branch of the if statement is executed, in this case causing the value "00" (0) to be assigned to result. The other branches are then ignored. However, if the condition was not true, then the next test is carried out. This test is the elsif condition, which tests for the case of a being less-than b. If this condition is true, then the second branch is executed, resulting in the value "11" (−1) being assigned to the result. Finally, if this condition is not true, then the final branch – an unconditional else branch – will be executed, resulting in the value "01" (+1) being assigned to the result.

There can be any number of elsif branches to an if statement, and they can of course be omitted entirely. There may be only one else branch to the statement, and if present it must be the last branch. The else branch can also be omitted.

Each branch of an if statement can contain any number of statements, it is not limited to a single statement as in this example. An if statement must always be terminated with an end if.

The hardware interpretation of an if statement is a multiplexer. The simplest form is the if statement with an else branch.

An example of this simplest form is:

```
library ieee;
use ieee.std_logic_1164.all, ieee.numeric_std.all;
entity compare is
  port (a, b : in unsigned (7 downto 0);
        equal : out std_logic);
end;
architecture behaviour of compare is
begin
  process (a, b)
  begin
    if a = b then
      equal <= '1';
    else
      equal <= '0';
    end if;
  end process;
end;
```

This example tests the equality of two signals of type unsigned – representing two 8-bit unsigned integers – and gives a result of type std_logic. The example shows how the built-in equality operator, which gives a boolean result, has effectively been converted to give a result of type std_logic by this if statement structure.

The resulting circuit is shown in Figure 8.1. Note that, as with all other examples, this circuit is diagrammatic and in practice the synthesiser will optimise away the inefficiencies in the circuit – in this case the constant inputs to the multiplexer – to give a minimal solution.

A multi-branch if statement is modelled by a multi-stage multiplexer. Consider the following multi-way if statement:

```
process (a, b, c, sel1, sel2)
begin
  if sel1 = '1' then
    z <= a;
```

```
  elsif sel2 = '1' then
        z <= b;
  else
        z <= c;
  end if;
end process;
```

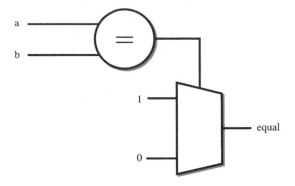

Figure 8.1   Multiplexer interpretation of if statement.

This results in the circuit shown in Figure 8.2.

You may recognise this circuit from Chapter 3, in the description of the conditional signal assignment. This is the if statement equivalent of:

```
z <= a when sel1 = '1' else
     b when sel2 = '1' else
     c;
```

The conditions in the successive branches of an if statement are evaluated independently. In this case, the conditions involved the two signals sel1 and sel2. There can be any number of conditions, each of which will be independent of the others. However, the structure of the if statement ensures that the earlier conditions are tested first. In this example, the test on sel1 was made before the test on sel2. This prioritisation is reflected in the hardware, where the multiplexer controlled by sel1 was first (nearest the output) and the multiplexer controlled by sel2 was second.

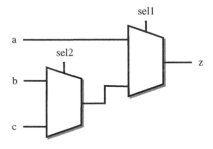

Figure 8.2   Multi-branch if statement.

This prioritisation of the conditions is important to remember so that redundant tests can be eliminated from the conditions. Consider the following, inefficient, example that has the same behaviour as the example above:

```
process (a, b, c, sel1, sel2)
begin
  if sel1 = '1' then
    z <= a;
  elsif sel1 = '0' and sel2 = '1' then
      z <= b;
  else
      z <= c;
  end if;
end process;
```

The extra test for sel1 = '0' is redundant since this elsif condition will only be tested if the first condition on the if statement was false. It is good practice to avoid such redundancies. There is no guarantee that a synthesiser will detect them and eliminate them for you. More importantly, such redundant expressions tend to obscure the logic of the conditions and make the model less readable.

With multiple-branch if statements, each condition can be, and will normally be, dependent on different signals and variables. If every branch is dependent on the same signal, then you probably really need a case statement. Case statements are dealt with in the next section.

All the examples of if statements so far have been complete. In other words, the target signal gets a value under all possible conditions. This gives straightforward combinational logic as illustrated by the 2-way and 3-way examples above. However, there are situations where a signal does not receive a value under all conditions.

There are two different situations where a signal does not receive a value: where there is a missing else part to the if statement and where the signal is not assigned to in some branches of the if statement. In both cases the interpretation is the same. In the conditions where the signal does not receive a value, the previous value is preserved.

This opens up the question: what is the previous value? If the signal has an earlier assignment, prior to the if statement, then the value comes from that assignment. If not, then the previous value comes from the previous execution of the process, leading to feedback in the circuit. Feedback will be covered in Section 8.6 on latch inference.

The first case is illustrated by the following VHDL:

```
process (a, b, enable)
begin
  z <= a;
  if enable = '1' then
    z <= b;
  end if;
end process;
```

In this case, the if statement is incomplete because it is lacking an else part. Thus, the signal gets a value in the if statement if the condition enable = '1' is satisfied, but remains unassigned if the condition is false. In this case, the previous value comes from the unconditional assignment before the if statement.

This is equivalent to:

```
process (a, b, enable)
begin
  if enable = '1' then
    z <= b;
  else
    z <= a;
  end if;
end process;
```

If there is no previous assignment, then there is feedback from the output of the circuit to an input. In other words, if the if statement is incomplete and there is no previous assignment, then feedback is produced. This is because the value of the signal from the previous execution of the process is preserved and becomes the value in this execution of the process.

One of the most common pitfalls in writing VHDL for logic synthesis is the accidental introduction of feedback in the circuit due to an incomplete if statement. The problem is that it is not possible for the synthesiser to check for these errors since it is quite legitimate to have incomplete if statements to describe latches. It is therefore up to the designer to check for these errors.

It is not just the structure of the if statement that is at issue here: if a circuit is meant to be purely combinational, then you must ensure that every signal assigned to in the process (an 'output' of the process) receives a value under every possible combination of conditions. There are two ways of doing this in practice: either be careful to make sure that every signal assigned to in an if statement is assigned to in every branch and that there is always an else part; or initialise signals with an unconditional assignment before the if statement.

This style of combinational process is a common one: there are many situations where the best structure is to start the process with a set of default assignments, then selectively override them in conditional statements.

Some illustrations of this pitfall are shown in the following examples:

```
process (a, b, c)
begin
  if c = '1' then
    z <= a;
  else
    y <= b;
  end if;
end process;
```

In this example, although the if statement looks complete, different signals are being assigned a value in each branch of the if statement. Thus, both signals z and y will be latched.

Another example is one in which there is a redundant test for a condition that must be true:

```
process (a, b, c)
begin
  if c = '1' then
    z <= a;
  elsif c /= '1' then
    z <= b;
  end if;
end process;
```

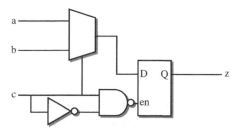

Figure 8.3 Incomplete if statement.

In this case, although the if statement looks complete, each of the conditions in an if statement is synthesised in isolation from the other conditions. The synthesiser will therefore not necessarily detect that this second condition is redundant. Thus, the if statement will be synthesised as a 2-way multiplexer, with a latched output to implement the missing else condition that requires storage of the previous value. The circuit for this example is shown in Figure 8.3.

So far, all the discussion about the use of if statements has centred around the use of signal assignments with conditionals. In fact, the same rules apply when using variables. Like a signal, if a variable is assigned to in some branches of the if statement but not others, then the previous value is preserved in exactly the same way, by latching the variable's value.

8.5   Case Statements

Case statements are like if statements in that they provide a way of branching on a condition. Unlike if statements, the condition of a case statement does not need to be boolean, and there can be many branches from one condition.

Case statements could also be described as the sequential equivalent of the selected signal assignment, although they are not exactly equivalent.

The condition in a case statement can be a signal, variable or expression of just about any type. The only restrictions are that the condition must be a discrete type (in other words, an integer or enumeration type) or a character array type such as bit_vector or std_ logic_vector. These restrictions are compatible with the restrictions due to the interpretation of types by logic synthesis.

As an example, consider the problem of sequencing through the values of an enumeration type representing the states of a traffic light. The state type is an enumeration type:

```
type light_type is (red, amber, green);
```

The process that controls the combinational part of the sequencer is best described using a case statement:

```
process (light)
begin
  case light is
    when red =>
      next_light <= green;
```

```
      when amber =>
        next_light <= red;
      when green =>
        next_light <= amber;
    end case;
  end process;
```

The condition in this case is a signal called light which is of type light_type. The case statement contains three branches, in this case one branch for each possible value of the type. The when parts are called the choices and these control the branching of the case statement. Each branch can contain any number of VHDL statements, although in this case, each branch only contains a single statement: a signal assignment to next_light.

The case statement choices must cover every possible value of the type or subtype of the condition. A convenient shortcut is the keyword others that acts as a mopping up choice that matches every value not covered in previous choices. The others choice must be the last choice in the case statement.

```
    case light is
      when red =>
        next_light <= green;
      when amber =>
        next_light <= red;
      when others =>
        next_light <= amber;
    end case;
```

It is also possible to group choices as either multiple choices or range choices. Furthermore, these can be combined into a multiple choice of ranges.

Some examples:

```
    when 0 to integer'high =>
    when red | amber =>
    when 0 to 1 | 3 to 4 =>
```

The vertical bar '|' separates multiple choices and can be read as 'or', so the second example reads 'when red or amber do'. The ranges (such as 3 to 4) choose all values in the specified range.

In hardware terms, a case statement is implemented as a multiplexer structure, much like the if statement. The difference is that the choices all depend on the same input and are mutually exclusive. Also, all the choices have the same priority, so there is no skewed priority tree in the circuit. This can lead to some optimisation of the control conditions compared with a multi-branch if statement where no dependencies between the conditions will be assumed.

8.6   Latch Inference

If a signal or variable is assigned only under some conditions and not others, usually due to the use of an incomplete if statement, then the previous value of the signal or variable is preserved. In a registered process, the previous value is the value stored in the register, so the feedback is synchronous. This will be discussed further in Chapter 9 on Registers. In a combinational process, the previous value is an output of the combinational logic and so the feedback is asynchronous. This asynchronous feedback is implemented as a latch.

The technique used by synthesisers to convert asynchronous feedback to latches is known as *latch inference*. The reason it has a special name is because it is a special technique – once again, there is no direct mapping from the VHDL to the latch circuit, so an interpretation has to be made which distinguishes between an `if` statement that describes a multiplexer and an `if` statement that describes a latch.

Latches are inferred by first detecting which signals are to be latched – that is, the signals that do not receive a new value under all possible conditions. The second stage is to extract the set of conditions that cause each signal to receive a value and using the `or` of these conditions as the enable control of the latch. Each signal is analysed independently, so a process can contain a mixture of combinational and latched outputs. Furthermore, each latch's enable control can have different logic.

An example of the use of latch inference is the following process that describes a 4-bit latch, using a signal of the type `std_logic_vector`. The latch inputs and outputs are modelled by vectors, whilst the enable condition depends on a single-bit `std_logic` signal. The signal declarations are:

```
signal input, output : std_logic_vector(3 downto 0);
signal enable : std_logic;
```

The latched process is described using an incomplete `if` statement:

```
process (enable, input)
begin
  if enable = '1' then
    output <= input;
  end if;
end process;
```

Note that the latched process is still a combinational process, so the sensitivity list must include all of the signals that are inputs to the process. This includes the `enable` signal.

To understand why this models a latch, consider what happens with this circuit when `enable` is held at `'1'`. Every time the input changes, the process is re-executed, causing the output to be assigned the new value of the input. The output therefore follows the input and the process acts as a latch in its transparent mode. If `enable` is held at `'0'`, however, the `if` statement bypasses the assignment, so the output does not follow the input. Since signals in VHDL preserve their last-assigned values, the output will keep the value last assigned to it indefinitely. The process is now modelling a latch in its hold mode.

It should be clear from this simple example that the process models a latch circuit. In this case, four latch gates will be synthesised since the signal being latched is a four-bit signal. The circuit is illustrated in Figure 8.4.

In principle, latch inference can be applied to a process of any complexity. If there are nested `if` and `case` statements, then the overall structure of the process can be analysed for missing assignments to any of the signals that are outputs of the process. In practice, this analysis becomes extremely difficult when there are many levels of nesting, so practical synthesisers will usually have a built-in limit. This limit will vary from synthesiser to synthesiser. The

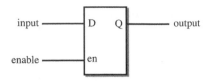

Figure 8.4   Latch inference.

recommended way to write latched circuits that are guaranteed to lead to correct latch inference
is to describe the latch behaviour only at the outermost level of the conditional. In other words,
have an if statement to implement the latch, then put all other circuitry inside that if statement.
Within the outer if statement, make all conditionals complete so that they describe combinational logic. The simple examples used so far in this section have followed this convention,
since they only have one level of conditional anyway. Here is a two-level conditional that
describes a latched multiplexer in such a way that the latch behaviour is modelled at the
outermost level of conditional:

```
process (en, sel, a, b)
begin
  if en = '1' then
    if sel = '0' then
      z <= a;
    else
      z <= b;
    end if;
  end if;
end process;
```

In this example, the outer if statement has no else part, so a latch is inferred. The inner if
statement is complete and therefore describes a multiplexer. The resulting circuit is shown in
Figure 8.5.

   This way of describing a latch circuit is also the most natural. A latched process is a
combinational process with latches on the outputs. The latches will always be inferred on the
outputs of the process because it is these outputs that need to be preserved for the next execution
of the process. Therefore, writing the latched process as a conditional (in this case an if
statement) modelling the latch, then enclosing the remaining purely combinational behaviour
of the process within that if statement, reflects the hardware structure.

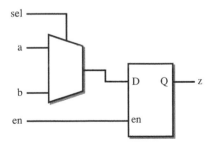

Figure 8.5   Latched multiplexer.

The same functionality could be achieved with a single-level, two-branch if statement:

```
process (en, sel, a, b)
begin
  if en = '1' and sel = '0' then
    z <= a;
  elsif en = '1' and sel = '1' then
    z <= b;
  end if;
end process;
```

This style can be clearer for some examples (although not this one). It is therefore a common alternative style for describing latch circuits. The missing else clause means that the signal z will not always get a new value under all conditions, so a latch will be inferred from the description. The enable control of the latch will be the or of the conditions of the if statement, which reduces to the condition en = '1'. The remainder of the circuit is implemented as a multiplexer – it is treated as if the if statement was complete by converting the final branch to an else clause. It can be seen that this is equivalent to the previous description.

8.7   Loops

A loop is a mechanism for repeating a section of VHDL code, for example to process every element in an array in the same way.

There are three types of loop in VHDL – the simple loop, the while loop and the for loop. The simple loop will continue looping indefinitely. The while loop will continue looping an unspecified number of times until a condition becomes false. The for loop continues looping a specified number of times. All loops can be ended by the exit statement (covered later in this section) and this is the only way of ending a simple loop.

However, the synthesis interpretation of loops is to replicate the hardware described by the contents of the loop statement once for each pass round the loop, and this is only possible with the for loop. This is the only kind of loop where the number of iterations is known and therefore the only kind where the number of replications of hardware required to implement it is known by the synthesiser.

For this reason, only the for loop will be described here.

8.7.1   For Loops

A for loop is a loop that repeats a fixed number of times. An example of how a for loop is used is shown below. This is typical of the most common use of for loops in that it is being used to apply the same operation to all elements of an array.

```
library ieee;
use ieee.std_logic_1164.all;
entity match_bits is
  port (a, b : in std_logic_vector(7 downto 0);
        matches : out std_logic_vector(7 downto 0));
end;
```

```
architecture behaviour of match_bits is
begin
  process (a, b)
  begin
    for i in 7 downto 0 loop
      matches(i) <= a(i) xnor b(i);
    end loop;
  end process;
end;
```

This example creates a set of bitwise equalities (implemented by the xnor functions), creating a result that has a '1' wherever the bits of a and b match and '0' otherwise.

There are a number of points to note. First, there is a loop constant called i that controls the loop execution. The loop constant only exists within the loop, so cannot be accessed once the loop has finished. It is not a variable and is therefore not declared in the process declarative part. It has been specified with a descending range, so counts downwards. It is given the value 7 in the first time round the loop, 6 in the second and so on until the end of the loop range is reached at 0. This is known as the loop constant because the value of i is treated as a constant value inside the loop; it can only be read and cannot be written to. The loop constant in this case is an integer, because the type has not been defined.

The full form of the loop range specification is similar to that for subtypes. This example could have been written:

```
for i in integer range 7 downto 0 loop
```

The equivalent circuit is created by replicating the circuit represented by the statements in the loop once for each value of the loop constant and replacing the loop constant in any statements by its value for that replication. For example, in the first replication, i will be replaced by 7. The resulting equivalent process is:

```
process (a, b)
begin
  matches(7) <= not(a(7) xor b(7));
  matches(6) <= not(a(6) xor b(6));
  matches(5) <= not(a(5) xor b(5));
  matches(4) <= not(a(4) xor b(4));
  matches(3) <= not(a(3) xor b(3));
  matches(2) <= not(a(2) xor b(2));
  matches(1) <= not(a(1) xor b(1));
  matches(0) <= not(a(0) xor b(0));
end process;
```

In this example, the ordering of the loop range was irrelevant, since there was no connection between the replicated logic blocks. The ordering of the loop range becomes important where there is a connection from one replicated block to another. This connection is usually created by a variable that stores a value in one iteration of the loop that is then read in another iteration of the loop. It is usually necessary to initialise such a variable prior to entering the loop.

The following example shows such a circuit:

```
library ieee;
use ieee.std_logic_1164.all, ieee.numeric_std.all;
entity count_ones is
  port (vec : in std_logic_vector(15 downto 0);
        count : out unsigned(4 downto 0));
end;
architecture behaviour of count_ones is
begin
  process (vec)
    variable result : unsigned(4 downto 0);
  begin
    result := "00000";
    for i in 15 downto 0 loop
      if vec(i) = '1' then
        result := result + 1;
      end if;
    end loop;
    count <= result;
  end process;
end;
```

This example is a combinational logic block that counts the number of bits in vec that are set to '1'. The result is initialised to zero using a string value ("00000" in this case). The result is then accumulated during the execution of the process in a variable called result and then assigned to the output signal count at the end of the process. The result needs to be five bits long since it must be capable of storing values in the range 0–16.

In synthesis, the process will be interpreted by unrolling the loop.

```
process (vec)
  variable result : unsigned(4 downto 0);
begin
  result := "00000";
  if vec(15) = '1' then
    result := result + 1;
  end if;
  if vec(14) = '1' then
    result := result + 1;
  end if;

  . . .

  if vec(0) = '1' then
    result := result + 1;
  end if;
  count <= result;
end process;
```

The contents of the loop – an if statement containing an assignment – represents a multiplexer-adder structure that will be replicated by the synthesiser once for each value of the loop constant. The result output of one replicated circuit block becomes the result input of the next circuit block. Figure 8.6 illustrates the circuit produced.

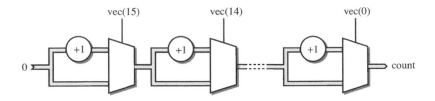

Figure 8.6   Interpretation of a for loop.

In this example, the bounds of the `for loop` were specified as the descending range `15 downto 0`. In practice, this explicit form of loop bounds is very rarely used when accessing an array. In practice, the array attributes are used to specify the loop bounds.

There are four possible permutations, depending on whether the array being accessed has an ascending or descending range and the order in which the loop is expected to visit the array elements.

If the loop is required to visit the elements of an array from left to right regardless of whether the array has a descending or ascending range, the `range` attribute is used:

```
for i in vec'range loop
```

If the loop is required to visit the elements of the array from right to left (the reverse of the above), then the `reverse_range` attribute is used:

```
for i in vec'reverse_range loop
```

On the other hand, if the loop is required to visit the elements of an array from the lowest index to the highest regardless of whether the array has a descending or ascending range, then the `low` and `high` attributes are used:

```
for i in vec'low to vec'high loop
```

Finally, if the reverse of the above is required, that is, the elements are visited from the highest to the lowest, then the `high` and `low` attributes are used the other way round:

```
for i in vec'high downto vec'low loop
```

It becomes particularly important to choose the correct loop bounds when writing subprograms, where it is not known in advance (when writing the subprogram) whether an array variable or signal passed to the subprogram as a parameter is going to have a descending range or an ascending range. This will be discussed in detail in Chapter 11. For array types being used to represent integer values, such as the types `signed` and `unsigned` in `numeric_std`, the convention is that the leftmost bit is the most-significant bit and the rightmost the least-significant bit, regardless of the array range. This means that the correct way to access such an array is to use the `range` attribute or `reverse_range` attribute.

Specifically, to access an array representing an integer from the m.s.b. to the l.s.b., use the `range` attribute:

```
for i in vec'range loop
```

To access the array from the l.s.b. to the m.s.b., use the `reverse_range` attribute:

```
for i in vec'reverse_range loop
```

Because of the hardware interpretation of a `for loop`, the bounds of the loop must be constant. This means that the bounds cannot be defined using the value of a variable or signal. All the examples above used the size of the array, not its value, to constrain the loop. The need to specify a constant loop count is a constraint that makes some circuits difficult to describe. For example, consider a case where you want to count the number of trailing zeros on a value. The natural way to do this is to count the zeros until a one is reached and then stop. This would be easy to describe as a `while` loop, but that would be unsynthesisable. The description of this kind of circuit using the `for loop` is facilitated by the `exit` and `next` statements.

Note: a common mistake is to declare a variable in the mistaken belief that this will then be used as the loop counter. This variable will be ignored and the loop will simply define a new loop constant with the same name that overrides it. After the loop has exited, the variable will not contain the last value of the loop counter, it will still be uninitialised.

8.7.2   Exit Statement

The `exit` statement allows the execution of a `for` loop to be stopped. That is, the `for` loop is exited, even though it hasn't completed all its iterations.

An application of the `exit` statement is the example just given, a circuit for counting the number of trailing zeros on a `std_logic_vector`. The solution is:

```
library ieee;
use ieee.std_logic_1164.all, ieee.numeric_std.all;
entity count_trailing_zeros is
   port (vec : in std_logic_vector(15 downto 0);
         count : out unsigned(4 downto 0));
end;
architecture behaviour of count_trailing_zeros is
begin
  process (vec)
    variable result : unsigned(4 downto 0);
  begin
    result := to_unsigned(0, result'length);
    for i in vec'reverse_range loop
      exit when vec(i) = '1';
      result := result + 1;
    end loop;
    count <= result;
  end process;
end;
```

This loop will visit every element in the array `vec`, starting at the l.s.b. and for each bit check whether it is a `'1'`. If it is, the loop exits and the current value in `result` is the count of the

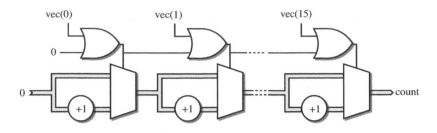

Figure 8.7   Exit statement.

trailing zeros. If the current element is zero, the value of `result` is incremented and the loop goes round again. The loop will stop anyway when every element has been visited, because it is a `for` loop.

This would seem to break the rule for synthesis that the number of iterations must be known to replicate the hardware. However, it is implemented by replicating the hardware once for every possible value of the loop constant. The `exit` statement is represented by a multiplexer that bypasses the incrementer for the current iteration and for all the remaining iterations if the `exit` condition becomes true. The circuit representing this example is shown in Figure 8.7.

It can be seen from this example that the hardware representation of an `exit` statement in a loop is similar to that of an `if` statement within a loop. It also shows how much hardware can be generated by a small piece of VHDL. This circuit requires 16 5-bit full adders and 16 5-bit wide multiplexers to implement a loop that required only four lines of VHDL to describe.

8.7.3   Next Statement

The `next` statement is closely related to the `exit` statement. Rather than exit the loop completely, the `next` statement skips any statements remaining in the current iteration of the loop and moves straight onto the next iteration. The loop carries on executing after a `next` statement.

A `next` statement can be a good substitute for an `if` statement for conditionally executing a group of statements. The hardware required to implement the `next` statement is the same as the hardware required to implement the equivalent `if` statement. The choice of whether to use a `next` statement or an `if` statement is then down to a question of style or readability.

As an example, consider again the first example used to illustrate a `for` loop with an `if` statement. Here it is again, rewritten to use a `next` statement:

```
library ieee;
use ieee.std_logic_1164.all, ieee.numeric_std.all;
entity count_ones is
  port (vec : in std_logic_vector(15 downto 0);
        count : out unsigned(4 downto 0));
end;
architecture behaviour of count_ones is
```

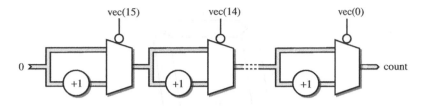

Figure 8.8 Next statement.

```
begin
  process (vec)
    variable result : unsigned(4 downto 0);
  begin
    result := to_unsigned(0, result'length);
    for i in vec'range loop
      next when vec(i) = '0';
      result := result + 1;
    end loop;
    count <= result;
  end process;
end;
```

This version of the model works by skipping over the increment statement if the value of the current element is zero, so the incrementer only counts the ones. Figure 8.8 illustrates the circuit produced. The only difference between this circuit and the original version of the circuit in Figure 8.6 is that the logic is inverted for controlling the multiplexers. This will have no impact on the circuit that will be synthesised; the two circuits are equivalent.

8.8 Worked Example

The problem that will be solved in this example is to design a BCD to 7-segment decoder with zero blanking. This circuit is used to drive the LED displays found in calculators and instrument panels. It is a purely combinational circuit.

The block diagram of the display decoder is shown in Figure 8.9.

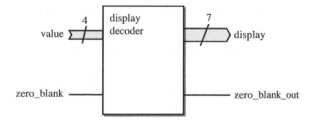

Figure 8.9 BCD to 7-segment decoder.

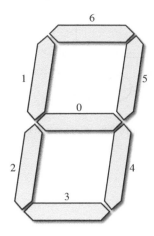

Figure 8.10   Segment positions.

Figure 8.11   Segment encodings.

The circuit decodes the four-bit BCD input value into a seven-bit output corresponding to the display segments. When the value is zero and the zero-blank input is high, then the display is blanked and the zero-blank output is set high. In all other cases the zero-blank output is low.

The segments are numbered from 0 to 6, corresponding to the segment positions shown in Figure 8.10.

The segments to be lit for each decimal digit are illustrated by Figure 8.11. The undefined values above 9 are mapped onto an arbitrary display pattern, in this case the letter E to represent an error.

The first stage is to write the entity according to the block diagram. Type unsigned will be used for value, since it represents a numeric value. Type std_logic_vector will be used for the display output, since it is just a pattern of bits with no numeric interpretation.

```
library ieee;
use ieee.std_logic_1164.all, ieee.numeric_std.all;
entity display_decoder is
  port (value          : in unsigned (3 downto 0);
        zero_blank      : in std_logic;
        display         : out std_logic_vector (6 downto 0);
        zero_blank_out : out std_logic);
end;
```

The architecture contains a single process, containing a `case` statement that decodes the values.

```
architecture behaviour of display_decoder is
begin
  process (value, zero_blank) begin
    display <= "1001111";
    zero_blank_out <= '0';
    case value is
      when "0000" =>
        display <= "1111110";
        if zero_blank = '1' then
          display <= "0000000";
          zero_blank_out <= '1';
        end if;
      when "0001" => display <= "0110000";
      when "0010" => display <= "1101101";
      when "0011" => display <= "1111001";
      when "0100" => display <= "0110011";
      when "0101" => display <= "1011011";
      when "0110" => display <= "1011111";
      when "0111" => display <= "1110000";
      when "1000" => display <= "1111111";
      when "1001" => display <= "1110011";
      when others => null;
    end case;
  end process;
end;
```

This process uses the combinational process style where all outputs are assigned a default value at the start and then this default value is overridden selectively in the conditional statements. The default display is an E for error. Note how it is necessary to have a `when others` clause to catch all the metalogical values on the inputs, because the VHDL rules require case statements to be complete, but it is unnecessary to have any assignments in that branch. VHDL requires at least one statement in each branch of a conditional, so the `null` statement is used.

Note also how the zero blanking has been handled. The `zero_blank` signal is initially assigned the value '0' outside the `case` statement. This is then overridden, along with the `display` output, only in the case where the zero blank output is to be set, which is done in the `if` statement in the zero branch of the decoder.

9

Registers

This chapter covers the VHDL required to describe registers. Throughout this chapter and the rest of the book, the term register will be used to refer to a flip-flop or a bank of flip-flops with common controls.

This chapter explains how a register is modelled, how this model is mapped onto flip-flops by a synthesiser. It then describes how to model other behaviour, such as resettable registers, gated registers. All of these models use synthesis templates to ensure the correct mapping from VHDL model to hardware.

9.1 Basic D-Type Register

The best way to explain how to describe a register in VHDL is with an example. The example is a complete VHDL design, with an entity and architecture. This is just to show the context; it is not a requirement that each register is described as a separate design unit and it is rare to do so.

The only recommended way to describe a register for logic synthesis is with a process. There are other ways of obtaining similar behaviour for simulation, but they will not necessarily be synthesisable.

```
library ieee;
use ieee.std_logic_1164.all;
entity Dtype is
  port (d, ck : in std_logic;
        q      : out std_logic);
end;
architecture behaviour of Dtype is
begin
  process
  begin
    wait until rising_edge(ck);
    q <= d;
  end process;
end;
```

VHDL for Logic Synthesis, Third Edition. Andrew Rushton.
© 2011 John Wiley & Sons, Ltd. Published 2011 by John Wiley & Sons, Ltd.

The register is described by the process statement. The process is recognised as a register because it matches a *template* built into the synthesiser. This is an example of a VHDL construct that does not have a direct mapping to hardware, so this template method must be used to identify the register. The template that is recognised is the `wait` statement that contains the `until` expression `rising_edge(ck)`. This expression must be used if the process is to be interpreted as a register, although there are other templates with similar behaviour that may be optionally used instead of this one. The full set of register templates is described in Section 9.4.

9.2   Simulation Model

This explanation will be based on the basic register example, the process of which is repeated here:

```
process
begin
  wait until rising_edge(ck);
  q <= d;
end process;
```

In this example, the `wait` statement is the first statement in the process, so when simulation is started, execution of the process will suspend at this point. The `wait` condition is:

```
wait until rising_edge(ck);
```

The condition is normally in two parts, the on clause and the `until` clause. The on clause contains a list of signals known as the sensitivity list. The `wait` statement is activated if an event occurs on any of the signals in the on clause. In this case, there is no on clause at all, so there is an implicit on clause containing all of the signals mentioned in the `until` clause. In this example, the only signal mentioned in the `until` clause is the signal `ck`, so there is an implied on clause containing just this signal. This makes the full form of the wait statement as follows:

```
wait on ck until rising_edge(ck);
```

The on clause means that the `wait` statement will only be activated if an event happens on signal `ck`. It will not be activated if an event occurs on `d`.

Once the `wait` statement is activated, the `until` condition is tested. The `until` condition is a `boolean` expression and so must evaluate to `true` or `false`. If it evaluates to `false`, the `wait` statement is deactivated and the process remains suspended. If on the other hand, the condition evaluates to `true`, then process execution will resume.

In this example, the `until` condition is the function call `rising_edge(ck)`. This function will return true if an event has just happened on `ck` and the current value of `ck` is `'1'`. The only way in which an event can happen on `ck` that leaves it with the current value `'1'` is a rising edge, so the overall effect of this `wait` statement is that the process only resumes when a rising edge happens on the clock signal `ck`.

Once the process has resumed, the statements in it are executed. The only statement in this process is:

```
q <= d;
```

This results in a transaction being generated for q. As with concurrent signal assignments (signal assignments outside a process – see Chapter 3), signal q does not get updated immediately, but a transaction is queued for the next delta cycle at the current simulation time. If the signal value is changed by the transaction, that change will be seen at the next delta cycle. When the process reaches its end, it loops back to the start where the wait statement is reached again and the process suspends. It will stay suspended until the next activating event, namely the next rising edge on signal ck. Event processing resumes only when the process is suspended at the wait statement.

In summary, after a rising edge on the clock signal ck, the process executes once, resulting in an update transaction on q being queued. This transaction results in q being updated on the next delta cycle, in other words at the same simulation time as the clock edge. This update transaction will only cause an event if the value of signal q is changed. The value of q is then preserved until the next activating event for the process, which will not happen until the next rising edge on signal ck causes the process to execute again.

It can be seen that this behaviour is the behaviour of an edge-triggered register.

9.3   Synthesis Model

The model just described has a simulation behaviour that is equivalent to an edge-triggered register (or flip-flop if you prefer). A synthesiser recognises it as such by recognising the specific syntax of the wait statement.

There is no hardware mapping of arbitrary processes, and synthesisers can only map those processes that meet the specific rules or templates. For registers, the specific rules require a wait statement of the kind in the example. No other conditions may be added to the until clause and no other signals may be added to the on clause. The exception to this rule is the asynchronous reset, dealt with in Section 9.9.

This kind of synthesis-specific rule is referred to as a *template*. It is important to know when a rule results from the definition of the VHDL language and when it results from a synthesis template. This is because errors arising from VHDL rules will be caught during the simulation phase of the design cycle, whereas errors arising from synthesis templates will not be caught until the synthesis phase of the design cycle. Furthermore, templates can vary between synthesis systems, whereas VHDL rules are constant between simulators. The register model described here is a synthesis template, so it is good practice to check the specific rules for the synthesiser you will be using and to test your understanding of them with some trial syntheses before committing to a large design project. In this case the template is standardised and should be recognised by all synthesis tools.

The register model is implemented by the synthesiser as a block of combinational logic representing the sequential code in the process, followed by a register on every output of the combinational logic block and every feedback within it. The outputs of the block are the set of all signals that are assigned to in the registered process. Thus, *all* signal assignments in a registered process will result in registers on those target signals.

Put more simply, registered and combinational processes are interpreted in similar ways, with the same combinational logic being generated from the contents. However, registered processes then have registers added to all the process's outputs and feedback paths. There is no latch inference, since feedback is registered.

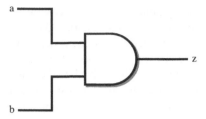

Figure 9.1   Simple combinational circuit.

To illustrate this, the following example shows a trivial combinational process:

```
process
begin
  wait on a, b;
  z <= a and b;
end process;
```

This is interpreted as the hardware shown in Figure 9.1.

The same process contents, but now contained in a process conforming to a register template, gives the same circuit but with a register on the output:

```
process
begin
  wait until rising_edge(ck);
  z <= a and b;
end process;
```

This is interpreted as the hardware in Figure 9.2.

The clock signal must be a one-bit type, which is nearly always of type std_logic.

This example illustrates the way in which a synthesiser interprets a registered process: the contents of the process, excluding the register template part, is interpreted as a combinational process and then a register is added to every output. Where latches

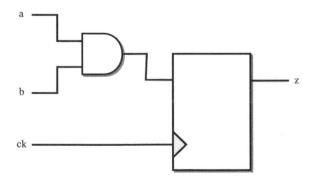

Figure 9.2   Registered circuit.

would be inferred in a combinational process, gated registers are inferred in the registered process.

9.4   Register Templates

So far, all the discussion has concerned one template for a registered process. There are in fact several different templates. The four most common templates will be described here, although there are others possible – this is not a comprehensive list but a safe list that should be supported by all synthesis tools. These example templates have two slightly different modes of operation in simulation, but are exactly equivalent in synthesis.

9.4.1   Basic Template

The template that has already been used is what will be referred to as the basic template:

```
process
begin
  wait until rising_edge(ck);
  q <= d;
end process;
```

This is the most compact form of the registered process. It only works for types that have an associated `rising_edge` function. The most common logic type, `std_logic`, has this function defined in the `std_logic_1164` package.

This template can optionally have an explicit on clause:

```
process
begin
  wait on ck until rising_edge(ck);
  q <= d;
end process;
```

This template is sometimes used in a modified form that avoids the use of a function call by using a simplified expression to recognise the clock edge. This results in what will be referred to as the basic template with an event expression:

```
process
begin
  wait until ck'event and ck = '1';
  q <= d;
end process;
```

This used to be the preferred form, but has been supplanted by the template using the edge function that was described earlier.

Note that there is some redundancy in this `wait` statement. The `wait` statement is activated only by events on signal `ck` anyway, therefore the test for `ck'event` is unnecessary. However, a synthesis system usually requires this test for an event in order to recognise that the process is a register template and not a combinational process, so it should be included even though it is redundant.

9.4.2   If Statement Template

This template has a slightly different mode of operation in simulation. To illustrate, here's what will be referred to as the `if` statement template:

```
process
begin
  wait on ck;
  if rising_edge(ck) then
    q <= d;
  end if;
end process;
```

The `wait` statement in this template has no `until` clause. This also means that it must have an explicit `on` clause. The `on` clause means that the process will be activated every time there's an event on clock signal `ck`, regardless of the event. The `if` statement then filters out just one kind of event – the rising edge in this example. If signal `ck` has just changed to `'0'`, then the `if` statement skips over the assignment; so signal `q` is not assigned to and retains its value. This means that the falling edge has no effect on the value of `q`. However, if signal `ck` has just changed to `'1'`, then the assignment is executed and so the output gets a new value, so this process still models a rising edge-triggered register.

This template can also use the event expression to test for the clock edge:

```
process
begin
  wait on ck;
  if ck'event and ck = '1' then
    q <= d;
  end if;
end process;
```

9.4.3   Sensitivity-List Template

The final template will be referred to as the sensitivity-list template. A process can have a sensitivity list instead of a `wait` statement to specify the set of signals that cause activation of the process. The form of the sensitivity-list template is:

```
process(ck)
begin
  if rising_edge(ck) then
    q <- d;
  end if;
end process;
```

In this example, the process remains suspended until an event happens on a signal in the sensitivity list which in this case is `(ck)`. This activates the process that executes once and then suspends again. In this case, the process will be activated on any event on `ck`, just like the `if` template. Inside the sensitivity-list process is an `if` statement that filters out all events but the rising edge of signal `ck`.

This process is functionally equivalent to the `if` template, with one subtle difference. In simulation, a design is initialised by running each process until the first `wait` statement is reached. However, in the sensitivity-list form of process, which cannot have a `wait` statement (this is a rule of VHDL), the whole process is executed once.

Therefore, the exact functional equivalence is:

```
process
begin
  if rising_edge(ck) then
    q <= d;
  end if;
  wait on ck;
end process;
```

In fact, the `if` template can also be written like this.

Like the `if` template, the condition can be expressed using the event expression:

```
process(ck)
begin
  if ck'event and ck = '1' then
    q <= d;
  end if;
end process;
```

The sensitivity-list template is clumsy in comparison with the basic template, but comes into its own when dealing with asynchronous resets, which use a variant of it – asynchronous resets will be covered later in Section 9.9. Many designers therefore choose to use the sensitivity-list template because it is easy to convert it into a resettable form or back again. The result is that this form is often used for all registers, resettable or not.

9.4.4   Positioning the Wait Statement

As the previous example shows, the sensitivity-list template is equivalent to the `if` statement template, but with the `wait` statement moved to the end of the process. This means the two templates are equivalent in synthesis but subtly different in simulation. This difference can be exploited such that simulations will initialise correctly without any impact on the synthesis results. The general rule is that the `wait` statement can be placed either at the beginning or at the end of the process for it to be recognised as a register template.

Processes with the `wait` statement at the end will be executed in their entirety at the start of simulation whereas processes with the `wait` statement at the beginning will not be executed at the start of simulation.

Note also the section on register initial values in Section 9.12, since this also has an impact on the results of simulation but not synthesis.

9.4.5   Specifying the Active Edge

So far, all the examples have been sensitive to the rising-edge of the clock signal. It is just as easy to specify a register sensitive to the falling-edge of the clock.

The edge function template for a falling-edge sensitive register is:

```
process (ck)
begin
  if falling_edge(ck) then
    q <= d;
  end if;
end process;
```

All the other templates can be used to specify falling-edge sensitive registers too.

One common concern is that the event expression way of testing for an edge is naive when using std_logic, since it relies on only detecting the value at the end of the edge and doesn't check the value at the start of the edge. Thus, a transition from 'X' to '1' would be regarded as a rising edge. Some designers prefer to filter out these doubtful edges and only consider '0' to '1' transitions as rising edges. This leads to a more complicated form of the event expression being used:

```
process (ck)
begin
  if ck'event and ck = '1' and ck'last_value = '0' then
    q <= d;
  end if;
end process;
```

This template also tests the last value of ck, namely the value it had before the event. This is a rather paranoid approach, but some designers insist on it. In fact, the edge functions (rising_edge and falling_edge) implement this expression, so using the edge function template gives you this extra checking already.

It is in any case good practice to ensure that the clock signal only ever takes real values during simulation. This should be simple, since the clock distribution circuit should be a simple circuit. This would seem to be good design practice anyway, since synchronous design requires the clock to be reliable and glitch-free, which should show itself as a reliable series of real-value transitions during simulations.

The main source of non-real values is elaboration. All signals are initialised with the leftmost value of the type unless specified otherwise. This means that std_logic clocks will initialise to the metalogical value 'U' by default. To avoid undesirable behaviour that causes false triggering of registers during simulation, simply give the clock a real initial value when it is declared:

```
signal ck : std_logic := '0';
```

or

```
signal ck : std_logic := '1';
```

The use of the edge functions is probably the simplest solution to the specification of the active clock edge and is far more readable than the other forms. It is also standardised and widely regarded as best practice.

9.5   Register Types

The register model is not limited to one-bit signal types. Indeed it is very rare to only register a single bit. The signal being registered can be of any synthesis-supported type including integer types that are used in counters, enumeration types that are used in state machines, array types as used to describe buses and record types for buses split into fields.

The following example shows a register for an 8-bit `signed` signal.

```
library ieee;
use ieee.std_logic_1164.all;
use ieee.numeric_std.all;
entity Dtype is
  port  (d : in signed(7 downto 0);
         ck : in std_logic;
          q : out signed(7 downto 0));
end;
architecture behaviour of Dtype is
begin
  process (ck)
  begin
    if rising_edge(ck) then
      q <= d;
    end if;
  end process;
end;
```

The register model is not limited to registering one signal. Any number of signals can be registered in the same registered process. All signals that are the targets of assignments in the process will be registered.

```
process (ck)
begin
  if rising_edge(ck) then
    q0 <= d0;
    q1 <= d1;
    q2 <= d2;
  end if;
end process;
```

In practice, register processes will be mixed with other logic, so an architecture will contain a number of concurrent assignments and processes representing combinational logic, intermixed with processes representing registers. Separating registers into separate design units simply makes the model unnecessarily complicated and obscures the design. It has only been done in these examples to show the processes in context.

9.6   Clock Types

So far, all the examples have used a clock that is of type `std_logic`. This is not a requirement of the template. It is possible to use any single-bit logical type for the clock signal. This includes the standard types `bit`, `boolean` and `std_logic`. It also includes any user-defined logical types.

However, it is good practice to use `std_logic` for clock signals, so all the examples in this book follow that practice.

9.7   Clock Gating

So far, all the register examples have been ungated. That is, they have taken a new value on every clock cycle. In practical circuit design it is very rare for a register to be ungated. Generally, some form of gating will be used.

From a circuit design viewpoint, there are two methods of register gating: clock gating and data gating. This section describes clock gating, the next section describes data gating.

In clock gating, the clock signal itself is switched on or off by some other control signal. However, it is rarely used and even considered bad practice to use clock gating in a synthesisable design. There are two very good reasons why clock gating is considered bad design practice for logic synthesis.

The first reason for not using clock gating is the advent of test synthesis tools. Test synthesis combines automatic overlay of scan paths with built-in test-pattern generation to give a complete test solution for a design, provided that it is a synchronous design. In particular, the scan overlay techniques require that the scan control circuitry can directly control the clocks in the design.

The second reason for not using clock gating is that the algorithms used in logic synthesis for logic minimisation cannot guarantee glitch-free logic. Indeed, this is not the intention, since the main thrust of logic minimisation is area saving and speed optimisation. It is therefore possible for a synthesis tool to create logic that is prone to generate glitches. Glitches in the control signal used to gate the clock would be disastrous to the correct functioning of the circuit.

As with many rules, there are exceptions. There are legitimate uses for clock gating, for example in the design of low-power circuits where clock gating is used to effectively disable a register, register bank or even a complete subsystem in a way that consumes no (or at least negligible) power. Indeed, simply charging and discharging the clock line itself can require significant power, so disabling the clock line to a region of a circuit that is not changing can give a significant power saving.

A good example of the application of clock gating would be a large register bank. In read mode, there is no need to clock the register bank – the registers need only be clocked in write mode. The clock could therefore be gated by the write enable control signal.

For this reason, clock gating is explained here with the warning that, if used, it is up to you to check that the synthesised clock circuits are safe. The simple rule for minimising the risk of glitches on gated clocks is to restrict the gating circuitry to be as simple as possible, with the gating signal driven through a minimum of control logic from a stable source – either from a register or a primary input that is itself guaranteed to be glitch-free.

The following example shows the use of a gated clock in a modified version of the basic D-type introduced at the start of the chapter.

```
library ieee;
use ieee.std_logic_1164.all;
entity GDtype is
  port (d, ck, en : in std_logic;
        q          : out std_logic);
end;
```

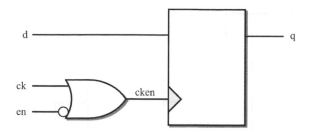

Figure 9.3 Clock gating circuit.

```
architecture behaviour of GDtype is
  signal cken : std_logic;
begin

  cken <= ck when en = '1' else '1';

  process (ck)
  begin
    if rising_edge(cken)
      q <= d;
    end if;
  end process;

end;
```

The resulting circuit is illustrated in Figure 9.3

Note that the clock enable disables the clock by holding it high. This is based on the assumption that the enable is driven by another register output triggered by the same clock edge as this register, namely on the rising edge of the clock, so any change in the enable signal will happen immediately after a rising edge when the clock is high. Any glitches are likely to happen in this early part of the clock cycle, so choosing to disable by holding the clock high effectively makes the circuit immune to these glitches. Of course, once the clock goes low the circuit will become sensitive to glitches on the enable signal again, but provided the propagation time of the enable signal is less than the high period of the clock, this will be a safe design.

9.8 Data Gating

Data gating is the preferred, glitch-resistant, way of providing an enable control on a synthesisable register. Data gating is so-called because it controls the data input of the register instead of the clock input. Therefore, the register is continuously clocked. Data gating works by feeding back the register output to its input whilst the enable is in its inactive state. The basic circuit is shown in Figure 9.4

A common concern is that the multiplexer is a large overhead. In practice most technologies have data-gating registers that have been optimised to eliminate most of this overhead and therefore have an area similar to a clock-gated register, but with the benefit that there are no special timing problems or testing problems with this circuit.

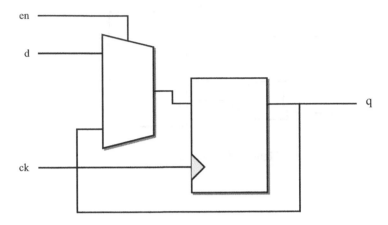

Figure 9.4   Data gating circuit.

The VHDL for a data-gated register based on the previous example used for clock gating is:

```
library ieee;
use ieee.std_logic_1164.all;
entity GDtype is
  port (d, ck, en : in std_logic;
          q         : out std_logic);
end;
architecture behaviour of GDtype is
begin

  process (ck)
  begin
  if rising_edge(ck) then
    if en = '1' then
        q <= d;
      end if;
    end if;
  end process;

end;
```

The combinational part of this register model looks exactly like a latch circuit. However, within a register model it becomes a data-gated register.

The reason this models a gated register is that, in simulation, the value of q is preserved until a new value is assigned to it. In this case, the assignment is bypassed as long as the enable signal is in its inactive state so the value of q is preserved even though the register is being clocked. In hardware terms, this is equivalent to the register output being connected back to the input. In fact, this is exactly what the synthesiser does. During the synthesis of the registered process, the structure of the if statement is analysed to see if there are any circumstances that would lead to the registered signal *not* getting a new value. If any such conditions are detected, they are implemented as feedback of the previous value. The conditions under which those signals require feedback become the control signals for the multiplexer.

A common pitfall when using the `if` template for registers is to try to combine the clock `if` statement with the enable `if` statement:

```
process (ck)
begin
  if rising_edge(ck) and en = '1' then
    q <= d;
  end if;
end process;
```

This will work perfectly in simulation, but may not be recognised as a register template during synthesis, because it does not correspond to any of the common register templates.

Note: this style of process has now been added to the VHDL synthesis standard and so some tools do allow it. However, adoption of the synthesis standard has been slow and patchy, so check the documentation to see if your synthesis tool does. If you are interested in cross-tool portability, avoid it.

The safe rule, which is probably still worth keeping to until all synthesisers conform to the new standard, is that the register template must be kept separate from its contents – in other words, the `if` statement that detects the clock edge must be kept separate from the `if` statement that implements the register enable. The safe and portable form of this process is:

```
process (ck)
begin
  if rising_edge(ck) then
    if en = '1' then
      q <= d;
    end if;
  end if;
end process;
```

9.9   Asynchronous Reset

It is sometimes useful to be able to reset a register to a predefined value. This is particularly true of wide, multi-bit registers.

There are two forms of register reset; asynchronous and synchronous. It is important to distinguish between the two forms and to use the correct one for the circumstances. This section describes asynchronous resets and the next section describes synchronous resets.

Asynchronous resets override the clock and so act immediately to change the value of the register. They are generally used to implement global system-level resets enforced from off-chip and therefore should only ever be controlled by primary inputs to a circuit. They are very sensitive to glitches on the reset signal because asynchronous reset controls are in effect always active. There is no guarantee that synthesised circuits are glitch free and there probably never will be such a guarantee. If they are used in a circuit, very careful design is needed to ensure safe operation.

In general, an asynchronous reset signal should be driven by a primary input as part of a system-wide asynchronous reset scheme. Internal resets should usually be synchronous.

Asynchronous resets require an alternative register template which is recognised as such by a synthesis tool. As with the basic D-type register, there are many different ways to write models

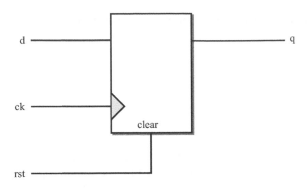

Figure 9.5   Asynchronous reset.

that act like asynchronous resets during simulation, but only those that conform to the synthesis templates will be synthesisable.

A basic resettable register based on the original D-type from the beginning of the chapter is shown here.

```
process(ck, rst)
begin
  if rst = '1' then
    q <= '0';
  elsif rising_edge(ck) then
    q <= d;
  end if;
end process;
```

The circuit represented by this example is shown in Figure 9.5. Note how the asynchronous clear input of a register has been used to implement the asynchronous reset.

The difference between this template and the basic register template is the addition of the `reset` signal to the sensitivity list of the process and the addition of the reset branch of the `if` statement before the clocked branch. This means that, during simulation, the process will execute due to any event on either the clock or the reset signal. The two branches of the `if` statement have conditions on them that filter out the relevant behaviour for both asynchronous reset and clock activation.

The reset condition must come as the first branch of the `if` statement because asynchronous resets override the clock behaviour. As long as the reset signal is in its active state, the reset value is asserted. When the reset is deactivated, the second branch of the `if` statement detects rising clock edges.

The reset value can be any *constant* value. This is particularly useful with multi-bit types, such as arrays, records, integers and enumeration types. Consider the following example where both q and d are 4-bit unsigned numbers:

```
process(ck, rst)
begin
  if rst = '1' then
```

```
        q <= to_unsigned(10, q'length);
    elsif rising_edge(ck) then
        q <= d;
    end if;
end process;
```

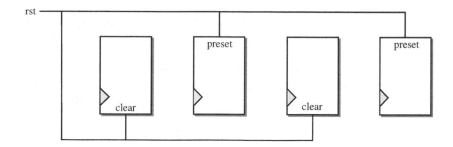

Figure 9.6   Asynchronous reset to a value.

The resulting reset circuitry is shown in Figure 9.6. Only the reset wiring has been shown to illustrate the way the reset signal is connected to either the preset or clear inputs of the individual registers to make up the bit-pattern for the reset value (equivalent to the unsigned integer value 10).

Because the asynchronous reset is implemented by using the registers' preset and clear inputs, the reset value must be a constant. It is not possible to asynchronously reset to the value of a signal.

Another consequence of the synthesis rules for asynchronous resets is that it is not generally possible to have both asynchronous preset and clear controls on a synthesisable register. This is far from obvious since it is easy to model such a register. Here is a VHDL model for a register with preset and clear controls:

```
process(ck, preset, clear)
begin
    if preset = '1' then
        q <= '1';
    elsif clear = '1' then
        q <= '0';
    elsif rising_edge(ck) then
        q <= d;
    end if;
end process;
```

However, this model is *not* generally synthesisable, although some synthesisers support it. The reason is that synthesis does not work only on single-bit registers (in other words, it does not work only on single flip-flops). Rather, most registers are multi-bit registers.

With multi-bit registers the concept of preset and clear is not very useful. What is far more useful is the concept of reset-to-value. In other words, the reset scheme for register synthesis is arranged so that the reset can set the register to any value, not just the limited values of all-zeros or all-ones.

As a consequence, synthesis vendors generally do not allow multiple reset controls, since that raises the possibility of enormous complexity in decoding the reset logic into preset and clear lines for each flip-flop in the register, with the possibility of introducing glitches on those control lines. Glitches in asynchronous reset circuits are absolutely fatal to a design because the reset lines act immediately and override the clock. Therefore, any such decoding is bad design practice. It is reasonable to disallow it.

It would be possible to recognise the special cases where no decoding logic was necessary (which happens when the values in each reset branch are mutually exclusive), but this is not generally done.

To give another full example, the following is an asynchronously resettable counter using type unsigned to count from 0 to 15 and round to 0 again. It exploits the characteristic of the arithmetic operators for the numeric types that wrap round on overflow. The asynchronous reset sets the counter to zero.

```
signal ck, rst : std_logic;
signal count : unsigned(3 downto 0);
...
process(ck, rst)
begin
  if rst = '1' then
    count <= to_unsigned(0, count'length);
  elsif rising_edge(ck) then
    count <= count + 1;
  end if;
end process;
```

9.9.1 Simulation Model of Asynchronous Reset

It is worth pausing at this point to look again at the simulation behaviour of the registered process when there is an asynchronous reset present. This is because the behaviour is rather more complex than the basic registered process described earlier.

Here is the simplest asynchronously resettable register model again:

```
process(ck, rst)
begin
  if rst = '1' then
    q <= '0';
  elsif rising_edge(ck) then
    q <= d;
  end if;
end process;
```

Note that this is similar in form to the sensitivity-list template for the basic register. The difference is that there are now two signals in the sensitivity list, the reset signal and the clock signal. This means that the process will be activated when an event occurs on either signal.

Consider what happens when the reset signal rst goes high, activating the asynchronous reset. The event on rst causes the process to execute. Since the reset signal has gone high, the first branch of the if statement will be executed and the signal q will have a transaction posted on the transaction queue that sets it to '0'. The process then finishes execution and suspends. During process suspension, the signal q will be updated with will go to '0' (or stay at '0' if it is already at that value).

If clock events happen whilst the reset is activated, those events will trigger the process again, but this will simply cause the first branch of the if statement to be re-executed. This first branch has no test for an event on rst, it is simply selected by the condition rst = '1', so the fact that events are happening on the clock signal does *not* cause the second branch of the if statement to be executed. This means that, as expected, the reset signal overrides the clock and simply causes the reset condition to be reasserted.

There is a little inefficiency in this model during simulation, in that it does cause the reset condition to be reasserted due to events on the clock. These cause unnecessary extra transactions on q. However, none of these transactions cause events to be generated because q is being held at a constant value.

When the rst signal goes low again, this will cause the process to be triggered yet again. However, since the value of rst is now '0' the first branch of the if statement will not be selected. Since there has been no event on ck, the second branch will also not be selected. As a consequence, the falling edge of rst has no effect on the value of q.

On the next clock edge, the process will be executed again. The first branch of the if statement will not be selected because the value of rst is '0'. The second branch of the if statement will be selected if the clock edge is a rising edge. This causes the signal q to be assigned the value of input d.

It can be seen that the process now acts like a conventional registered process and will continue to do so as long as the reset signal remains inactivated.

In summary, the reset signal overrides the clock signal and acts as a level-sensitive control signal. The reset is activated immediately and is not synchronised to the clock. The clocked part of the model acts like a conventional register when the reset is inactivated.

There are many other ways in which this behaviour could be modelled in VHDL. However, this particular model has been chosen for its relative simplicity. Remember that synthesisers recognise this as an asynchronously resettable register by matching it to a template. If a different model is used, it may simulate correctly but it will not match the template during synthesis.

9.9.2 Asynchronous Reset Templates

When discussing simple registers, four standard templates were discussed. Not all of these templates can be converted into asynchronously resettable registers. In fact, there are only two templates that have resettable versions: these are the sensitivity-list template and the if-statement template.

The sensitivity-list is the template that has just been used in the previous sections. Its basic form is:

```
process(ck, rst)
begin
  if rst = '1' then
    q <= '0';
  elsif rising_edge(ck) then
```

```
      q <= d;
    end if;
end process;
```

The if-statement template is similar to the above template, but uses a `wait` statement instead of the sensitivity list:

```
process
begin
    wait on ck, rst;
    if rst = '1' then
      q <= '0';
    elsif rising_edge(ck) then
      q <= d;
    end if;
end process;
```

Both templates use the same structure of `if` statement, it is just the way in which the processes are sensitised to their control signals that differs between the two templates.

9.10   Synchronous Reset

A synchronous reset control uses a very similar structure to the data-gating circuit described in Section 9.8. The only difference is that, instead of feeding back the register output to its input, a reset value is fed to the register input when the reset signal is in its active state.

In contrast to asynchronous resets, synchronous resets take effect on the active edge of the clock and so fit the synchronous design philosophy. As such, they require no special handling at all and are simply a special case of combinational logic with a registered output. Synchronous resets can be controlled by any data signal in a circuit and are insensitive to glitches. All resets driven by logic within the circuit should generally be synchronous resets.

The VHDL for a synchronously resettable process is shown below.

```
signal d, ck, rst : std_logic;
signal q : std_logic;
...
process (ck)
begin
    if rising_edge(ck) then
      if rst = '1' then
        q <= '0';
      else
        q <= d;
      end if;
    end if;
end process;
```

The circuit created for this example is shown in Figure 9.7.

The example shows the register being reset to '0'. There is no restriction on the value to which the register is reset, a feature that becomes particularly useful with other types, such as arrays, records, integers and enumeration types.

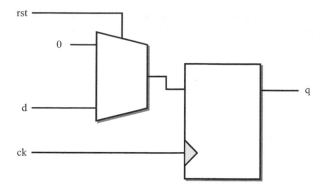

Figure 9.7   Synchronous reset.

Consider the following example where both q and d are 4-bit signals of type unsigned:

```
signal d : unsigned (3 downto 0);
signal ck, rst : in std_logic;
signal q : out unsigned (3 downto 0);
...
process (ck)
begin
  if rising_edge(ck) then
    if rst = '1' then
      q <= to_unsigned(10, q'length);
    else
      q <= d;
    end if;
  end if;
end process;
```

The resulting reset circuitry is shown in Figure 9.8. Only the reset wiring has been shown to illustrate the way the reset signal controls the individual registers to make up the bit-pattern for the reset value (equivalent to the unsigned integer value 10 in this example).

In fact, the synchronous reset is a special case of a multiplexed register and the 'reset' value needn't even be a constant value. It could be another signal.

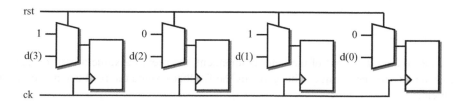

Figure 9.8   Synchronous reset to a value.

To give another example, the following is a description of a resettable counter using types unsigned to count from 0 to 15 and then wrap round back to 0. It is the synchronously resettable version of the example in the previous subsection.

```
signal ck, rst : std_logic;
signal count : unsigned(3 downto 0);
...
process (ck)
begin
  if rising_edge(ck) then
    if rst = '1' then
      count <= to_unsigned(0, count'length);
    else
      count <= count + 1;
    end if;
  end if;
end process;
```

Note the way in which the reset if statement is nested inside the register template if statement. The argument here is exactly the same as for data gating – the register template part must be kept separate from the combination part of the model.

9.11   Registered Variables

It is also possible to create registers using variables. Remember that a registered process is interpreted by synthesis as a combinational circuit and then placing a register on every signal assigned to in the registered process and every feedback path. This usually means that variables do not get registered. However, if there is feedback of a previous variable value, then this feedback must be via a register to make the process synchronous.

Here is an alternative way of writing a counter, using the type unsigned that implements integers with wrap-around:

```
process (ck)
  variable count : unsigned (7 downto 0);
begin
  if rising_edge(ck) then
    if rst = '1' then
      count := to_unsigned(0, count'length);
    else
      count := count + 1;
    end if;
    result <= count;
  end if;
end process;
```

In this case, in the else part of the inner if statement, the previous value of count is being read to calculate the next value. This gives us the feedback since the read occurs before the assignment.

Note that this example actually creates two registers. According to the feedback rules, variable count will be registered. However, signal result will also have a register on it, because all signals assigned to in a registered process get registered. The extra register will always contain the same value as the register for `count`. This redundant register will be optimised away by the synthesiser, so can be ignored.

9.12   Initial Values

There has already been a discussion of initial values in Section 8.3. However, it is worth having a reminder of the pitfalls associated with initial values at this stage since the most common problems with initial values are found in registered logic.

Every signal in a VHDL model is given an initial value during the pre-simulation period known as elaboration. The initial value can be user defined in the signal declaration or it defaults to the leftmost value of the signal type or subtype. This means that the simulation is guaranteed to start in a known state for every simulation on any simulator. Specifically, it means that counters and state machines will always start at a known state during simulation.

There is, however, no hardware equivalent of elaboration and so initial values are ignored by synthesis. This means that registers, and specifically counters and state machines, will start in an unknown state. It is the designer's responsibility to allow for this by ensuring that a circuit can be put into a known state either by designing in an initialisation scheme or by providing an external reset. Either scheme can be tested in simulation by giving registered signals a 'bad' initial value and then ensuring that the circuit initialises correctly.

10

Hierarchy

Hierarchy should be used extensively in a design to divide-and-conquer the design problem. Furthermore, third-party components can be incorporated into a design more easily this way and this leads to a higher degree of confidence in the integrity of the design.

The natural form of hierarchy in VHDL, at least when it is used for RTL design, is the component. Do not be tempted to use subprograms as a form of hierarchical design! Any entity/architecture pair can be used as a component in a higher-level architecture. Thus, complex circuits can be built up in stages from lower-level components.

It is possible to design parameterised circuits using generics. The combination of generics with synthesis, which allows the design of technology-independent circuits, is a very powerful combination.

10.1 The Role of Components

There are a number of reasons for using hierarchy.

Each subcomponent can be designed and tested in isolation before being incorporated into the higher levels of the design. This testing of intermediate levels is much simpler than system testing and consequently is usually more thorough. Using components for hierarchy is compatible with the Test-Driven Design paradigm since it makes testing of subsystems easier. This means that the designer can have a high degree of confidence in the subcomponents and this also contributes to the overall integrity of the design.

It is good practice to collect useful subcomponents together into reusable libraries so that they can be used elsewhere in the same design and later on in other designs. One of the great gains of logic synthesis is that such modules are technology independent and so can be reused in a wide variety of different projects. This means that, over a period of time, the level of reuse in each design increases as more components become available.

It is important to have a strategy to collect useful components together into libraries for reuse. Many reuse policies, both in software and hardware design, fail due to excessive bureaucracy, which discourages users from submitting reusable components to be added to the library. A realistic reuse policy should therefore be based on the minimum of bureaucracy to encourage the generation and use of such components.

VHDL for Logic Synthesis, Third Edition. Andrew Rushton.
© 2011 John Wiley & Sons, Ltd. Published 2011 by John Wiley & Sons, Ltd.

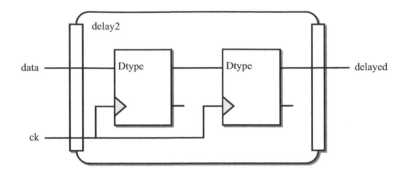

Figure 10.1   Target circuit.

The first official version of VHDL, VHDL-1987, was very clumsy in its handling of components. The original language designers concentrated on making the component mechanism as flexible as possible, and succeeded in this aim, but the result was cumbersome in use and the full flexibility was rarely used. This form of component is known as "indirect binding". A simpler form of components was added in VHDL-1993 to make them more usable. This form of component is known as "direct binding".

The use of components will be illustrated first with a complete example using indirect binding. Then, this example will be converted to use the simpler direct binding. Direct binding is the form that will then be used throughout the rest of the book – indirect binding is included for completeness.

The example is a trivial one so that the focus is on the components rather than their contents. It is a simple two-register pipeline. The problem is illustrated by Figure 10.1, and is to create a circuit containing two register components connected in series.

It will be assumed that the register component that will be used already exists and has the following interface:

```
library ieee;
use ieee.std_logic_1164.all;
entity Dtype is
   port (D, Ck : in std_logic; Q, Qbar : out std_logic);
end;
```

10.2   Indirect Binding

Due to the flexibility of the indirect binding mechanism in VHDL, there is more than one way that this could be described using components, depending on the level of detail that is given. Some of the information required to create component instances is optional and, if omitted, default values will be used. This is covered in detail in the following subsections. The target circuit in its *fullest* form is:

```
library ieee;
use ieee.std_logic_1164.all;
entity delay2 is
```

```
   port (data, ck : in std_logic; delayed : out std_logic);
end;
architecture behaviour of delay2 is

   -- component declaration
   component dtype
     port (d, ck : in std_logic; q, qbar : out std_logic);
   end component;

   --  configuration specification
   for all : dtype use entity work.Dtype(behaviour)
     port map (D => d, Ck => ck, Q => q, Qbar => qbar);

   signal internal : std_logic;

begin

   -- component instances
   d1 : dtype
     port map (d => data, ck => ck, q => internal, qbar => open);
   d2 : dtype
     port map (d => internal, ck => ck, q => delayed, qbar => open);

end;
```

Note that the entity has just three port declarations, since the target circuit only has a delayed output, compared with the dtype component that has both q and qbar outputs.

The architecture has been called behaviour, despite the fact that it contains no behavioural VHDL, because it is part of a synthesisable design. The architecture name should reflect the *usage* of the architecture, not its contents.

The architecture contains three sections related to the use of components. These have been labelled with comments giving the names of the three parts: component declaration, configuration specification and component instance. It is the component instances that make up the circuit. The component declaration and configuration specification are declarations that control the selection and means of connection of the component instances.

The following three subsections will discuss the three parts of this architecture.

10.2.1 Component Instances

Component instances are concurrent statements. In this sense, they model hardware components closely, because hardware is inherently concurrent.

In the above example, there are two component instances, which have been labelled d1 and d2. These labels are for reference only but must be present and must be unique within the architecture. Each component instance in this example creates a subcircuit using the component dtype and the connections to this component.

It should be noted that the component instance is an instance of the component declaration and not the entity. The relationship between the component declaration and the entity is controlled by the configuration specification.

In the example, the distinction between the entity and the component declaration has been made by giving the entity initial capitals on its name and the name of its ports. The component instance and the internal signals in the target circuit have been given lowercase names. Put another way, everything declared within the target circuit has lowercase names, whereas everything outside the target circuit – namely the entity being used as a component – have initial capitals. This distinction is made just for clarity and is not of any other significance, since VHDL is not sensitive to the case of names anyway.

The component instance part of the example is reproduced here:

```
d1 : dtype
   port map (d => data, ck => clock, q => intern, qbar => open);
d2 : dtype
   port map (d => intern, ck => clock, q => delayed, qbar => open);
```

A component instance has two parts, the component name and its port map. The component name part is fairly self-explanatory – it labels the instance (in this case d1) and then gives the name of the component that is to be used for the instance (in this case dtype). Note that this is the name of a component declaration, not the entity. This component must be declared before it can be instanced, either in the architecture (see later) or in a package (see Section 10.4).

The port map describes how signals in the architecture are to be connected to the ports of the component. There are two ways of specifying the port map: using named association or positional association. The form shown is named association.

With named association, each port is listed by name (remember that these are the component ports, not the entity ports). The port name is then followed by the "=>" symbol, known as *finger* and pronounced 'connected to' and then the signal that it is to be connected to. This signal must, of course, be declared as either a signal in the architecture or a port in the entity of the target circuit. Output ports on the component can be left unconnected by using the keyword open.

With positional association, the port names and fingers are omitted, and the signals are listed in the same order as the port declarations in the component. Thus, the above example could be rewritten using positional association:

```
d1 : dtype port map (data, clock, internal, open);
d2 : dtype port map (internal, clock, delayed, open);
```

Positional association is clearly simpler and shorter, but named association has the advantage of acting as a reminder of which port each signal is being connected to. This means that the component instance is more self-explanatory to a reader of the model. With named association it is also possible to change the ordering of the ports if a different ordering seems more natural to show the data flow in the design. Finally, the named association gives a VHDL analyser an extra opportunity for checking the consistency of the design.

10.2.2 Component Declarations

The component declaration defines a component that can then be instanced. It is in effect saying that, somewhere, there is an entity that can be used as a component. However, the component declaration itself does not specify where that entity is or even what it is called or what its ports are called. This is the job of the configuration specification, which is described in the next section.

In practice, the component declaration usually matches the entity that it represents exactly. That is, the name and the port names and their order are taken directly from the entity. The only difference is that the component has a slightly different syntax. In the following two examples, the entity and its representative component are shown with their differences underlined.

The entity for the `Dtype` was declared as:

```
entity Dtype is
   port (D, Ck : in std_logic; Q, Qbar : out std_logic);
end;
```

The component `dtype` is declared as:

```
component dtype
   port (d, ck : in std_logic; q, qbar : out std_logic);
end component;
```

The need for a component declaration is probably the clumsiest part of the use of indirect binding. However, the need for components declarations can be reduced by using component packages. There is no need for component declarations at all when using direct binding.

10.2.3   Configuration Specifications

The configuration specification closes the gap between entities and their component declarations. The reason for separating the entity and its components is a feature of the language that allows the association (technically known as *binding*) between component and entity to be made as late as possible in the simulation process. In fact, binding is not carried out during the analysis phase at all, it is deferred until elaboration – the start of simulation. This means that a hierarchical design can be compiled in any order – with the extreme cases being top-down or bottom-up.

The configuration specification defines where the entity is to be found, what it is called and how the component ports relate to the entity ports. The full form of a configuration specification is long-winded, but all aspects of it have sensible defaults and indeed, the configuration specification is itself optional and can be omitted completely to get all the default options.

First, an explanation of the parts of a configuration specification. The configuration from the example is reproduced here:

```
for all : dtype use entity work.Dtype(behaviour)
   port map (D => d, Ck => ck, Q => q, Qbar => qbar);
```

The configuration specification has three parts.

The first part specifies the components that the configuration applies to. In this case, the components are selected by the keyword `all`, so all the components called `dtype` are configured by this specification. It would have been possible to have separate configuration specifications for each component separately:

```
for d1 : dtype ...
for d2 : dtype ...
```

The second part of the configuration selects the entity to use for the component. This is known as the entity binding. The entity binding must specify which library the entity is in as well as its name. In this case, the entity is specified as being in the same library as the target circuit, so is referred to by the reserved library name work.

The entity binding also specifies an architecture to use for the component. Multiple architectures are rarely used in RTL design, so the ability to select an architecture is rarely needed. In the example, the architecture behaviour is specified. To omit the architecture, omit the parentheses as well. The default architecture is, strangely, the most recently compiled one! For single-architecture entities, though, this means that the only architecture present is automatically selected.

There are other forms of binding, but these are not generally supported by synthesisers so will not be covered here.

The third part of the configuration specification defines the port bindings. That is, it associates the entity ports with the component ports. This allows port names to be different in the entity and the component, although it is hard to imagine any reason for doing this.

The port binding is optional, and if omitted, the component ports are bound to entity ports of the same name. If present, then the port binding, enclosed in parentheses, is a list of entity ports associated with their corresponding component ports. In other words, each entry in the port binding is the entity port name, finger, then the component port name. The entity port name and finger can be omitted, in which case the component ports are bound to the entity in the order defined on the entity.

10.2.4   Default Binding

If the configuration specification is missing completely, then the component is bound to an entity of the same name as the component, in the work library, and to the architecture most recently analysed, and the ports are bound to the entity ports with the same names. Since this is the most commonly desired binding, configuration specifications would appear to be unnecessary.

However, there is one case where configuration specifications are necessary. This is where the component is to be bound to an entity in a different library. Also, some VHDL systems do not implement the default entity binding, so some form of configuration declaration must be included even with an entity in library work. These systems do not implement default entity binding because, so the vendors argue, the VHDL standard does not require it. This is a moot point and default binding has become the accepted practice, so these vendors are out of line with most users' expectations.

Binding could be achieved by making all of the entities in the library visible with a library and use clause. For example, if the component was in a library called basic_gates, then the following library and use clauses could be added to the architecture to make all the entities in that library visible, making a configuration specification unnecessary:

```
library basic_gates;
use basic_gates.all;
architecture behaviour of delay2 is
  component dtype
    port (d, ck : in std_logic; q, qbar : out std_logic);
  end component;
  signal internal : std_logic;
```

```
begin
  d1 : dtype port map (data, clock, internal, open);
  d2 : dtype port map (internal, clock, delayed, open);
end;
```

The problem with this approach is that all the entities in that library become visible, regardless of whether they are going to be used, causing what is known as *clutter* in the number of names visible at once. This could result in name clashes, especially if more than one library is made visible in this way. A better approach that avoids clutter is to use the minimum configuration specification, which just binds the entity and uses the defaults for everything else:

```
library basic_gates;
architecture behaviour of delay2 is
  component dtype
    port (d, ck : in std_logic; q, qbar : out std_logic);
  end component;
  for all : dtype use entity basic_gates.Dtype;
  signal intern : std_logic;
begin
  d1 : dtype port map (data, clock, intern, open);
  d2 : dtype port map (intern, clock, delayed, open);
end;
```

Note that the `library` clause is still needed in order to be able to make the entity binding.

Due to the difference in interpretation of the default binding rules of VHDL, it is recommended that at least a minimum configuration is used, even for creating components of entities that are in library `work`; don't rely on the default entity binding.

10.2.5   Summary of Indirect Binding Process

The best way to think of the relationship between a component instance, the component declaration and the entity being used as a component is as a two-layer binding process.

The outer layer defines the connections between the signals in the target circuit and the ports of the component and is described by the component instance.

The inner layer defines the binding between the component and the entity and is described by the configuration specification.

This two-layer model is illustrated by Figure 10.2.

10.3   Direct Binding

A simpler form of component instantiation is direct binding. In direct binding, the component instance is directly bound to the entity without the need for an intermediate component declaration or a configuration specification. Another way of putting this is that direct binding has only one layer, unlike the two layers of indirect binding.

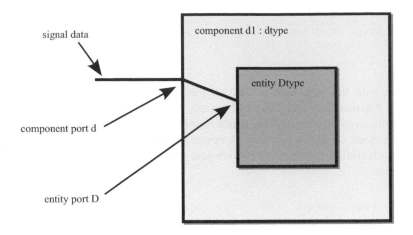

signal data

component d1 : dtype

entity Dtype

component port d

entity port D

Figure 10.2   The two layers of indirect binding.

The syntax of direct binding is like a combination of the component instance and the configuration specification. This allows the above example to be reduced to a very simple architecture indeed:

```
library basic_gates;
architecture behaviour of delay2 is
  signal internal : std_logic;
begin
  d1 : entity basic_gates.dtype(behaviour)
    port map (data, clock, internal, open);
  d2 : entity basic_gates.dtype(behaviour)
    port map (internal, clock, delayed, open);
end;
```

Direct binding is the recommended form for simple hierarchy within a design. However, indirect binding does have its uses as the next section shows.

10.4   Component Packages

It can be convenient to collect together a set of components within a project or a component library and write a package containing all of their component declarations in one place. This package can then be used to instantiate those components and, because it is in one place, act as a user reference as well. It is good practice to do this if you are responsible for a reusable component library for example. This approach makes use of indirect binding with the component declaration moved into a separate package.

Once you have such a package, it is only necessary to have a use clause for the package to make use of the component. Note that this is only worthwhile for commonly used components, since it is still necessary to write the component declaration once – in the package! For simple hierarchy within a design, use direct binding.

To illustrate the use of component packages, the previous indirect binding example will be used. The package containing the component declaration would look like:

```
package basic_gates is

  component dtype
    port (d, ck : in bit; q, qbar : out bit);
  end component;

  ... other components

end;
```

Only the component declaration can be placed in a package; the configuration specification is part of the binding process and so must be part of the specific design, by placing it in the architecture where the binding is to take place.

Assuming that this package is analysed into the current work library, the component would be used in the target circuit as follows:

```
use work.basic_gates.all;
architecture behaviour of delay2 is
  signal intern : std_logic;
  for all : dtype use entity work.Dtype;
begin
  d1 : dtype port map (data, clock, intern, open);
  d2 : dtype port map (intern, clock, delayed, open);
end;
```

This example uses indirect binding with the minimum configuration specification, as recommended in Section 10.2.

The package can be analysed into any library in this way. It need not be in the same library as the entity that its component declarations represent. However, it is good practice to analyse the package into the same library as the entities. Such component packages are quite commonplace, especially for reusable libraries. These libraries contain whole families of related circuits described as entity/architecture pairs. Then, all the entities are declared as component declarations in a single package for ease of use. Often, the package is given the same name as the library to make its purpose clear.

If the entities are in a different VHDL library from your design, then a configuration specification is always required. Taking the same example as before, and again assuming that the Dtype entity and its component package are in the library basic_gates rather than work, the architecture now looks like:

```
library basic_gates;
use basic_gates.basic_gates.all;
architecture behaviour of delay2 is
  signal intern : bit;
  for all : dtype use entity basic_gates.Dtype;
begin
  d1 : dtype port map (data, clock, intern, open);
  d2 : dtype port map (intern, clock, delayed, open);
end;
```

10.5   Parameterised Components

It is possible to design parameterised circuits using *generics*. The combination of generics with
synthesis, which allows the design of technology-independent circuits, is a very powerful
combination. Generics are especially powerful where there is an established reuse policy, so
useful models are written in a generic style with reuse in mind. Indeed, it is good practice to
consider the reusability of subcomponents of a design during the early stages of the design
cycle, possibly as a part of the design review process, so that subcomponents recognised as
reusable are earmarked for parameterisation.

Probably the most common use of generics is to parameterise the port width of a component,
so the same ALU model, for example, can be used as an 8-bit ALU in one design and then as
a 16-bit ALU in another design. Other uses are to parameterise the number of pipeline stages in
a design or to include/exclude features such as output registers.

10.5.1   Generic Entity

To illustrate the use of generics, a simple example will be used. The example is a shifter with
a parallel load. Once again, this example does not really represent typical usage since such
a simple circuit would be embedded in a bigger design and not have an entity and architecture of
its own. A simple example is used so that the focus of attention is on the VHDL, not the circuit.

The interface of the shifter is:

```
library ieee;
use ieee.std_logic_1164.all;
entity shifter is
   generic (n : natural);
   port (ck      : in   std_logic;
         load    : in   std_logic;
         shift   : in   std_logic;
         input   : in   std_logic_vector(n-1 downto 0);
         output  : out std_logic);
end;
```

This entity has an extra field – the generic clause – that defines the circuit's parameters. In this
case, there is just one parameter, n, which specifies the width of the shifter. This has then been
used to define the width of the `input` port. Simple calculations can be used in the definition of
port widths, provided they can be evaluated as a constant value once the generic parameter is set
to an actual value. Generally, this restricts you to using built-in operators on integer, as in this
case where the range of an n-bit signal is defined as (n-1 downto 0), which uses built-in
integer subtraction.

Before going into the writing of a generic model, the way in which this entity would be used
as a component will be illustrated.

10.5.2   Using Generic Components

A generic entity is used as a component in the same way as a non-generic entity. The only
difference is that a value must be specified for the generic parameter for each instance of
the component. This value must be a constant so that the synthesiser can calculate the size of

the ports and is usually given directly as a numeric value in the component instance's `generic map`.

For example, to use the `shifter` parameterisable entity to make an 8-bit shifter component, the following would be used:

```
library ieee;
use ieee.std_logic_1164.all;
entity shifter8 is
  port (ck     : in  std_logic;
        load   : in  std_logic;
        shift  : in  std_logic;
        input  : in  std_logic_vector(7 downto 0);
        output : out std_logic);
end;

architecture behaviour of shifter8 is
begin
  shift : entity work.shifter
    generic map (n => 8)
    port map (ck, load, shift, input, output);
end;
```

The component instance contains an extra field – the generic map – which gives the value to be used for the generic parameter. This instance of the `shifter` circuit will be built with this value substituted throughout in place of the parameter n.

The generic map can alternatively be written using positional association:

```
    shift : entity work.shifter
      generic map (8)
      port map (ck, load, shift, input, output);
```

The rule in synthesis, remember, is that all signals must be of a known size for the synthesiser to be able to calculate the size of the bus to implement. With parameterised circuits, this calculation is carried out *after* generic parameters have been replaced with their actual values. Therefore, a generic can be treated as a constant value.

In the original entity definition the port `input` was defined as:

```
    input : in std_logic_vector(n-1 downto 0);
```

When an instance is created, the value n will be replaced by its actual value. In the example above, the value 8 was used. This means that the port declaration becomes:

```
    input : in std_logic_vector(7 downto 0);
```

So the synthesiser creates an 8-bit bus for this port.

10.5.3   Parameterised Architecture

Now that the use of a generic entity as a component has been discussed, it is time to get down to actually writing the generic circuit.

Within an architecture, a generic parameter acts as if it was a constant value, so can be used wherever a constant value would be used.

The architecture for the ripple carry adder is:

```
architecture behaviour of shifter is
  signal store : std_logic_vector(n-1 downto 0);
begin

  process
  begin
    wait until rising_edge(ck);
    if load = '1' then
      store <= input;
    elsif shift = '1' then
      store <= store(store'left-1 downto 0) & store(store'left);
    end if;
  end;

  output <= store(store'left);

end;
```

Note how the store signal that is then registered by the register process is defined in terms of the generic parameter n. Then, within the architecture, the bounds of this store are referred to using attributes rather than constant values, so that they can adjust to stores of different sizes. For example, the shift is implemented using a slice of the store:

```
store <= store(store'left-1 downto 0) & store(store'left);
```

The rule with slices is their bounds must be constant to be synthesisable. In this case, this will evaluate to a constant once the attribute has been substituted. For the case where $n = 8$, the store has a range of (7 downto 0):

```
store <= store(7-1 downto 0) & store(7);
```

This is now expressed in terms of built-in integer operators and can be evaluated as a constant slice, concatenated with a single bit to give a result that is the same size as the store:

```
store <= store(6 downto 0) & store(7);
```

This could alternatively have been written in terms of the generic parameter instead:

```
architecture behaviour of shifter is
  signal store : std_logic_vector(n-1 downto 0);
begin

  process
  begin
    wait until rising_edge(ck);
    if load = '1' then
      store <= input;
    elsif shift = '1' then
```

```
        store <= store(n-2 downto 0)
      end if;
    end;

    output <= store(n-1);

  end;
```

Again, all slices and indices are expressed as calculations involving integer and its built-in operations, so this generic architecture is synthesisable.

10.5.4   Generic Parameter Types

In simulation VHDL, generic parameters can be of any type. However, synthesisers are limited in the types of generic parameter supported. The only types that can generally be used are integer types, although some synthesisers also allow enumeration types, of which the most useful are probably `boolean`, `bit` and `std_logic`.

The shifter circuit conformed to these restrictions, since the generic parameter type was `natural`, a subtype of `integer`. The use of subtype `natural` tells users not to try to create a shifter with a negative number of bits.

All the bit-array types used for synthesis are indexed by integer. This is true of `bit_vector`, `std_logic_vector` and the `signed`, `unsigned`, `sfixed`, `ufixed` and `float` synthesisable types. Therefore, integer generics can be used to constrain array ports of any of these types.

Generic parameters of other types can make other features parameterisable. For example, type `boolean` can be used to specify conditional features. An example of this will be deferred until Section 10.6, because conditional features are most easily described using `if generate` statements which are covered in that section.

10.6   Generate Statements

Generate statements are used to create replicated or conditional hardware structures. Generate statements are concurrent statements that can be used in any architecture, but they are described here because they really come into their own when used with generics in parameterised circuits.

There are two types of generate statement: the `for generate` statement for replicated structures and the `if generate` statement for conditional structures.

10.6.1   For Generate Statements

The `for generate` statement replicates (copies) its contents a specified number of times. It is in effect a concurrent form of `for loop`. However, being a concurrent statement, it can be used to replicate any other concurrent statements, including processes, concurrent signal assignments, component instances and other generate statements. This means that `for generate` statements can replicate structures that are impossible to replicate using `for loops`. The most obvious example is a register: since a register must be a process, one way of replicating a register to get a register bank is with a `for generate` statement replicating a register process.

To illustrate the use of the `for generate` statement, a simple parameterisable register bank will be written using multiple register processes. This circuit has an array of enable inputs, one per register. The generic register bank:

```
library ieee;
use ieee.std_logic_1164.all;
entity register_bank is
  generic (n : natural);
  port (ck     : in  std_logic;
        d      : in  std_logic_vector(n-1 downto 0);
        enable : in  std_logic_vector(n-1 downto 0);
        q      : out std_logic_vector(n-1 downto 0));
end;

architecture behaviour of register_bank is
begin
  bank: for i in 0 to n-1 generate
    process
    begin
      wait until rising_edge(ck);
      if enable(i) = '1' then
        q(i) <= d(i);
      end if;
    end;
  end generate;
end;
```

The `for generate` is similar in appearance to the `for loop`. Note that it has a label (`bank:`) and that this is required. The `for loop` is a sequential statement, whereas the `for generate` is a concurrent statement. In many other respects they are very similar. There is a generate constant, which in the above example is called `i`, which controls the generation. This is known as the generate constant because the value of `i` is treated as a constant value inside the loop. The rules for writing the conditions of a `for generate` statement are the same as for a `for loop`. The range of the generate must have a constant range, because the synthesiser must know how many times to replicate the structure, so it is not possible to use a signal value to define the range.

The order of execution of the generation is irrelevant, since there is no order to the execution of the concurrent statements being executed. It has been specified in this case with an ascending range, so counts upwards. Concurrent statements conceptually are all executed simultaneously, so reversing the order of the generation can't possibly make any difference to the resultant circuit. It is important to realise that, although the example has been written with a top to bottom signal flow, particularly with regard to the carry path, in fact this is a concurrent description and so could be written in any order. The top to bottom flow is for clarity to the human reader.

Notice that the d, q and `enable` signals are array signals that are then indexed by the generate constant in the generate statement. The equivalent circuit is created by replicating the circuit represented by the statements in the generate statement once for each value of the generate constant and replacing the generate constant in any statements by its value for that replication. For example, in the first replication, `i` will be replaced by 0.

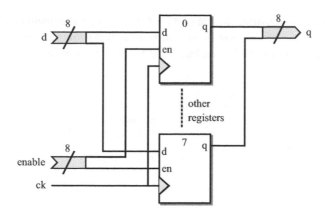

Figure 10.3 For-generate circuit.

The equivalent architecture after the generation for an 8-bit instance is:

```
architecture behaviour of register_bank is
begin
  process
  begin
    wait until rising_edge(ck);
    if enable(0) = '1' then
      q(0) <= d(0);
    end if;
  end;
  ... other processes
  process
  begin
    wait until rising_edge(ck);
    if enable(7) = '1' then
      q(7) <= d(7);
    end if;
  end;
end;
```

The resultant circuit for this example is shown in Figure 10.3.

10.6.2 If Generate Statements

The `if generate` statement allows optional structures to be described. For example, an optional output register could be added to a general-purpose component and controlled by a `boolean` generic parameter. They are not so commonly used as `for generate` statements, but will occasionally have a use.

An `if generate` statement is controlled by a `boolean` expression. This could be the value of a `boolean` generic parameter or it could be the test for an integer being a certain value. Either way, the result is a `boolean` and must be a constant. It is not possible to use a signal in the condition.

Again, the best way to illustrate the use of the `if generate` statement is with an example, in this case showing the description of just such an optional output register. The example consists of just the register in isolation, but could be incorporated into any generic entity.

```
library ieee;
use ieee.std_logic_1164.all;
entity optional_register is
   generic (n : natural; store : boolean);
   port (a : in std_logic_vector (n-1 downto 0);
         ck : in std_logic;
         z : out std_logic_vector (n-1 downto 0));
end;
architecture behaviour of optional_register is
begin

   gen: if store generate
     process
     begin
       wait until rising_edge(ck);
       z <= a;
     end process;
   end generate;

   notgen: if not store generate
     z <= a;
   end generate;

end;
```

If the generic parameter `store` is set to `true`, then a registered process is generated, so the output z will be a registered version of input a. However, if the generic parameter store is set to `false`, then a concurrent signal assignment is generated, so the output z will be directly connected to a.

It is important, as this example shows, to consider both the possible conditions of the `if generate` statement, which means that `if generate` statements often appear in pairs, like this example, with opposite conditions. It is not clear why there is no `else generate` statement in VHDL-1993, since there is an obvious need for one, but this is effectively what is being done in this example.

Note: VHDL-2008 does have an `else generate`. So when this functionality makes its way into VHDL synthesis and simulation tools, you will be able to write a much clearer version of this:

```
gen: if store generate
  process
  begin
    wait until rising_edge(ck);
    z <= a;
  end process;
else generate
  z <= a;
end generate;
```

There is also an `elsif generate` statement in VHDL-2008 to allow multiple conditions to be tested:

```
gen: if registered generate
  process
  begin
    wait until rising_edge(ck);
    z <= a;
  end process;
elsif latched generate
  process (a, enable)
  begin
    if enable = '1' then
      z <= a;
    end if;
  end process;
else generate
  z <= a;
end generate;
```

However, check this is available in both simulator and synthesiser before using it.

10.6.3   Component Instances in Generate Statements

There is a common problem when using indirect binding of components inside generate statements. It is one of the most common pitfalls in using generate statements.

The problem is that the generate statement is seen as a separate sub-block of the main architecture, and so components within the generate block cannot be configured by a configuration specification in the architecture. That is, configuration specifications are required to be in the same *scope* – the term for a sub-block – as the component instances themselves, but the generate statements are considered to be a separate level of scope from the architecture.

To illustrate the problem, here is an *incorrect* architecture with a configuration specification in the architecture. This example creates a generic word-width register built from the `dtype` component defined in a package earlier:

```
library ieee;
use ieee.std_logic_1164.all;
entity word_delay is
  generic (n : natural);
  port (d : in std_logic_vector(n-1 downto 0);
        ck : in std_logic;
        q : out std_logic_vector(n-1 downto 0));
end;
use basic_gates.basic_gates.all;
architecture behaviour of word_delay is
  for all : dtype use entity work.Dtype;
begin
  gen: for i in 0 to n-1 generate
    d1 : dtype port map (d(i), ck, q(i), open);
  end generate;
end;
```

The problem is that the error in this example is obscured by the rules of VHDL – no compilation error will be produced by this example. This is because the configuration specification uses the selection all, which can legitimately match with no component instances at all. Indeed in this example the configuration specification does not bind any components and the component instance is unbound.

The error would show itself if the configuration selection was made explicit:

```
for d1 : dtype use entity work.Dtype;
```

This form of configuration specification will cause an error in compilation, since there is no component instance in the architecture with the label d1.

If indirect binding is still required, then the solution is to add the declaration to the generate statement itself:

```
architecture behaviour of word_delay is
begin
  gen: for i in 0 to n-1 generate
    for all : dtype use entity work.Dtype;
  begin
    d1 : dtype port map (d(i), ck, q(i), open);
  end generate;
end;
```

The alternative solution is to use direct binding to eliminate the configuration specification completely:

```
architecture behaviour of word_delay is
begin
  gen: for i in 0 to n-1 generate
    d1 : entity work.dtype port map (d(i), ck, q(i), open);
  end generate;
end;
```

10.7 Worked Examples

10.7.1 Pseudo-Random Binary Sequence (PRBS) Generator

This is a simple example that shows a number of useful features of VHDL for creating parameterised circuits. It also illustrates the use of look-up tables in a parameterised design.

A pseudo-random binary sequence (PRBS) is a sequence of bits that are pseudo-random. That is, they are not really random but they can be used where a good approximation to random values is required. Their most common application is to create digital white noisc. They are also used in built-in self-test circuits to create test vectors.

One of the desirable characteristics of pseudo-random sequences is that they are evenly distributed, so when creating white noise they give a flat frequency distribution.

The PRBS generator is a very simple circuit. It is a shift-register in which the shift input is generated by feedback of the exclusive-or of two or more tap points in the shift register. For this reason it is also known as a linear feedback shift register or LFSR. The term PRBS describes what the circuit *does*, whilst the term LFSR describes how it is *implemented*.

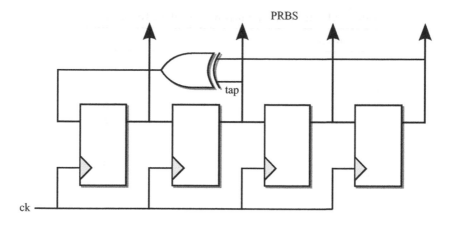

Figure 10.4   Four-bit PRBS generator.

A PRBS generator goes through a sequence of states such that the states are non-consecutive (this is the pseudo-random part). The number of states determines the repeat frequency of the generator. The more states, the larger the circuit but also the more random the sequence appears to be.

The maximum possible number of states is one less than the total number of permutations of the bits. In other words, for an n-bit register, the number of states is $2^n - 1$. The only state not covered is the all-zeros state because that state has no exit (in other words the generator gets stuck in this state). This limitation becomes part of the design requirement – the PRBS generator must be designed to avoid or escape from the all-zero state.

Figure 10.4 shows an example of a 4-bit PRBS generator with taps at the outputs of the fourth and the second bit of the register.

The entity of the PRBS generator is generic with a parameterised output bus that is the PRBS output. It also has a clock and a reset input. The reset can set the register to any value except the all-zeros state, which would cause the generator to stick.

The entity declaration is:

```
library ieee;
use ieee.std_logic_1164.all;
entity PRBS is
  generic (bits : integer range 4 to 32);
  port (ck, rst : in std_logic;
        q : out std_logic_vector(bits-1 downto 0));
end;
```

The definition of the generic parameter constrains the value of the generic to the range 4–32, so enforces the built-in limits. Only PRBS generators from 4 to 32 bits can be requested. The advantage of placing the range constraint on the generic parameter is that it is then visible to any user of the component, who will rely on reading the entity to see what interface is available to them. It therefore makes the entity self-documenting. It is also visible to the compiler, which will generate an error if a value outside the range is used.

Table 10.1   Tap points for maximal-length PRBS generators

PRBS size (n)	Tap point (t)
4	3
5	3
6	5
7	6
9	5
10	7
11	9
15	14
17	14
18	11
20	17
21	19
22	21
23	18
25	22
28	25
29	27
31	28
33	20

The implementation problem is that, although LFSRs can be created of any length with any combination of tap points, not all such generators can visit all the possible states. Some only go through a subset of the possible set, but there is a special class that do and these are called maximal-length LFSRs. In order to meet the design requirements, the generic component must implement a maximal-length LFSR and not any other kind.

Furthermore, there is a special set of maximal-length LFSRs which only require two tap points. These are the easiest to implement and are suitable for the implementation of this parameterisable circuit.

Table 10.1 gives a sample set of two-tap maximal-length LFSRs. This data is taken from (Horowitz and Hill, 1989). This table is by no means complete, but gives a selection of possible tap points. One tap is always at the end of the shift-register (bit n), the table gives the second tap point (t).

Note that there are gaps in the table. There are some shift-register lengths for which there is no possible two-tap maximal-length LFSR. In these cases, the next larger shift register length will be used.

It can be seen that neither the shift-register lengths nor the tap points follow any regular pattern. The design objective is to create a parameterisable circuit that implements the next largest two-tap maximal-length LFSR to the size required by the user, up to a maximum of 32 bits. In other words, if the user specifies a PRBS generator of at least 32 bits, the next largest LFSR, a 33-bit circuit with a tap at bit 20, will be generated. The user is insulated from the implementation and does not need to know that, when requesting a 32-bit generator, in fact a 33-bit shift register is used.

The shift-register size and tap points will have to be implemented as look-up tables since they do not follow any regular pattern. A look-up table is implemented as just an array of integer constants.

Here is the type definition for the look-up table:

```
type table is array (natural range <>) of integer;
```

Here is the look-up table for the shift-register lengths:

```
constant sizes : table(4 to 32) :=
    (              4,   5,   6,   7,   9,
      9,  10,  11,  15,  15,  15,  15,  17,
     17,  18,  20,  20,  21,  22,  23,  25,
     25,  28,  28,  28,  29,  31,  31,  33);
```

Finally, here is the look-up table for the tap points:

```
constant taps  : table(4 to 32) :=
    (              3,   3,   5,   6,   5,
      5,   7,   9,  14,  14,  14,  14,  14,
     14,  11,  17,  17,  19,  21,  18,  22,
     22,  25,  25,  25,  27,  28,  28,  20);
```

Note that the tables have a range of 4–32. The idea is to use the generic parameter, which is constrained to this range, to index these constant arrays. Thus, if the generic parameter is 32, then element index 32 in the sizes array will give the value 33 and element 32 in the taps array will give the value 20, as required.

The trick in using these arrays is to remember that generic parameters are replaced by their constant values for each instance and that, furthermore, constants are evaluated before synthesis takes place. The look-up tables are declared as constants and will be indexed by the generic parameter that is also treated as a constant, so the table look-up will take place prior to synthesis.

This becomes clear when the signal declaration for the shift register is examined. The signal declaration is:

```
signal shifter : std_logic_vector(sizes(bits) downto 1);
```

The signal is declared with an unconventional range ending at 1 because the data we have available uses this convention – the tap points are numbered from 1 to the length of the register rather than from 0. We could adjust the values by subtracting 1 from each tap point, but that would introduce possible errors. It is better to use the supplied data as given rather than make transformations. It is then easy to check the data in the implementation against the data sheet.

Prior to synthesis, the generic parameter will be replaced by its value. Using the example value of 32 again, this signal declaration will simplify to:

```
signal shifter : std_logic_vector(sizes(32) downto 1);
```

The array index `sizes(32)` is a constant index into a constant array. It can therefore be evaluated prior to synthesis. This gives a further simplification:

```
signal shifter : std_logic_vector(33 downto 1);
```

The signal is now a known size, so can be synthesised.

The rest of the architecture is quite simple. It consists of a registered process that describes the shift register and its feedback path. The register incorporates a synchronous reset to the all-ones state (one solution to avoiding the all-zeros state). The linear feedback is a simple 2-input `xor` function:

```
process
begin
  wait until ck'event and ck = '1';
  if rst = '1' then
    shifter <= (others => '1');
  else
    shifter <=
      shifter(shifter'left-1 downto 1)
      &
      shifter(shifter'left) xor shifter(taps(bits));
  end if;
end process;
q <= shifter(bits downto 1);
```

The last line connects the appropriate bits of the shift register to the output of the PRBS generator. The shifter can be longer than the output bus, for example, a 32-bit PRBS generator uses a 33-bit shifter, so a slice is used to select the relevant part of the shifter. It also normalises the result, since the output signal has the range `(bits-1 downto 0)`.

Note that the tap for the feedback also uses the look-up table technique:

```
shifter(shifter'left) xor shifter(taps(bits));
```

Substituting the generic parameter gives:

```
shifter(shifter'left) xor shifter(taps(32));
```

Evaluating constants gives:

```
shifter(33) xor shifter(20);
```

Putting it all together, here's the complete architecture:

```
architecture behaviour of PRBS is
  type table is array (natural range <>) of integer;
  constant sizes : table(4 to 32) :=
    (              4,  5,  6,  7,  9,
       9, 10, 11, 15, 15, 15, 15, 17,
      17, 18, 20, 20, 21, 22, 23, 25,
      25, 28, 28, 28, 29, 31, 31, 33);
```

```
      constant taps  : table(4 to 32) :=
        (              3,  3,  5,  6,  5,
          5,  7,  9, 14, 14, 14, 14, 14,
         14, 11, 17, 17, 19, 21, 18, 22,
         22, 25, 25, 25, 27, 28, 28, 20);
      signal shifter : std_logic_vector(sizes(bits) downto 1);
    begin
      process
      begin
        wait until ck'event and ck = '1';
        if rst = '1' then
          shifter <= (others => '1');
        else
          shifter <=
            shifter(shifter'left-1 downto 1)
            &
            shifter(shifter'left) xor shifter(taps(bits)));
        end if;
      end process;
      q <= shifter(bits downto 1);
    end;
```

This is clearly much simpler than having a library full of non-generic versions, one for each size of PRBS generator, and the use of look-up tables means that the critical design data is clearly defined in one place.

10.7.2 Systolic Processor

This is a larger but slightly more obscure example. It is unlikely that you will ever design such a processor, but it is a good example in that it contains several examples of the use of generics in a hierarchical design.

A systolic processor is a form of pipelined processor that is well suited to regular matrix-type or signal-processing problems. This example is a single processor unit (or systole) of a systolic processor that is used to multiply a matrix by a vector. This systole can be used to build a complete systolic processor.

The external interface of the systole is shown in Figure 10.5.

All the inputs are integers of the same precision and will be represented by type signed from numeric_std, with their size controlled by a generic parameter. In addition to the inputs and outputs shown in the figure, a clock input will be required.

The entity for this interface is:

```
  library ieee;
  use ieee.std_logic_1164.all;
  use ieee.numeric_std.all;
  entity systole is
    generic (n : natural);
    port (left_in, top_in, right_in : in signed(n-1 downto 0);
          ck : in std_logic;
          left_out, right_out : out signed(n-1 downto 0));
  end;
```

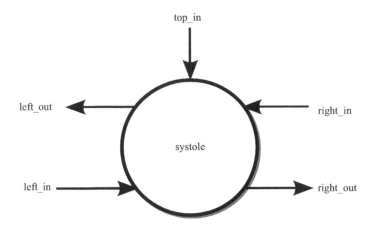

Figure 10.5 Systole interface.

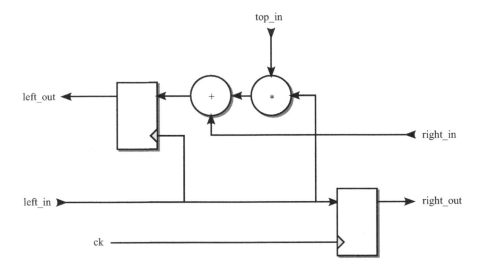

Figure 10.6 Internal structure of the systole.

The systole is a simple multiply-accumulate structure with a register to store the intermediate results. One of the inputs is fed through a register to the outputs directly. The internal architecture is shown in Figure 10.6.

All the intermediate signals are the same precision as the inputs and outputs and so should also be parameterised by the generic parameter.

The architecture for the systolic processor is:

```
architecture behaviour of systole is

    signal sum : signed(n-1 downto 0);
    signal product : signed(2*n-1 downto 0);
```

```
begin

  product <= left_in * top_in;

  sum <= right_in + product(n-1 downto 0);

  process
  begin
     wait until rising_edge(ck);
     left_out <= sum;
     right_out <= left_in;
  end process;

end;
```

Note how a double length `product` is created as in intermediate signal, since the output of the multiply operator in `numeric_std` gives a double length result. This is then truncated to the required length by the slice in the addition, so that only the lower half of `product` is kept. The resize function was not used for the truncation because it was required that overflow causes wrap-around and the resize function for `signed` preserves the sign (see Section 6.4 for an explanation of the problems with the `resize` function).

The most significant bits of the product are not used in the circuit, so synthesis will remove them. Thus, all the datapaths in the solution are of equal width, specified by the generic parameter n.

Now that the systolic element has been designed, it is possible to complete the systolic multiplier.

The multiplier is intended to multiply a 3 × 3 matrix by a 3-element vector. The systolic multiplier is illustrated by the block diagram in Figure 10.7. The data inputs are fed into the processor through a single 8-bit port, so part of the solution is to store the data and feed it to the array at the right time.

In fact, the b elements are fed in 2 cycles before the a elements so that b(1) meets a(1,1) in the middle systole. The first result will emerge from the left-hand end of the array another 2 cycles later.

The first stage in the solution is to find a solution to the data input scheduling. This requires that the data samples are stored prior to being fed into the systolic array. the most appropriate solution here is to use shift registers with an enable control. When enabled, they

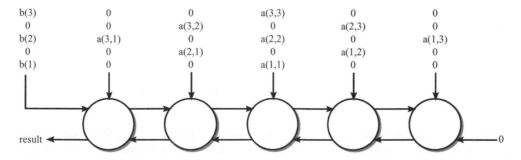

Figure 10.7 Data flow of the systolic multiplier.

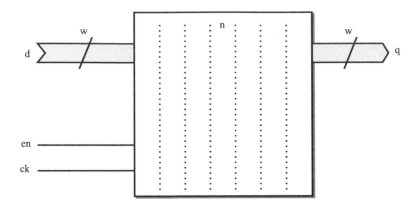

Figure 10.8 Interface to the shift register.

act as first-in, last-out shifters. When disabled, they hold their values and output a zero to the systolic array.

The shifter circuit has the interface shown in Figure 10.8.

The shift register is parameterised in both word-width (w) and number of stages (n). The design of the shift register is:

```
library ieee;
use ieee.std_logic_1164.all, ieee.numeric_std.all;
entity shifter is
  generic (w, n : natural);
  port (d : in signed(w-1 downto 0);
        ck, en : in std_logic;
        q : out signed(w-1 downto 0));
end;
architecture behaviour of shifter is
  type signed_array is array(0 to n) of signed(w-1 downto 0);
  signal data : signed_array;
begin
  data(0) <= d;
  gen: for i in 0 to n-1 generate
    process
    begin
      wait until ck'event and ck = '1';
      if en = '1' then
        data(i+1) <= data(i);
      end if;
    end process;
  end generate;
  q <= data(n) when en = '1' else (others => '0');
end;
```

Note how the generate statement has been simplified by making the data array one element larger than necessary and using data(0) as the register input. This element is not registered since it is never used as the target of an assignment in a registered process. Only elements data(1) to data(n) will be registered.

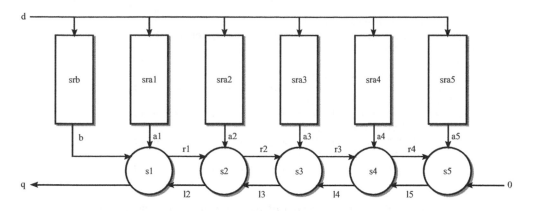

Figure 10.9   Internal structure of the systolic multiplier.

The next stage is to create the basic structure of the systolic multiplier, at this stage without the controller that will synchronise the activities of the systoles and the shift registers. The structure of the systolic multiplier is shown in Figure 10.9.

For this example, all the datapaths will be 16-bit `signed` numbers.

The interface to the systolic multiplier is described by the following entity:

```
library ieee;
use ieee.std_logic_1164.all;
use ieee.numeric_std.all;
entity systolic_multiplier is
  port (d : in signed (15 downto 0);
        ck, rst : in std_logic;
        q : out signed (15 downto 0));
end;
```

The architecture contains five component instances of the `systole` component and six instances of the `shifter` component, with different generic shift lengths reflecting the number of values to be stored in each shifter.

```
architecture behaviour of systolic_multiplier is
  signal b, a1, a2, a3, a4, a5, r1, r2, r3, r4, 12, 13, 14, 15, nil
    : signed (15 downto 0);
  signal enb, ena1, ena2, ena3, ena4, ena5 : std_logic;
begin
  nil <= (others => '0');

  s1 : entity work.systole
    generic map (16) port map (b, a1, 12, ck, q, r1);
  s2 : entity work.systole
    generic map (16) port map (r1, a2, 13, ck, 12, r2);
  s3 : entity work.systole
    generic map (16) port map (r2, a3, 14, ck, 13, r3);
  s4 : entity work.systole
    generic map (16) port map (r3, a4, 15, ck, 14, r4);
  s5 : entity work.systole
    generic map (16) port map (r4, a5, nil,ck, 15, open);
```

```
    srb : entity work.shifter
      generic map (16, 3) port map (d, ck, enb, b);
    sra1 : entity work.shifter
      generic map (16, 1) port map (d, ck, ena1, a1);
    sra2 : entity work.shifter
      generic map (16, 2) port map (d, ck, ena2, a2);
    sra3 : entity work.shifter
      generic map (16, 3) port map (d, ck, ena3, a3);
    sra4 : entity work.shifter
      generic map (16, 2) port map (d, ck, ena4, a4);
    sra5 : entity work.shifter
      generic map (16, 1) port map (d, ck, ena5, a5);
  end;
```

The final stage of the design is the design of the controller that synchronises the systolic multiplier. This will appear as processes in the structural description and controls the enable signals for the shift register components. The controller is basically a simple sequencer with decoding for the various enable signals. The synchronous reset input to the systolic multiplier will be used to simply reset the sequencer. There is no need for resets on any of the other components.

The controller is in two parts, a counter and a decoder. The counter is based on an enumeration type so that each control state can be given a meaningful name. The calculation proceeds in two phases: load the data into the shift registers; perform the calculation. It takes 12 clock cycles to load the shift registers and 9 clock cycles to perform the calculation, a total of 21 clock cycles. In principle, it should be possible to overlap the two phases, but for this example they will be kept separate for clarity. The results come out on the q output separated by zeros, with the first element of the result appearing on the fifth cycle of the calculation phase and the last on the ninth cycle.

In order to implement the counter, a new type and counter signal will need to be added to the architecture:

```
type state_type is
  (ld_a11, ld_a12, ld_a13,
   ld_a21, ld_a22, ld_a23,
   ld_a31, ld_a32, ld_a33,
   ld_b1, ld_b2, ld_b3,
   calc1, calc2, calc3, calc4, calc5, calc6, calc7, calc8, calc9);
signal state : state_type;
```

The next stage is to design a counter for this type. It must have an external synchronous reset and must also reset to the start of the sequence after the last state. The counter is:

```
process
begin
  wait until ck'event and ck = '1';
  if rst = '1' or state = state_type'right then
    state <= state_type'left;
  else
    state <= state_type'rightof(state);
  end if;
end process;
```

The final part of the solution is the decoding logic, implemented as a combinational process containing a case statement that implements the decoding of the current state:

```
process (state)
begin
  enb <= '0';
  ena1 <= '0';
  ena2 <= '0';
  ena3 <= '0';
  ena4 <= '0';
  ena5 <= '0';
  case state is
    when ld_a11 => ena3 <= '1';
    when ld_a12 => ena4 <= '1';
    when ld_a13 => ena5 <= '1';
    when ld_a21 => ena2 <= '1';
    when ld_a22 => ena3 <= '1';
    when ld_a23 => ena4 <= '1';
    when ld_a31 => ena1 <= '1';
    when ld_a32 => ena2 <= '1';
    when ld_a33 => ena3 <= '1';
    when ld_b1 => enb <= '1';
    when ld_b2 => enb <= '1';
    when ld_b3 => enb <= '1';
    when calc1 => enb <= '1';
    when calc2 => null;
    when calc3 => enb <= '1'; ena3 <= '1';
    when calc4 => ena2 <= '1'; ena4 <= '1';
    when calc5 => enb <= '1'; ena1 <= '1';
                 ena3 <= '1'; ena5 <= '1';
    when calc6 => ena2 <= '1'; ena4 <= '1';
    when calc7 => ena3 <= '1';
    when calc8 => null;
    when calc9 => null;
  end case;
end process;
```

Before the case statement, all the control signals are initially set to '0'. Within the case statement, these values are selectively overridden by a '1' for those control signals that are required to be set. This makes the case statement both simpler and clearer than setting all of the controls in all of the branches. Note the use of the null statements where there are no assignments in a branch of the case statement. For no particular reason, it is illegal in VHDL to have an empty branch, so the null statement overcomes this.

11

Subprograms

Subprograms in VHDL are very much like subprograms in software programming languages. They contain series of sequential statements and can be called from anywhere in a VHDL model to execute those statements.

There are two types of subprogram in VHDL: the function and the procedure. Operators (for example the add operator "+") are functions.

The use of subprograms in VHDL for logic synthesis is to carry out commonly repeated operations. Hierarchy is implemented using components (Chapter 10). This chapter describes how to write subprograms and when it is appropriate to do so.

11.1 The Role of Subprograms

In software languages, subprograms are the natural form of hierarchy. A task is broken down into subtasks, each of which is written as a subprogram and called from other subprograms.

However, in VHDL, the natural form of hierarchy is the entity/architecture pair, which is then invoked as a *component*. Designs should be partitioned into separate components, each of which can be simulated and tested in isolation. These components can then be used as instances (the term is *instantiated*) in a higher-level architecture.

It is easy for VHDL users with a software background to fall into the trap of using VHDL like a software language and partitioning a problem into subprograms. This is always a mistake. Bear in mind that only processes can model registers, and subprograms cannot contain processes because they contain only sequential statements, so subprograms must only contain combinational logic.

Subprograms are therefore usually restricted to small atomic operations.

11.2 Functions

A function is a subprogram that can be called in an expression. An expression can appear in many places in the VHDL language, but the most obvious examples are the source (right-hand side) of an assignment and the condition in an if or case statement.

VHDL for Logic Synthesis, Third Edition. Andrew Rushton.
© 2011 John Wiley & Sons, Ltd. Published 2011 by John Wiley & Sons, Ltd.

11.2.1   Using a Function

If there was a function called `carry` for the carry function of a full-adder, then an example of its use would be:

```
library ieee;
use ieee.std_logic_1164.all;
entity carry_example is
  port (a, b, c : in std_logic;
        cout : out std_logic);
end;
architecture behaviour of carry_example is
begin
  cout <= carry(a, b, c);
end;
```

This example shows the function being called in a concurrent signal assignment. It omits the declaration of the function itself.

Functions can be declared in a number of different places: in the declaration part of a process (before the `begin`), in the declaration part of an architecture and in a package are the most common places, but they can also be declared within other subprograms, within a package body (and therefore only usable within the package itself) and one or two other places that are rarely used in practice.

This example would be complete if the function was included in the declarative part of the architecture. The use of subprograms declared in packages will be dealt with in more detail in Section 11.6.

Functions can be called in either sequential VHDL or concurrent VHDL, although the function itself can only contain sequential statements. The best way of thinking of a concurrent function call is by its equivalent process. The equivalent process for the signal assignment in the example above is:

```
process (a, b, c)
begin
  cout <= carry(a, b, c);
end process;
```

This shows that the assignment is sensitive to changes on any of the function parameters, so any changes on any of the parameters results in the assignment statement being re-executed. In other words, a concurrent signal assignment containing a function call models combinational logic, just as a concurrent signal assignment without a function call would.

11.2.2   Function Declaration

To show how a function declaration is made up, the `carry` function used in the example above will be used. The following VHDL shows how such a carry function would be written:

```
function carry (bit1, bit2, bit3 : in std_logic)
return std_logic is
  variable result : std_logic;
begin
  result := (bit1 and bit2) or (bit1 and bit3) or (bit2 and bit3);
  return result;
end;
```

Focusing first on the structure of the function, the first part is the function declaration that gives the name of the function and details of its parameters and its return type. In this case, the function is called `carry`, takes three `in` parameters called `bit1`, `bit2` and `bit3` and returns a result of type `std_logic`. The result returned from the function can then be assigned to a signal or variable of type `std_logic` or even be used to build up an expression of type `std_logic`.

The parameters of a function must be of mode `in`. Since mode `in` is the default mode anyway, the mode is usually dropped for function declarations:

```
function carry (bit1, bit2, bit3 : std_logic)
return std_logic is
...
```

Think of the parameters of the function as its inputs and the return value as its single output. A function can only return one value, although it can be of any type. Therefore, functions are suitable for describing blocks of combinational logic with any number of inputs and one output, like the carry logic. The output can be an array or record and this is how to implement multiple outputs.

The second part of the function is the declaration part. In this case, the function contains one local variable named `result`:

```
variable result : std_logic;
```

Variables can be used to accumulate results or store intermediate values used in the calculation of the return value (the output). Unlike processes, variables declared inside functions do not preserve their values between executions, they are reinitialised every time the function is called, so they cannot be used to store internal state.

A function may also include declarations used only within the function, such as types, subtypes, constants and other subprograms. However, except for local constants, these are rarely useful.

The third part of the function is the statement part. In this case it contains two statements:

```
result := (bit1 and bit2) or (bit1 and bit3) or (bit2 and bit3);
return result;
```

The first statement is a straightforward variable assignment. Any sequential VHDL can appear here. The last statement is a `return` statement. Functions must finish on a `return` statement; it is an error if a simulator can reach the end of the function without a `return` statement being reached.

The `return` statement specifies the value to be returned by the function – the output of the circuit. In this case, the return value is the value stored in the internal variable `result`. A function must exit via a return statement.

In fact, the `return` statement can have an expression of any complexity to calculate the return value, so the `carry` function could have been written:

```
function carry (bit1, bit2, bit3 : in std_logic) return std_logic is
begin
   return (bit1 and bit2) or (bit1 and bit3) or (bit2 and bit3);
end;
```

Of course, this would not have demonstrated the declaration of local variables or the use of variables in a function.

11.2.3   Initial Values

Initial values assigned to variables in a function *are* synthesised. This is different from the interpretation of the initial values of signals or the initial values of variables when they are declared in a process.

The reason for this is to do with the way in which simulators interpret subprograms. When a subprogram is called, the variables are initialised at the time of the call and are reinitialised each call. This is not the same as elaboration at time zero, it is just a part of the statement sequence. It can be synthesised as if it was a variable assignment. For example, the following function takes a `boolean` parameter and returns a `bit` with the same logical value:

```
function to_bit (a : in boolean) return bit is
  variable result : bit := '0';
begin
  if a then
    result := '1';
  end if;
  return result;
end;
```

In this example, the variable `result` is initialised with the value `'0'`. Then this value is overridden by the value `'1'` only if the input parameter is `true`. This is equivalent to:

```
function to_bit (a : in boolean) return bit is
  variable result : bit;
begin
  result := '0'
  if a then
    result := '1';
  end if;
  return result;
end;
```

The re-initialisation of variables every time a function is called allows a variable to be initialised with the value of one of the parameters. To take the use of initial values to its absurd extreme, the same function can be written such that the functionality is entirely in the initial value:

```
function to_bit (a : in boolean) return bit is
  variable result : bit := bit'val(boolean'pos(a));
begin
  return result;
end;
```

In this example, the attributes `val` and `pos` are used to make the conversion. The parameter, which is of type `boolean`, is converted to an integer value by the `boolean'pos` expression. This integer value is then converted to type `bit` by the `bit'val` expression. This conversion takes advantage of the fact that the positional values of the two types match up, so that both `'0'` and `false` have the same positional value (0) and `'1'` and `true` have the same positional value (1). This method could not be used for conversions of `boolean` to `std_logic` for example, because the positional values do not match up.

This example also shows how the initial value of the variable depended on the value of the input parameter of the function. Once again, the initialisation of the variable can be thought of as being equivalent to the declaration of an uninitialised variable followed by an assignment in the function body, so the following code is equivalent to the example above:

```
function to_bit (a : in boolean) return bit is
  variable result : bit;
begin
  result := bit'val(boolean'pos(a));
  return result;
end;
```

11.2.4 Functions with Unconstrained Parameters

One of the most useful features of functions is the ability to define them with unconstrained arrays as parameters. It would appear that this breaks the rule for synthesis that all datapaths must be of a known size. However, such a parameter will in fact be constrained to fit the size of the variable or signal passed to it when the function is called. Effectively there is a family of functions, one for every possible array size, in one function. Everywhere in a design that a function is called a new copy is created and that copy is constrained to the size of its parameters prior to synthesis. This means that the parameter sizes themselves must be a known size (you cannot, for example, use a variable array slice as a parameter), but that is the only constraint. However, there are some pitfalls that make such unconstrained functions tricky to write correctly.

An example of a function with an unconstrained parameter is:

```
function count_ones (vec : std_logic_vector) return natural is
  variable count : natural;
begin
  count := 0;
  for i in vec'range loop
    if vec(i) = '1' then
      count := count + 1;
    end if;
  end loop;
  return count;
end;
```

In this example, the parameter vec has been defined as a std_logic_vector – an unconstrained array type – without giving any bounds. This function can therefore be used on any signal or variable that is a std_logic_vector.

The declaration part of the function contains the declaration for an intermediate variable called count, which is going to be used to accumulate the return value of the function. At the start of the statement part of the function the variable is initialised to 0, then in the for loop it is incremented every time a '1' is found in the input array vec.

The key point to the way the function has been written is that neither the size nor the direction of the input array are known. By using the range attribute in the for loop, the function is guaranteed to visit the leftmost bit first and to go through the array from left to right. This will be true, regardless of whether the array starts at 0 or at 100, whether it has an ascending range or a descending range, or what size it is.

This example is a simple one, since the range and direction of the input array does not have any effect on the algorithm. This is not always the case. A more complicated example will be used to illustrate other issues that arise with unconstrained parameters.

The reason for the need for flexibility in the design of a function is that it is not known at the time of writing the function, exactly what array range is going to be passed to it. When an array is passed to an unconstrained function parameter like this, the parameter takes on exactly the range of the array being passed to it. It is important to write general-purpose functions so that the user of the function has complete freedom in the way they use it. It is not good practice to assume that users will conform to the convention that bit-array types are descending ranges ending in zero. Furthermore, there are situations where, even if the convention is conformed to, the parameter still could have a different range. Take the following example. In this, the user wishes to count the number of ones in the top half of a 16-bit word:

```
library ieee;
use ieee.std_logic_1164.all;
entity count_top_half is
   port (vec : in std_logic_vector(15 downto 0);
         count : out natural range 0 to 8);
end;
architecture behaviour of count_top_half is
begin
   count <= count_ones(vec(15 downto 8));
end;
```

In this example, the top half of the vector has been passed to the function using the slice `vec(15 downto 8)`. Therefore, the parameter inside the function will have the range `(15 downto 8)`.

A final cause of problems is the string value. Remember that a constant value of a character array type such as `std_logic_vector` can be represented by a string value. For example:

```
count <= count_ones("00001111");
```

In this case, the VHDL language definition says that the range of the string value should be `(0 to 7)`, an ascending range. If the function was defined assuming a descending range, then it would not work with string values. For example, a loop that went from the `high` value to the `low` value would visit the elements from right to left instead of from left to right. Furthermore, different synthesis vendors have been inconsistent with their implementations of string values. A quick survey of three synthesis and simulation tools revealed three different interpretations: the first used the range `7 downto 0`; the second used the range `0 to 7` (the correct interpretation) and the third used the range `1 to 8`, all for the same 8-bit string. The result is that no assumptions can be made about the range of a string value, since it may vary from one VHDL tool to another, and indeed does.

This means that, to write a function that will work for all parameters on all systems, it is essential to make no assumptions at all about the range or direction of an unconstrained array parameter.

The main technique that is used to simplify the writing of functions with unconstrained parameters is a technique that will be referred to as *normalisation*. In this, the in parameters are immediately assigned to local variables that are exactly the same size as the parameters, but with ranges that conform to the common conventions, and these variables are then used instead of the parameters.

The following example shows how a function could be written that counts the number of matches between the bits in two std_logic_vector parameters:

```
function count_matches(a, b : std_logic_vector)
  return natural
is
  variable va : std_logic_vector (a'length-1 downto 0) := a;
  variable vb : std_logic_vector (b'length-1 downto 0) := b;
  variable count : natural := 0;
begin
  assert va'length = vb'length
    report "count_matches: parameters must be the same size"
    severity failure;
  for i in va'range loop
    if va(i) = vb(i) then
      count := count + 1;
    end if;
  end loop;
  return count;
end;
```

This function takes two parameters, a and b. These are both normalised by assignment to va and vb. The assignment has been incorporated into the variable declarations as the initial values of the variables. Note how the sizes of va and vb are calculated in terms of the sizes of the parameters, by using a'length and b'length in the range constraint of the two local variables.

The next feature of this function is the assertion. This is a simulation-only feature, ignored by synthesis, which checks that the two arrays are of equal size. The assertion is present because the rest of the function has been written assuming that the two arrays are of equal size. It is always good practice to use assertions to check such assumptions are being honoured. If anyone ever uses the function wrongly by passing different-sized parameters, the simulator will report the fact by printing the assertion message. There are some conditions (namely, when a is longer than b) where trying to continue after the assertion would cause a subsequent failure of the simulator, so the severity of the assertion is failure, which stops the simulator immediately.

The reason why the normalisation is necessary is that the for loop is controlled by the value i, which takes on the values in the range of va. This index is then used to access the elements of vb. This is only safe to do if vb has the same range as va.

It is generally good practice to always normalise unconstrained array parameters. This will avoid all the common pitfalls inherent in writing such functions.

11.2.5   Unconstrained Return Values

It is also possible to write functions that have an unconstrained array as their return type. Once again, there are some important guidelines to follow if the common pitfalls are to be avoided.

Consider the case where there is a function that returns a std_logic_vector, then the size of the std_logic_vector must be known by the user when writing the VHDL that uses the function in order that a signal or variable of the right size can be declared to receive the result. Furthermore, the size of the return value must be calculable by the synthesiser prior to synthesis in order that the right number of bits can be allocated to the bus. This also means that the size of the result must not vary during a simulation run since it would then be impossible to define a signal or variable to receive that value. What this effectively means is that the size of the result array must be either of a fixed size or entirely dependent on the sizes of the input parameters. The return size must not depend on the *values* of the parameters since that would cause it to vary. The exception to this is where the parameter values are constant literals.

In fact, the general rules of VHDL do allow the return of differing length results under some conditions (not in signal assignments though), but these do not apply to synthesis. The rule stated above is a requirement of synthesis VHDL, regardless of the circumstances. The reason is quite simple: the synthesiser must know how many bits are going to be required to implement the return value and this must be clear from the structure of the function with knowledge only of the size and not the values of the parameters.

For example, assume there is a function called matches, which returns the bitwise equality of two std_logic_vector values. This function is also based on an example from Section 8.7 where it was used to illustrate for loops:

```
function matches(a, b : std_logic_vector)
return std_logic_vector is
  variable va : std_logic_vector(a'length-1 downto 0) := a;
  variable vb : std_logic_vector(b'length-1 downto 0) := b;
  variable result : std_logic_vector(a'length-1 downto 0);
begin
  assert va'length = vb'length
    report "matches : parameters must be the same size"
    severity failure;
  for i in va'range loop
    result(i) := va(i) xnor vb(i);
  end loop;
  return result;
end;
```

In this example, the result is constrained to be the same size as the input parameter a. Furthermore, the assertion constrains b to be the same size as a. In use, then, this constraint will be known by the user (either from reading documentation or the function itself), so a signal or variable of the right size can be used to receive the result. Furthermore, it is known to the synthesiser since the constant expression a'length-1 used to constrain the result will be evaluated prior to synthesis of the function body.

Taking the original example from Section 8.7, which was a complete entity/architecture pair, the following code shows how this circuit could have been written with the aid of the function.

```
library ieee;
use ieee.std_logic_1164.all;
entity match_bits is
  port (a, b : in std_logic_vector (7 downto 0);
        result : out std_logic_vector (7 downto 0));
```

```
  end;
  architecture behaviour of match_bits is
  begin
    result <= matches(a, b);
  end;
```

This example is still incomplete, because nothing has been said about where the function itself is defined. This will be addressed in Section 11.6, so for now the example will be left in this slightly incomplete state.

The one exception to the general rule that the return size must not depend on the value of a parameter is when that parameter is always going to be given a constant value. In this case, the synthesiser can use the constant value to calculate the size of the return value. An example of this is the following simple sign-extension function that interprets a std_logic_vector as a signed integer and sign-extends the integer to a specified width:

```
  function extend (a : std_logic_vector; size : natural)
  return std_logic_vector is
    variable va : std_logic_vector(a'length-1 downto 0) := a;
    variable result : std_logic_vector(size-1 downto 0);
  begin
    assert va'length <= size
      report "extend: must extend to a longer length"
      severity failure;
    assert va'length >= 1
      report "extend: need at least a sign bit to sign extend"
      severity failure;
    result := (others => va(va'left));
    result (va'range) := va;
    return result;
  end;
```

In this example, the length of result is defined in terms of the input parameter size. This will only be legal if size is associated with a constant value, so synthesis will fail if it is ever associated with a variable or signal. The reason it will fail is that the signal result will not be of a known size when it comes to synthesis, although it will not necessarily fail in simulation.

The assertions enforce the usage rules for the function: in this case it is only legal to sign extend to a length greater than the current length of the input parameter and there must be at least one bit in the input parameter for the sign extension to work. The sign-extension algorithm itself works by pre-filling the result with the sign bit from the input parameter and then copying the input parameter into the subrange corresponding to its length.

In use, this function must be used with a constant size parameter. A simple example is the following entity/architecture pair that shows how the function could be used to sign extend an 8-bit input to create a 16-bit output:

```
  library ieee;
  use ieee.std_logic_1164.all;
  entity extend_example is
    port (a : in std_logic_vector(7 downto 0);
          z : out std_logic_vector(15 downto 0));
  end;
  architecture behaviour of extend_example is
```

```
begin
  z <= extend(a, 16);
end;
```

This will be synthesised by first taking a copy of the function, performing constant calculations within that function copy (including array-size calculations), then synthesising the result.

After parameter substitution, the function looks like:

```
function extend (a : std_logic_vector(7 downto 0); 16 : natural)
  return std_logic_vector
is
  variable va : std_logic_vector(7 downto 0) := a;
  variable result : std_logic_vector(15 downto 0);
begin
  assert 8 <= 16
    report "extend: must extend to a longer length"
    severity failure;
  assert 8 >= 1
    report "extend: need at least a sign bit to sign extend"
    severity failure;
  result := (others => va(7));
  result (7 downto 0) := va;
  return result;
end;
```

It is clear that the sizes of all the buses are now known. The assertions, once checked, can be ignored, leaving the basic function, which can be synthesised.

It is pushing the limits of what is possible in synthesis to write functions like this, and so it is highly recommended that the synthesiser documentation is checked and possibly some examples tried to establish what is possible.

11.2.6 Multiple Returns

There can be more than one return statement in a function. The only unbreakable rule is that the function must be exited by a return statement; it must not be possible to reach the end of the function without encountering one. However, it is not necessary for there to be a return statement physically at the end of the function.

Multiple return statements are generally used in two contexts: to break out of a for loop and return a value, or to return different values from different branches of a condition. Both of these common cases will be illustrated here.

Using a return statement to break out of a for loop and return a value is very similar to the use of an exit statement in a for loop. The following example shows a function that has two return statements:

```
function count_trailing (vec : std_logic_vector) return natural is
  variable result : natural := 0;
begin
  for i in vec'reverse_range loop
    if vec(i) = '1' then
      return result;
```

```
      end if;
      result := result + 1;
   end loop;
   return result;
end;
```

In this example, the result is incremented for every trailing '0' in the parameter. The value of the result is returned immediately from the function when a '1' is found. However, if a '1' is never found – because the parameter contains only '0's – then the final `return` statement catches this special case and ensures that the result is still returned.

The second use of multiple returns is to give different results depending on the value of a condition. This can give a clearer structure to a function than setting the value of an intermediate variable in the conditional and then returning the result.

The use of multiple returns comes into its own with type-conversion functions for types that cannot be converted directly. The following shows a function that converts a `std_logic` parameter to a `character`:

```
function to_character (a : std_logic) return character is
begin
   case a is
      when 'U' => return 'U';
      when 'X' => return 'X';
      when '0' => return '0';
      when '1' => return '1';
      when 'Z' => return 'Z';
      when 'W' => return 'W';
      when 'L' => return 'L';
      when 'H' => return 'H';
      when '-' => return '-';
   end case;
end;
```

Note that this function does not actually end in a `return` statement, but that the rule that it is impossible to reach the end of the function without encountering a `return` statement has been met.

In this example, there are two uses of each of the character literals 'U' through to '-'. This looks confusing but is quite clear on closer inspection. The case statement is branching on the value of the input parameter a, which is of type `std_logic`, so the values used as the choices of the `case` statement (in the `when` clauses) are the character literals defined in type `std_logic`. However, the return type of the function is type `character`, so the values used in the `return` statements are the character literals defined in type `character`.

A final point on the use of multiple returns is that, when returning an unconstrained array type, all the `return` statements must return an array of the same size. This is a direct consequence of the need to know the size of the return array prior to synthesis.

As an example, here is the `count_trailing` example rewritten to return an `unsigned` rather than an `integer` result. Since the return value is an unconstrained array it has an extra parameter specifying the size of the return subtype.

```
function count_trailing(vec : std_logic_vector, size : natural)
  return unsigned
is
  variable result : unsigned(size-1 downto 0) := (others => '0');
begin
  for i in vec'reverse_range loop
    if vec(i) = '1' then
      return result;
    end if;
    result := result + 1;
  end loop;
  return result;
end;
```

11.2.7   Function Overloading

It is not necessary to find unique names for all your functions. It is possible to reuse names by *overloading*. When a function is called in VHDL, it is not just the name that is used to determine which function is being called; the number of parameters and their types and also the return type are all used to identify the function. This means that a function is unique if one or more of these characteristics is unique. The process that a VHDL analyser goes through to identify an overloaded function is called *overload resolution*.

There are pitfalls in overloading functions excessively. There is not always sufficient information available for the analyser to know the types of all the parameters or the expected return type. The biggest problems seem to occur when functions are differentiated by return types alone. If the function call is itself a parameter to an overloaded function or operator, then the analyser cannot work out from the context which of the functions is intended.

It is therefore recommended that functions that differ in their return type but not their parameters are given different names. Indeed, other programming languages that allow function overloading, such as C++, disallow the overloading of functions distinguished only by their return type for exactly this reason. So even though VHDL allows it, it is strongly recommended that you don't do it.

One application of overloading is where two types are expected to have similar behaviour, although they are used in different circumstances. A good example of this is functions acting on the types `bit` and `std_logic`. Both types are logical types and they are almost interchangeable at the RTL level of modelling. Using type `bit` gives simpler modelling and simpler simulation, but using `std_logic` gives access to tristates. Therefore, when writing general-purpose utilities, it is common to provide both `bit` and `std_logic` versions of a function with the same name.

11.3   Operators

The operators in VHDL have been dealt with in some detail in Chapter 5. That chapter concentrated on the synthesis interpretation of the built-in operators. This section will look at operators as subprograms.

Operators are just functions with special rules that allow them to be used in-fix. That means, they can be used in expressions such as:

```
z <= a + b * c;
```

The source expression in this signal assignment (the right-hand side) contains two operators, "+" and "* ". These are in fact function calls, and the same signal assignment could have been written:

```
z <= "+"(a, "*"(b, c));
```

The two different ways of writing the expressions shown above are exactly equivalent. This example shows that operators are special functions with names corresponding to the operator symbol enclosed in double quotes to make it a string value.

When a type is defined, a set of operator functions are automatically defined by the language. These are the built-in operators. In addition, operators can be written by a user to either add operators to a type that the type does not have built-in, or to replace the built-in operator with a function with different behaviour. In both cases, the writing of user-defined operators is referred to as operator overloading. Examples of both kinds of operator overloading will be shown.

First, however, it is helpful to summarise the built-in operators that will be predefined for any type.

11.3.1 Built-In Operators

The set of built-in operators that a type has depends on the type. This section will summarise the relationship between operators and types.

Table 11.1 shows which types have which operators predefined. It only lists the synthesisable types from package `standard`, since these are the only built-in types.

The following six groupings of operators are used in the table to keep it simple. These are the same groups as used in earlier chapters when describing operators, but the comparison has been further split into equality and ordering operators:

```
boolean: not, and, or, nand, nor, xor, xnor
equality: =, /=
ordering: <, <=, >, >=
shifting: sll, srl, sla, sra, rol, ror
arithmetic: **, abs, *, /, mod, rem, sign +, sign -, +, -
concatenation: &
```

Table 11.1 Built-in operators for each type

Type	Boolean	Equality	Ordering	Shifting	Arithmetic	Concat
bit, boolean	√	√	√			
enumeration types		√	√			
integer		√	√		√	
record types		√				
arrays		√	√	√		√
arrays of bit, boolean	√	√	√	√		√

There are some notable special cases here, namely the built-in logical types `bit` and `boolean`. These have predefined logical operators, even though they are enumeration types. Furthermore, any array type defined in terms of these two logical types will also have bitwise logical operations predefined.

Other logical types, such as `std_logic`, are simply enumeration types, so only have the basic set of relational operators when first defined. This means that the author of a multi-valued logic package is responsible for overloading the other operators required for the type, and this has been done for `std_logic` to provide the logical operators.

Similarly, array types such as the types `signed` and `unsigned` will have non-numeric comparison and no arithmetic operators, so the authors of the `numeric_std` package had to add the arithmetic operators and replace the built-in comparison operators.

11.3.2   Operator Overloading

An operator is just a function, the name of which is the symbol for the operator enclosed in quotes. For example, the function declaration of the "+" operator for type `integer` is:

```
function "+" (l, r : integer) return integer;
```

Since operators are functions, they can be overloaded in the same way that functions can.

There are two situations where operator overloading is used. The first is to add operators to a type that hasn't got them. The second is to change the behaviour of existing operators for a type.

In order to overload the "+" operator for integer, a function with this name, parameter types and return type would be written. However, changing the behaviour of predefined types by overloading their operators is not good practice, because it can be very confusing. Operator overloading comes into its own when defining your own types.

The rules for operator overloading mean that it is only possible to overload operators that are already defined in the language. These are the set listed above. You couldn't, for example, define a new operator "@" because there is no such operator in the language.

Secondly, the number of parameters for the operator must be correct. Most operators take two arguments (binary operators), so these must be written as functions with two parameters. A few of the operators take only one argument (unary operators) so these must be written as functions with one parameter. Some operators have both unary and binary variants, such as the "−" operator: with one parameter you overload the minus sign, with two you overload subtraction.

Operators can be defined taking any combination of types as parameters and returning any type. The VHDL analyser is responsible for working out which of a number of operators with the same symbol is intended. The process of working out which operator to use is called *operator resolution*, and is identical to function overload resolution except it works on operator symbols rather than function names. It works by trying to match the operator name, the number of parameters, the parameter types and the return type with the context in which the operator has been used. However, excessive overloading of operators causes ambiguities that make operator resolution impossible, thus making the operators almost unusable. For example, if two operators take exactly the same parameters but return different types, then the VHDL analyser may not be able to work out from the context, which of the two operator functions to use. This problem is exhibited by the deprecated package `std_logic_arith` (Chapter 7), which excessively overloads the arithmetic operators.

To avoid such problems, it is good practice to provide only one operator of each kind for a particular type and to use type conversions to cover the other permutations. For example, to allow the addition of a `std_logic_vector` to an `integer`, provide an operator that can add two `std_logic_vector` parameters, giving a `std_logic_vector` result. Then require the user to type convert the `integer` to a `std_logic_vector` and then add that result to the other `std_logic_vector`.

This approach will avoid all of the common problems encountered with operator overloading.

When overloading operators, the following templates are the recommended way to overload each of the operators. In each case, the word *type* refers to the type that the operator is being overloaded for. In the array concatenation operators, the word *element* refers to the element type of the array.

The one-parameter functions corresponding to the unary operators are:

```
function "not" (r : type) return type;
function "-" (r : type) return type;
function "+" (r : type) return type;
function "abs" (r : type) return type;
```

The following are the reducing logic operators that are only allowed in VHDL-2008:

```
function "and" (l : type) return element;
function "or" (l : type) return element;
function "nand" (l : type) return element;
function "nor" (l : type) return element;
function "xor" (l : type) return element;
function "xnor" (l : type) return element;
```

The two-parameter functions corresponding to the binary operators are:

```
function "and" (l, r : type) return type;
function "or" (l, r : type) return type;
function "nand" (l, r : type) return type;
function "nor" (l, r : type) return type;
function "xor" (l, r : type) return type;
function "xnor" (l, r : type) return type;
function "and" (l : element; r : type) return type;
function "or" (l : element; r : type) return type;
function "nand" (l : element; r : type) return type;
function "nor" (l : element; r : type) return type;
function "xor" (l : element; r : type) return type;
function "xnor" (l : element; r : type) return type;
function "and" (l : type; r : element) return type;
function "or" (l : type; r : element) return type;
function "nand" (l : type; r : element) return type;
function "nor" (l : type; r : element) return type;
function "xor" (l : type; r : element) return type;
function "xnor" (l : type; r : element) return type;
function "=" (l, r : type) return boolean;
function "/=" (l, r : type) return boolean;
```

```
function "<" (l, r : type) return boolean;
function "<=" (l, r : type) return boolean;
function ">" (l, r : type) return boolean;
function ">=" (l, r : type) return boolean;
function "sll" (l : type; r : integer) return type;
function "srl" (l : type; r : integer) return type;
function "sla" (l : type; r : integer) return type;
function "sra" (l : type; r : integer) return type;
function "rol" (l : type; r : integer) return type;
function "ror" (l : type; r : integer) return type;
function "**" (l, r : type) return type;
function "*" (l, r : type) return type;
function "/" (l, r : type) return type;
function "mod" (l, r : type) return type;
function "rem" (l, r : type) return type;
function "+" (l, r : type) return type;
function "-" (l, r : type) return type;
function "&" (l, r : type) return type;
function "&" (l : element; r : type) return type;
function "&" (l : type; r : element) return type;
function "&" (l : element; r : element) return type;
```

These templates correspond to the predefined operators that are automatically defined for one or other of the standard types.

A final point to be aware of is that it is the basetype that is used in operator resolution, not the subtype. It is not possible to overload different functions for different subtypes. If an operator is defined with subtype parameters, then the operator will be applied to all signals and variables of the parameters' basetypes or any other of their subtypes, but there will be an additional constraint on the values of the type that may be used. For example, if an operator "+" was defined for `natural`, which is a subtype of `integer`, then the operator effectively replaces the `integer` operator. However, the range constraint of `natural` would mean that negative values could no longer be used with that operator.

When overloading operators, all the guidelines for writing functions apply. For example, to overload an operator for an unconstrained array type, use the normalisation technique described in Section 11.2. If the operator needs to return an unconstrained type, use the guidelines on unconstrained return types in the same section. Finally, use the source code of `std_logic_1164` and `numeric_std` as a reference.

11.4   Type Conversions

Built-in type conversions and user-defined type conversions are quite different, so it is not possible to overload the built-in type conversions to change their behaviour. However, user-defined type conversions are subprograms and can be written as functions.

11.4.1   Built-In Type Conversions

There are type conversion functions automatically available for conversion between what the language reference manual describes as 'closely related types'. This term needs some explanation.

All integer types are considered closely related, so it is possible to convert a value of any integer type to any other integer type. The type conversion is done by using the name of the target type as if it were a function. It is possible to use either a type or a subtype for the name of the type-conversion function. In the case of a subtype name being used, there is no difference in the type conversion compared with using the basetype, but the result will be checked against the constraints of the subtype during simulation. It is good practice to use subtype names for type conversions so that out-of-range values can be detected by the simulator during the type conversion.

For example, suppose you are using a type called short and wish to convert it to a natural. This would be carried out using the name natural as a function call.

```
signal sh : short;
signal int : natural;
...
int <= natural(sh);
```

Array types are considered closely related under the following conditions:

1. they have the same number of dimensions;
2. they are indexed by types which can be converted to each other;
3. the elements are of the same type.

The first condition must be true for a synthesisable model, because only one-dimensional arrays are allowed for synthesis. The second rule will always be true if the array is indexed by any integer type; indeed, most array types used in synthesis are indexed by natural. However, if you use an array type indexed by an enumeration type, it will only be convertible to other array types indexed by the same enumeration type, because enumeration types are not convertible to each other.

The final condition is generally the only constraining condition in synthesis VHDL.

The outcome of this is that most arrays of bit, for example, bit_vector and the types signed and unsigned defined in numeric_bit, can be converted between each other. Similarly, arrays of std_logic, for example, std_logic_vector and the types signed and unsigned in numeric_std can also be converted between each other.

11.4.2   User-Defined Type Conversions

It is not possible to define a function with the same name as a type to create a new type conversion. User-defined type conversions are simply functions that take a value of one type and return another type, presumably with the same value. However, there are some conventions for writing type conversions that are worth knowing and using.

The first convention is the name of the type-conversion function. The convention is to call the function to_*type* where *type* is the name of the target type of the type conversion and therefore also the return type of the function. This means that all type-conversion functions that convert to type integer, for example, will be called to_integer. The VHDL analyser will still have some function overloading to resolve, but experience has shown that overloading in this way, which ensures that the function input parameter is always a different type for the same name and that the function return type is always the same for the same name, rarely causes ambiguity and so the analyser will generally have no problem resolving which `to_integer` function is being referred to. Also, if this convention is always kept, there is no problem remembering the name of the type-conversion function and it is obvious when reading a VHDL model what the function is doing.

A simple example of the type-conversion function is a conversion from `boolean` to `bit`:

```
function to_bit (arg : boolean) return bit is
begin
  case arg is
    when true => return '1';
    when false => return '0';
  end case;
end;
```

A complete type conversion is not always possible and it is necessary to decide which values to discard. For example, converting from `std_logic` to `bit` requires the metalogical values to be discarded. The function for this is:

```
function to_bit (arg : std_logic) return bit is
begin
  case arg is
    when '1' => return '1';
    when others => return '0';
  end case;
end;
```

In this case the arbitrary decision has been made to map the metalogical values onto the `'0'` value. Alternatively, an assertion could be raised. In this case only a warning will be given because the change of value from, say, `'Z'` to `'0'` may not be significant in itself. It certainly doesn't warrant an error:

```
function to_bit (arg : std_logic) return bit is
begin
  case arg is
    when '1' => return '1';
    when '0' => return '0';
    when others =>
      assert false report "conversion from metalogical value"
        severity warning;
      return '0';
  end case;
end;
```

Note that a return value is still needed, since the assertion will not stop the simulation.

Extra parameters are sometimes needed to give the type-conversion function extra information on how to convert the type. The second convention when writing type-conversion functions is, where there is more than one parameter, always make the value to be converted the first parameter of the function.

One use of a second parameter is when converting from a non-array type to an array type. The parameter specifies how many bits to use in the conversion. For example, in the conversion from `integer` to `std_logic_vector`, it is necessary to tell the type-conversion function how many bits to use in the bitwise representation:

```
function to_std_logic_vector (arg : integer; size : natural)
  return std_logic_vector
is
  variable v : integer := arg;
  constant negative : boolean := arg < 0;
  variable result : std_logic_vector(size-1 downto 0);
begin
  if negative then
    v := -(v + 1);
  end if;
  for count in 0 to size-1 loop
    if (v rem 2) = 1 then
      result(count) := '1';
    else
      result(count) := '0';
    end if;
    v := v / 2;
  end loop;
  if negative then
    result := not result;
  end if;
  return result;
end;
```

A second use for an additional argument is to specify what to do with the extra values when converting between logical types that have metalogical values. For example, an alternative version of the function to convert from `std_logic` to `bit` is:

```
function to_bit (arg : std_logic; xmap : bit := '0') return bit is
begin
  case arg is
    when '1' => return '1';
    when '0' => return '0';
    when others => return xmap;
  end case;
end;
```

`Bit` has only two values, whereas `std_logic` has nine, of which two map onto the `bit` values directly. In this example, the extra parameter specifies the value of `bit` to map the other seven metalogical values of `std_logic` onto. It has been given a default value so that, if not specified, then the metalogical values will map onto '0', giving exactly the same functionality as the original version of the function. This extra parameter only effects simulation anyway; it will have no effect on synthesis, because the seven metalogical values are eliminated by the synthesiser, so this synthesises to exactly the same circuit as the original `to_bit`.

11.5   Procedures

Procedures are subprograms and, in that sense, are like functions. Indeed, they look very like functions in their declaration. However, they are used differently and have different rules.

Like functions, they can be declared in any declarative region: architectures, processes, within other subprograms, but most of all, in packages, where they are most useful.

The same general guidelines apply to the use of procedures as to functions: they should not be used as a means of partitioning a design; components should be used for this. The main use should be small, atomic operations and commonly needed routines.

11.5.1 Procedure Parameters

The first difference with procedures is that the parameters can be of mode in, mode out or mode inout. These modes should not be confused with the modes of the ports on entities that have a slightly different interpretation.

Procedures do not have return values, so the parameters are the only means by which values can be passed into and out of a procedure.

Mode in is comparable to the input parameters of a function. Parameters of mode in are used to pass values into the procedure but cannot be used to pass values back out again. Mode out is used to pass parameters out of a procedure but cannot be used to pass values in. In this sense, they are comparable to the return value of a function, except that there can be any number of out parameters. Finally, mode inout is used to pass a value into a procedure where it can be modified and then passed back out again. Mode inout does *not* model tristates or bidirectional signals.

To illustrate the use of procedures, consider the following example that describes a single full-adder. The reason for using a procedure here is that a full-adder has two outputs:

```
procedure full_adder (a, b, c : in std_logic;
                          sum, cout : out std_logic) is
begin
   sum := a xor b xor c;
   cout := (a and b) or (a and c) or (b and c);
end;
```

Notice that the assignments used in the procedure are variable assignments. This is because out parameters on a procedure are variables by default. However, signal parameters can also be specified, and these will be dealt with separately later in this section.

This procedure can only be used in a sequential procedure call, because the out parameters are variables and therefore can only be used to assign to variables. They cannot be used with signals.

An example of the use of this procedure in a sequential procedure call is shown in the following example that puts together four full-adders to make a four-bit adder with carry in and carry out.

```
library ieee;
use ieee.std_logic_1164.all;
entity adder4 is
   port (a, b : in std_logic_vector (3 downto 0);
         cin : in std_logic;
         sum : out std_logic_vector (3 downto 0);
         cout : out std_logic);
```

```
   end;
   architecture behaviour of adder4 is
   begin
     process (a, b, cin)
       variable result : std_logic_vector(3 downto 0);
       variable carry : std_logic;
     begin
       full_adder (a(0), b(0), cin, result(0), carry);
       full_adder (a(1), b(1), carry, result(1), carry);
       full_adder (a(2), b(2), carry, result(2), carry);
       full_adder (a(3), b(3), carry, result(3), carry);
       sum <= result;
       cout <= carry;
     end process;
   end;
```

There are a number of subtleties in this model. The first is that signals are used on the in parameters of the procedure. This is because, like with functions, there is no difference between passing a signal or a variable to a subprogram in parameter. However, the out parameters must be associated with variables, so two intermediate variables result and carry are used for this purpose. Notice how the carry is both the input and output of procedure calls add1 to add3. Because this is sequential VHDL, the value of carry is passed in first, then the procedure is executed, calculating a new carry output, which is then passed back out through the cout parameter and thus assigned back to the carry variable. Finally, these variables are assigned to the entity out parameters, which are signals.

11.5.2 Procedures with Unconstrained Parameters

Procedures can also be declared with unconstrained parameters in the same way as functions. For in parameters, the rules are exactly the same as for functions. The main difference is that it is possible to declare out parameters in this way as well.

The use of unconstrained out parameters has its own set of pitfalls similar to those of the use of unconstrained in parameters. The problem is that, when a procedure is called, the parameter inherits its range from the variable associated with the parameter in the call, just as with in parameters.

For example, consider the case of a procedure with the following interface:

```
   procedure add (a, b : in std_logic_vector;
                  sum : out std_logic_vector);
```

When the procedure is called, all three parameters take their ranges from the variables passed to them in the call. For example:

```
   library ieee;
   use ieee.std_logic_1164.all;
   entity add_example is
     port (a, b : in std_logic_vector(7 downto 0);
           sum : out std_logic_vector(7 downto 0));
   end;
   architecture behaviour of add_example is
```

```
begin
  process (a, b)
    variable result : std_logic_vector(7 downto 0);
  begin
    add(a, b, result);
    sum <= result;
  end process;
end;
```

In this case, the parameters all take on the range 7 downto 0. However, the writer of the subprogram cannot assume anything about the range of a parameter. Note that this is the same argument as used with function parameters, but here it is an out parameter that has also been constrained by the variable passed to it. In a sense, the range of the parameter has been passed in to the procedure, even though the parameter is of mode out.

The safe way of writing such procedures is to normalise all of the unconstrained parameters. Normalisation of out parameters is slightly different from that of in parameters. A local variable is created to be the temporary working variable for that output and it is assigned to the parameter at the end of the procedure instead of at the beginning.

To illustrate how this is done, the add procedure would be written:

```
procedure add (a, b : in std_logic_vector;
               sum : out std_logic_vector) is
  variable a_int : std_logic_vector(a'length-1 downto 0) := a;
  variable b_int : std_logic_vector(b'length-1 downto 0) := b;
  variable sum_int : std_logic_vector(sum'length-1 downto 0);
  variable carry : std_logic := '0';
begin
  assert a_int'length = b_int'length
    report "inputs must be same length"
    severity failure;
  assert sum_int'length = a_int'length
    report "output and inputs must be same length"
    severity failure;
  for i in a_int'range loop
    sum_int(i) := a_int(i) xor b_int(i) xor carry;
    carry := (a_int(i) and b_int(i)) or
             (a_int(i) and carry) or (b_int(i) and carry);
  end loop;
  sum := sum_int;
end;
```

The assertions check that the assumptions made in writing the procedure have been kept by the user of the procedure. In this case, the procedure has been kept simple by insisting that all three parameters are the same length. Notice how the length attribute has been used to find the length of the sum parameter, even though the sum parameter is an output and therefore unreadable. It is only the *values* of out parameters that are unreadable, not their *size* attributes.

11.5.3   Using Inout Parameters

An inout parameter is a parameter that can be modified by the procedure. That is, it can be read and then given a new value. Conceptually, it is both an in and an out parameter. It is worth stating again at this stage that this is not equivalent to a bidirectional or a tristate signal.

A simple example of the use of an `inout` parameter is:

```
procedure invert (arg : inout std_logic_vector) is
begin
  for i in arg'range loop
    arg(i) := not arg(i);
  end loop;
end;
```

This example simply inverts each element of the parameter and passes the result back. This example also shows that `inout` parameters can be unconstrained too.

The example does not show the normalisation of `inout` parameters, but this is simply a combination of the techniques used for in parameters and out parameters. The same example could have been written with normalisation:

```
procedure invert (arg : inout std_logic_vector) is
  variable arg_int : std_logic_vector(arg'length-1 downto 0);
begin
  arg_int := arg;
  for i in arg_int'range loop
    arg_int(i) := not arg_int(i);
  end loop;
  arg := arg_int;
end;
```

11.5.4   Signal Parameters

So far, all the examples of procedures have had `out` and `inout` parameters that were variables. This meant that all the examples could only be called in sequential VHDL – either in a process or from another subprogram.

The full story of parameters is more complex than these examples have shown, although they illustrated the most common usage of subprograms. However, for completeness, here's the whole story of parameters, with a practical example of the use of signal parameters.

A subprogram parameter can have three modes: `in`, `out` and `inout`. It can also have three classes: `constant`, `variable` and `signal`. Not all permutations are usable, for example, a `constant out` parameter does not make any sense, but there are still a number of legal combinations. Function parameters may only be of mode `in`, but they can still have any class.

Parameters of class `signal` must be associated with a signal. Parameters of class `variable` must be associated with a variable. However, a parameter of class `constant` can be associated with any expression – for example, a variable, a signal, another function call – but can only be of mode `in`.

Parameters of mode `in` are of class `constant` by default. This is the class of parameter that has been used so far in all the examples. This is why it is possible to use the same function with either variables or signals as the parameters.

It is possible to restrict the use of `in` parameters by making them class `variable` or `signal`, but this is rarely done in practice.

Parameters of modes `out` and `inout` are class `variable` by default. This is why they must be associated with variables when the procedure is used, thus restricting them to use within sequential VHDL. They cannot be of class constant, but they can be of class signal.

By making out and inout parameters class signal, they may be associated with signals and it is then possible to use the procedure in either sequential VHDL or concurrent VHDL. However, signal class parameters cannot then be associated with variables.

For example, a signal parameter version of the full-adder example shown before is:

```
procedure full_adder_s (a, b, c : in std_logic;
                        signal sum, cout : out std_logic) is
begin
  sum <= a xor b xor c;
  cout <= (a and b) or (a and c) or (b and c);
end;
```

There are a number of features of this procedure to highlight. The first is that only the out parameters have been changed to class signal. The in parameters are class constant and so can be associated with either signals or variables anyway. Within the procedure, the assignments to the out parameters have become signal assignments. Finally, the name of the procedure has been changed. This is because overload resolution does not take into account the class of a parameter when resolving procedure calls, only the types, so this procedure has exactly the same parameter profile as the original version and could not be distinguished from it if the name was the same.

To avoid problems with overload resolution, it is good practice to use a notation to distinguish between procedures written with class variable parameters and procedures written with class signal parameters. The recommended notation is to use the suffix '_s' or even '_signal' on the names of procedures defined with class signal parameters.

In use, there is now no need to declare intermediate variables to accumulate the result as was the case before. However, it is necessary to declare intermediate signals for the carry path. The same 4-bit adder example now looks like:

```
library ieee;
use ieee.std_logic_1164.all;
entity adder4 is
  port (a, b : in std_logic_vector (3 downto 0);
        cin : in std_logic;
        sum : out std_logic_vector (3 downto 0);
        cout : out std_logic);
end;
architecture behaviour of adder4 is
  signal c : std_logic_vector (2 downto 0);
begin
  full_adder_s (a(0), b(0), cin, sum(0), c(0));
  full_adder_s (a(1), b(1), c(0), sum(1), c(1));
  full_adder_s (a(2), b(2), c(1), sum(2), c(2));
  full_adder_s (a(3), b(3), c(2), sum(3), cout);
end;
```

In this solution, the procedures have been used as concurrent procedure calls, so there are no processes.

A concurrent procedure call will be re-evaluated every time one of its inputs changes. The inputs are the set of mode `in` and mode `inout` parameters. Therefore it is equivalent to a combinational logic block. It is in fact equivalent to a process containing the procedure call in sequential VHDL, so the first procedure call in the above example is equivalent to:

```
process (a(0), b(0), cin)
begin
   full_adder_s (a(0), b(0), cin, sum(0), c(0));
end process;
```

It is legal in VHDL to have `wait` statements in a procedure (but not in a function), but `wait` statements are disallowed for synthesis so that the procedure can be implemented as combinational logic using this equivalence.

11.6   Declaring Subprograms

Most of the examples so far have been incomplete! They have shown a subprogram and then an example of how to use it without really addressing the issue of where the subprogram is declared.

This section covers the rules for placement of subprogram declarations and then discusses the use of packages for subprograms.

11.6.1   Local Subprogram Declarations

Subprograms can be declared locally in architectures, processes and within other subprograms. In these cases, the subprograms can only be used within that structure. For example, a subprogram declared in a process can only be used in that process.

The only reason for having local declarations is to clarify a model by breaking it down into manageable parts. However, for hardware modelling for synthesis, it is better practice to break down a model into components. Nevertheless, there will be situations where this is appropriate, so the rules are covered here.

The main rule has already been stated: a subprogram declared locally can only be used locally. The only place that subprograms can be declared and then used elsewhere is the package, which will be discussed in the next section.

An example of the declaration and use of a local subprogram is shown in the following example, which puts together a function used in an earlier example and its use.

```
library ieee;
use ieee.std_logic_1164.all;
entity carry_example is
  port (a, b, c : in std_logic;
        cout : out std_logic);
end;
architecture behaviour of carry_example is
   function carry (bit1, bit2, bit3 : in std_logic) return std_logic
   is
   begin
     return (bit1 and bit2) or (bit1 and bit3) or (bit2 and bit3);
```

```
    end;
  begin
    cout <= carry(a, b, c);
  end;
```

This function is local to the architecture and so can be used anywhere in the architecture.

Alternatively, a function can be declared in a process. The same architecture could have been written with a single combinational process:

```
  architecture behaviour of carry_example is
  begin
    process (a, b, c)
      function carry (bit1, bit2, bit3 : in std_logic)
        return std_logic is
      begin
        return (bit1 and bit2) or (bit1 and bit3) or
               (bit2 and bit3);
      end;
    begin
      cout <= carry(a, b, c);
    end process;
  end;
```

11.6.2   Subprograms in Packages

The main reason for writing subprograms is to model common operations that will be useful elsewhere. These subprograms can be collected together in a package. From there, they can be used throughout a design or even in other designs by sharing that package.

A package is in two parts, the package header (also known as just the package) and the package body. The package header contains the declarations of the subprograms, whilst the package body contains the subprogram bodies.

Normally, a package will be used to collect together a set of closely related subprograms. Often, although not always, the subprograms will be associated with a new type that is declared in the package – subprograms acting on that type will be stored in the same package as the type, so that they are always available to users of the type. This is particularly true of operators, for which it is always good practice to have the operators with the type they act on.

As an example, consider the earlier examples of the carry function and the full_adder procedure, and add to these a sum function. Finally, put them together into a package called std_logic_vector_arith and the package would look something like the following example:

```
  library ieee;
  use ieee.std_logic_1164.all;
  package std_logic_vector_arith is

    function carry (a, b, c : std_logic) return std_logic;
    function sum (a, b, c : std_logic) return std_logic;
```

```
      procedure full_adder (a, b, c : in std_logic;
                            s, cout : out std_logic);

   end;
```

The package header declares the subprograms, but not the subprogram bodies. The subprogram bodies are defined in the package body:

```
   package body std_logic_vector_arith is

     function carry (a, b, c : std_logic) return std_logic is
     begin
       return (a and b) or (a and c) or (b and c);
     end;

     function sum (a, b, c : std_logic) return std_logic is
     begin
       return a xor b xor c;
     end;

     procedure full_adder (a, b, c : in std_logic;
                           s, cout : out std_logic) is
     begin
       s := sum (a, b, c);
       cout := carry (a, b, c);
     end;

   end;
```

The separation of the package header and the package body is similar to the separation of an entity and an architecture. It means that the interface is separate from the contents. This makes it possible to make modifications to the contents of the package body without affecting the interface. Probably more significantly for users of packages, it makes the package more readable.

Note the difference between the subprogram declaration and the subprogram body. The subprogram declaration is:

```
   function carry (a, b, c : std_logic) return std_logic;
```

Note the semi-colon immediately after the return type. For procedures, the semi-colon follows the close parenthesis:

```
   procedure full_adder (a, b, c : in std_logic;
                         s, cout : out std_logic);
```

The subprogram bodies are distinguished by the fact that the semi-colon is replaced with the keyword is:

```
   function carry (a, b, c : std_logic) return std_logic is ...
   procedure full_adder (a, b, c : in std_logic;
                         s, cout : out std_logic) is ...
```

11.6.3   Using Packages

Having declared a package like that above, it can be used by placing a use clause before the design unit where the package is needed. In fact, there are a number of places where the use clause can go, but the only common placing is before the design unit. Thus, to finish off the first example of the use of the carry function in this chapter, here's the same VHDL but with the use clause added:

```
library ieee;
use ieee.std_logic_1164.all;
entity carry_example is
  port (a, b, c : in std_logic;
        cout : out std_logic);
end;
use work.std_logic_vector_arith.all;
architecture behaviour of carry_example is
begin
  cout <= carry(a, b, c);
end;
```

In this case, the use clause is before the architecture, because the architecture was where the package was needed. If there was a type defined in the package and that type was to be used in the interface, then the use clause would have been placed before the entity.

The make-up of the use clause needs some clarification. The example has a three-part use clause work.std_logic_vector_arith.all.

The first part of the use clause, in this example work, refers to the library that the package is to be found in. The keyword work refers to the current working library – that is, the library that the architecture itself is being compiled into. Library work can always be referred to in a use clause, but other libraries can only be referred to if there is also a library clause declaring the existence of another library. In fact, there is, in effect, an implicit library clause for library work before every design unit. The library clause looks like:

```
library work;
```

Since this is implicit, there is never any need for a library clause for the work library, but all other libraries must be declared like this.

The second part of the use clause is the name of the package, std_logic_vector_arith. This is fairly self-explanatory. It makes that package available for use in the design unit.

The third part of the use clause, in this case all, is the item or items in that package that are to be made available for use in the design unit. The keyword all makes everything in the package available for use. It would be possible to specify individual subprograms, but this is rarely done in practice and it is very rare indeed to see any other form of use clause.

11.7   Worked Example

To give a realistic illustration of operator overloading, consider the situation where a package defining a new type is to be written, which models complex numbers as arrays of std_logic.

The complex number representation will use the left half of the array as the real part and the right half as the imaginary part. The array will be constrained to have a length that is even such that the two parts are always the same length.

The first stage is to create a skeleton package declaration containing the type definition:

```
library ieee;
use ieee.std_logic_1164.all;
package complex_std is
   type complex is array (natural range <>) of std_logic;
end;
```

This defines an unconstrained array of std_logic called complex. Note that this is not the same as std_logic_vector, it is a completely new and separate type.

By declaring the type, a set of built-in operators will automatically become available for use on this type. The built-in operators for this type, referring back to Table 11.1, are the equality, ordering and concatenation operators. That is, the following operators will automatically be generated by the VHDL analyser:

```
equality: =, /=
ordering <, <=, >, >=
concatenation &
```

More specifically, the following functions will be automatically declared:

```
function "=" (l, r : complex) return boolean;
function "/=" (l, r : complex) return boolean;
function "<" (l, r : complex) return boolean;
function "<=" (l, r : complex) return boolean;
function ">" (l, r : complex) return boolean;
function ">=" (l, r : complex) return boolean;
function "&" (l, r : complex) return complex;
function "&" (l : std_logic; r : complex) return complex;
function "&" (l : complex; r : std_logic) return complex;
function "&" (l : std_logic; r : std_logic) return complex;
```

However, the boolean and arithmetic operators will not be predefined:

```
boolean: not, and, or, nand, nor, xor
arithmetic: **, abs, *, /, mod, rem, sign +, sign -, +, -
```

The concatenation operators are fine as they are, since they can be used to build up a complex number from its component parts. There are in fact four concatenation operators, one to concatenate a complex with a complex, one to concatenate a complex with a std_ logic, one to concatenate a std_logic with a complex and one to concatenate two std_logic. All of them create a complex.

The problem areas lie with the predefined equality and ordering operators and with the lack of boolean and arithmetic operators.

The problem with the equality operators is that the predefined equality for arrays do not give the correct ordering for any numeric types, including the complex type. The problem lies with arrays of different lengths, which are not equal according to the rules for array comparisons, regardless of their values. This means that leading zeros are significant in the comparison. To resolve this problem, the equality and the inequality operators will have to be overloaded for type complex.

The ordering operators also cause problems. There is no sensible definition of the ordering operators for type complex, and yet they are implicitly defined by the language. The best response is to overload them with operators that raise errors if called.

Also, the type does not have boolean and arithmetic operators, so these will also be added. All the boolean operators will be added in this way, plus the set of generally synthesisable arithmetic operators, these being the sign "-", abs, "+", "-" and "*" operators.

Finally, it will be useful to be able to break complex numbers up into real and imaginary parts and to put the parts back together again. It will also be useful to be able to change the precision of a complex number, so that, for example, a 16-bit complex number can be converted into a 24-bit complex number. A set of utility functions will be defined to do these jobs.

All the internal behaviour is going to be defined in terms of the integer arithmetic defined in the numeric_std package. It is always good practice to build up in layers using existing packages in this way, rather than re-inventing basic operations, in the same way as it is good practice to reuse hardware components. The utility functions for assembling and disassembling type complex will use type signed in numeric_std to represent the real and imaginary parts.

The first stage is to add the function declarations of those operators that are to be overloaded to the complex_std package. These should be declared after the complex type, but can appear in any order. Also, the utility function declarations can be added.

```
library ieee;
use ieee.std_logic_1164.all;
use ieee.numeric_std.all;
package complex_std is

    type complex is array (natural range <>) of std_logic;

    function "not" (r : complex) return complex;
    function "and" (l, r : complex) return complex;
    function "or" (l, r : complex) return complex;
    function "nand" (l, r : complex) return complex;
    function "nor" (l, r : complex) return complex;
    function "xor" (l, r : complex) return complex;

    function "=" (l, r : complex) return boolean;
    function "/=" (l, r : complex) return boolean;
    function "<" (l, r : complex) return boolean;
    function "<=" (l, r : complex) return boolean;
    function ">" (l, r : complex) return boolean;
    function ">=" (l, r : complex) return boolean;

    function "-" (r : complex) return complex;
    function "abs" (r : complex) return complex;
    function "+" (l, r : complex) return complex;
    function "-" (l, r : complex) return complex;
    function "*" (l, r : complex) return complex;
```

```
function real_part (arg : complex) return signed;
function imag_part (arg : complex) return signed;
function create (R, I : signed) return complex;
function resize (arg : complex; size : natural) return complex;

end;
```

The library and use clause make the packages std_logic_1164 and numeric_std available for use in the interface to this package. It also makes the packages available for use in the package body, since package bodies inherit use clauses from their own package header.

The final stage is to write the package body, containing the function bodies for these operators. To keep things clear, the function bodies have been separated out from the package body so that each can be discussed individually, but in practice they would appear where the ellipsis (...) is.

```
package body complex_std is
   ...
end;
```

The first function to be examined is a purely local function for calculating the maximum of two integers. This function will be used inside the package for calculating the lengths of results. The purpose of this function will become clear when it is used.

```
function max (l, r : natural) return natural is
begin
  if l > r then
    return l;
  else
    return r;
  end if;
end;
```

The first set of functions to be examined are the utilities create, real_part, imag_part and resize since these will be used by the operators.

The create function takes two elements of type signed and concatenates them to form a complex. To make a complex, both real and imaginary parts must be of equal length. There is a design choice here as to what happens if the two arguments are not the same size. Either an error could be raised, or the arguments could be normalised to the same size. The more flexible solution is the latter, so this is the one that has been used. This means that the complex returned from the function is twice the length of the longer of the real and imaginary arguments to the function. In practice, the most common usage will be with arguments of the same size, so the result will be the simple concatenation of the arguments.

```
function create (R, I : signed) return complex is
   constant length : natural := max(R'length,I'length);
   variable R_int, I_int : signed (length-1 downto 0);
begin
   R_int := resize(R, length);
   I_int := resize(I, length);
   return complex(R_int
end;
```

The `resize` functions here are resizing type `signed` from `numeric_std`, then the result is formed by concatenating the two `signed` numbers (using the concatenation operator for `signed` again) and then type converting the result to `complex`. In fact it can be written as a single expression:

```
function create (R, I : signed) return complex is
   constant length : natural := max(R'length,I'length);
begin
   return complex(resize(R, length) & resize(I, length));
end;
```

The `real_part` and `imag_part` functions perform simple slice operations on the argument. They also check the rule that complex numbers are of even length. Finally, they convert the result to type `signed`.

```
function real_part (arg : complex) return signed is
   variable arg_int : complex (arg'length-1 downto 0) := arg;
begin
   assert arg'length rem 2 = 0
      report "complex.real_part: argument length must be even"
      severity failure;
   return signed(arg_int(arg_int'length-1 downto arg_int'length/2));
end;

function imag_part (arg : complex) return signed is
   variable arg_int : complex (arg'length-1 downto 0) := arg;
begin
   assert arg'length rem 2 = 0
      report "complex.imag_part: argument length must be even"
      severity failure;
   return signed(arg_int(arg_int'length/2-1 downto 0));
end;
```

Notice how the argument in each case is normalised in order to make the rest of the function simpler to write. However, it does mean that the return value of the `real_part` function is offset, which might be undesirable. It is good practice to stick to the normalisation convention for all functions, so the function needs a normalised intermediate variable to store the result.

```
function real_part (arg : complex) return signed is
   variable arg_int : complex (arg'length-1 downto 0) := arg;
   variable result : signed(arg'length/2-1 downto 0);
begin
   assert arg'length rem 2 = 0
      report "complex.real_part: argument length must be even"
      severity failure;
   result :=
      signed(arg_int(arg_int'length-1 downto arg_int'length/2));
   return result;
end;
```

The `resize` function changes the size of a `complex` argument and returns the result. It is defined in terms of the `numeric_std resize` function and the three utilities just defined.

```
function resize (arg : complex; size : natural) return complex is
begin
  assert size rem 2 = 0
    report "complex.resize: size must be even"
    severity failure;
  return create (resize(real_part(arg),size/2),
                 resize(imag_part(arg),size/2));
end;
```

This function turns out to be quite simple, since it uses existing functions from both package `complex_std` and package `numeric_std`. It creates a `complex` from two `signed` numbers, each of which is created by `resize` from `numeric_std`, acting on the real and imaginary parts of the argument. The assertion enforces the rule that `complex` numbers must be of even length by preventing the user from specifying an odd `size`. The length of the `complex` argument will be checked when it is passed to the `real_part` and `imag_part` functions so does not need an additional assertion.

Bear in mind once again that the `resize` function from `numeric_std` has unexpected behaviour when truncating in that the sign is preserved, whereas the normal convention is to simply truncate by discarding the most significant bits and allowing the result to wrap round if the truncation causes overflow. If the normal convention is the desired behaviour, then the function should be rewritten without the use of the `resize` function calls.

The boolean operators are quite simple, in that they are implemented by simply applying the operator to each element in turn. The common convention, used in both `std_logic_1164` and the numeric packages, is that boolean operators take two arguments of the same size and return a result of that size. They do not allow different length arguments. In the case of the `not` operator, there is only one argument of course. In this case, I will follow these conventions but add an extra rule, which is that the return value will have a normalised range.

The `not` operator is:

```
function "not" (arg : complex) return complex is
  variable result : complex(arg'length-1 downto 0) := arg;
begin
  for i in result'range loop
    result(i) := not result(i);
  end loop;
  return result;
end;
```

The `not` operator used within the loop is that for type `std_logic`, so the `not` of a type `complex` is defined in terms of the logical behaviour of its element type – again this is good practice.

The binary operators are similar, but there are two parameters and there must be an assertion to enforce the rule that they must be of equal length.

```
function "and" (l, r : complex) return complex is
  variable l_int : complex(l'length-1 downto 0) := l;
  variable r_int : complex(r'length-1 downto 0) := r;
  variable result : complex(l'length-1 downto 0);
```

```
begin
  assert l'length = r'length
    report "complex_std: ""and"" arguments are different lengths"
    severity failure;
  for i in result'range loop
    result(i) := l_int(i) and r_int(i);
  end loop;
  return result;
end;
```

All the other boolean operators are written in exactly the same way, with the appropriate element operator used inside the loop. All of the boolean operators could have an extra assertion to check the even-length rule if it was considered desirable.

Many of the other operators are layered onto numeric_std in the same way. For example, the equality function is true if both the real parts and the imaginary parts are equal. The parts of the complex number are of type signed, so the equality for signed will be used for the comparison. The equality operator for type signed is written to compare operands of different lengths and to take into account leading zeros, so suits our purposes exactly.

```
function "=" (l, r : complex) return boolean is
begin
  return real_part(l) = real_part(r)
         and
         imag_part(l) = imag_part(r);
end;
```

The inequality function is best defined in terms of the equality function:

```
function "/=" (l, r : complex) return boolean is
begin
  return not (l = r);
end;
```

The ordering operators are required to give errors if called. This is done with an assertion that will always fail.

```
function "<" (l, r : complex) return boolean is
begin
  assert false
    report "complex_std:""<"" illegal operation"
    severity failure;
  return false;
end;
```

The return statement after the assertion is still needed, even though the assertion will stop a simulation, because most VHDL analysers will insist that a function has at least one return statement in it. If the return statement was missing, the package would not even compile successfully.

All the other ordering operators are written in the same way.

Finally, the arithmetic operators for sign "-", abs, "+", "-" and "*" are defined using the arithmetic operators from numeric_std and will use the same conventions. This means that the sign "-" and abs operators will return a result that is the same length as the argument, the "+" and "-" operators will return a result that is the length of the longer operand, and the "*" operator will return a result that is the sum of the lengths of the arguments.

```
function "-" (r : complex) return complex is
begin
   return create(-real_part(r), -imag_part(r));
end;
function "abs" (r : complex) return complex is
begin
   return create(abs real_part(r), abs imag_part(r));
end;
function "+" (l, r : complex) return complex is
begin
   return create(real_part(l)+real_part(r),
                 imag_part(l)+imag_part(r));
end;
function "-" (l, r : complex) return complex is
begin
   return create(real_part(l)-real_part(r),
                 imag_part(l)-imag_part(r));
end;
function "*" (l, r : complex) return complex is
begin
   return
     create(
       real_part(l)*real_part(r)-imag_part(l)*imag_part(r),
       real_part(l)*imag_part(r)+imag_part(l)*real_part(r));
end;
```

An important point about this package is that, nowhere in it are there definitions of how to actually perform arithmetic, or even how to perform bitwise logic. The predefined and, presumably, well tested packages std_logic_1164 and numeric_std have been used to provide the boolean and arithmetic operations throughout.

12

Special Structures

This chapter is a mopping-up chapter to cover a few hardware structures that are important but that haven't fitted into the earlier chapters. These special structures have been collected together here.

The first special structure is the tristate driver, how to model this in simulation using a synthesis template and how to model tristate buses.

Then finite state machines (FSMs) are covered. These are often used to implement controllers. A variety of templates are described that provide Moore machines, Mealy machines and FSMs with either combinational or registered outputs.

The next section covers memories, which can be implemented as register banks or converted into RAMs using RAM inference. A number of different templates are described that allow different types of RAM to be inferred.

Finally, decoders are described, which can be converted into ROMs using ROM inference.

12.1 Tristates

There are two aspects to the modelling of tristate systems in VHDL. One covers the modelling of tristate drivers. The other is the modelling of tristate buses.

Tristate drivers are hardware structures that have to be recognised as templates, since there is no direct mapping from VHDL to tristate drivers. In other words, a synthesiser has to perform *tristate inference*, just as it has to perform latch inference and register inference.

The most common tristate driver template uses sequential VHDL to model tristate behaviour. This means that tristate drivers must be modelled as processes. It is possible to write a concurrent signal assignment (using conditional assignments) that has the behaviour of a tristate driver, but most synthesisers will not recognise it as a tristate driver because it does not match the template.

Tristate buses require the use of a multi-valued logic type, which can model a high-impedance value. The logic type should also be able to handle multiple drivers driving the same bus. In VHDL terms, this means that the type must be *resolved*, that is, capable of being the target of more than one signal assignment. In simulation, resolution is implemented as a *resolution function* that is called with all the values being driven onto the signal from all the tristate drivers. The resolution function then decides what value the signal should have.

VHDL for Logic Synthesis, Third Edition. Andrew Rushton.
© 2011 John Wiley & Sons, Ltd. Published 2011 by John Wiley & Sons, Ltd.

Resolution only applies to signals, so tristate buses must be modelled by signals and cannot be modelled by variables.

The IEEE standard type `std_logic` models high-impedance values as the value `'Z'`. Furthermore, it is resolved in a way that models tristate drivers. For example, if one driver is driving the value `'1'` and another `'Z'`, then the signal gets the value `'1'`. It is not necessary to understand any more than this about the operation of a resolved type if you are using VHDL for synthesis. It is sufficient to know that the subtype `std_logic` is resolved for use as a tristate bus, whereas its basetype `std_ulogic` is not.

The tristate driver template uses a combinational process containing an `if` statement. An example is shown in the following VHDL code, which shows just the tristate driver process:

```
process (d, en)
begin
  if en = '1' then
    q <= d;
  else
    q <= 'Z';
  end if;
end process;
```

It can be seen that this process is a combinational logic process, apparently containing a two-way multiplexer modelled by the `if` statement. However, since the assignment to q in one branch of the `if` statement is the value `'Z'`, this will be interpreted as a tristate driver. The equivalent circuit is shown in Figure 12.1.

The tristate driver is implemented in a similar way to the register. The process is implemented as if it was a combinational process (ignoring the tristate part) and then a tristate driver is added to the outputs. Because of this similarity with the register template, it is not possible to combine the two templates. In other words, it is not possible to describe a register with a tristateable output as a single process. Instead, it should be modelled by separate processes; a registered process to model the register and a combinational process to model the tristate driver on its output.

The exact rules for writing the tristate driver will vary between synthesisers. The safest template to use is as illustrated by the example above which is the simplest standardised form. Assume that the whole process will be implemented as a tristate driver and that the `if` statement must be in the form shown, with two branches: one containing the high-impedance assignment and the other containing any other combinational logic. Most synthesisers offer

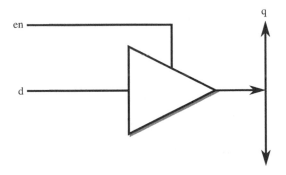

Figure 12.1   Tristate driver.

more flexibility than this, but using that extra flexibility may be non-portable between synthesisers. In any case, the use of the template makes the intention of the circuit clearer both to the original designer and to anyone else who subsequently has to work on it, so the use of this template is recommended, even if it is not obligatory.

The tristate bus is simply a signal of a subtype, known as a resolved subtype, which can model tristates. This is always `std_logic` or any one of its array types, since this subtype already has a resolution function that models tristates and is the only standard type that does. However, the subtype itself is not enough to make a signal a tristate – after all the subtype `std_logic` is generally used for all signals in a design. The other condition is that a signal is treated as a tristate bus if it has tristate drivers driving it. If any one driver of a signal is a tristate driver, then all the drivers of that signal must be tristate drivers. Furthermore, if one element of an array signal is a tristate signal, then all elements must be tristate signals.

The only other special handling of tristate signals is when they are used as ports to an entity. Tristate ports should be modelled using mode `inout`, so that the synthesiser knows that they must be implemented as tristate buses. This rule should be followed even if the bus is being driven by a tristate driver but is not being read. An example of this is the above example of the basic driver, shown in context in an entity/architecture pair.

```
library ieee;
use ieee.std_logic_1164.all;
entity tristate is
  port (d : in std_logic;
        en : in std_logic;
        q : inout std_logic);
end;
```

In this entity, port q has been identified as a tristate bus and therefore it should be driven by a tristate driver.

```
architecture behaviour of tristate is
begin
  process (d, en)
  begin
    if en = '1' then
      q <= d;
    else
      q <= 'Z';
    end if;
  end process;
end;
```

Generally, tristate drivers will be mixed with other logic in an architecture. This example showed the tristate driver as the only component in the architecture for clarity, to show the relationship between the driver process and the port modes. Despite the fact that the process only writes to the port and doesn't read it, the port is `inout` mode.

Note: there are a lot of examples available on the Internet that show tristate buses modelled as `in` or `out` parameters. Modern synthesisers don't need the `inout` port class to know that a port is a tristate bus and to synthesise it correctly. However, it is still recommended practice to use `inout` ports for all tristate buses because it makes the interface of the entity more self-explanatory.

Array drivers can be created simply by using any array of `std_logic`, such as `std_logic_vector` and the synthesis types `signed` and `unsigned` from `numeric_std`.

Note: the fixed-point and floating-point types from the VHDL-1993 compatibility versions of fixed_pkg and float_pkg cannot be used for tristates because they use std_ulogic. This problem is fixed in the VHDL-2008 versions of the packages so these types will also be usable as tristate buses when the VHDL-2008 versions become available. In the meantime, tristate buses should be implemented using std_logic_vector and the fixed and floating-point types converted to and from this type using the bit-preserving type conversions described in Section 6.7.

For tristate arrays. the assignment of the value 'Z' is changed into an array assignment of a string of 'Z' values

```
library ieee;
use ieee.std_logic_1164.all
use ieee.numeric_std.all;
entity tristate_vec is
  port (d : in signed(7 downto 0);
        en : in std_logic;
        q : inout signed(7 downto 0));
end;

architecture behaviour of tristate_vec is
begin
  process (d, en)
  begin
    if en = '1' then
      q <= d;
    else
      q <= "ZZZZZZZZ";
    end if;
  end process;
end;
```

An alternative that is simpler to type, especially for large buses, and the only form that works for generic-sized buses, is to use an aggregate with an others clause:

```
q <= (others => 'Z');
```

To illustrate the use of tristates in a slightly larger example, a tristateable multiplexer will be written using the synthesisable tristate template.

To keep the example clear, the design is placed in the context of a separate entity and architecture.

The entity is:

```
library ieee;
use ieee.std_logic_1164.all;
entity tristate_mux is
  port (a, b, sel, en : in std_logic;
        z : inout std_logic);
end;
```

There are two possible solutions to this problem: either two tristate drivers driving the same output or a multiplexer followed by a single tristate driver. Both of these two solutions will be shown.

The block diagram of the first possible implementation circuit is shown in Figure 12.2.

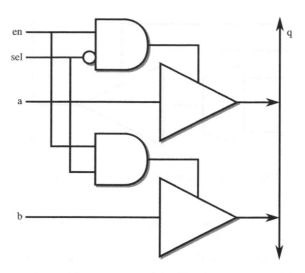

Figure 12.2   Tristate multiplexer using two drivers.

Inputs a and b are the data inputs, sel selects between a (sel = 0) and b (sel = 1). Input en enables the output driver: when en = 0 the output is high impedance.

The architecture for this two-driver solution is:

```
architecture behaviour of tristate_mux is
begin

  t0: process (en, sel, a)
  begin

    if en = '1' and sel = '0' then
      z <= a;
    else
      z <= (others => 'Z');
    end if;
  end process;

  t1: process (en, sel, b)
  begin
    if en = '1' and sel = '1' then
      z <= b;
    else
      z <= (others => 'Z');
    end if;
  end process;

end;
```

The solution contains two copies of the template for a tristate driver. The drivers are connected straight to the tristate port z. Note that no attempt has been made to combine the two drivers into one process: this is a common mistake. The design is clearer and simpler and therefore far less error-prone if each part of the design is expressed separately in this way. There is a one-to-one correspondence with the original block diagram.

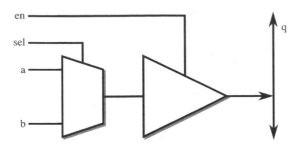

Figure 12.3 Tristate multiplexer using one driver.

The alternative solution uses a single tristate driver with a multiplexer on its input. This solution is illustrated by Figure 12.3.

Since the tristate driver process can contain other logic in the combination branch (that is, the branch without the high-impedance assignment) of the `if` statement, this can be expressed as a single process:

```
process (en, sel, a)
begin
  -- tristate driver
  if en = '1' then
    -- multiplexer
    if sel = '0' then
      z <= a;
    else
      z <= b;
    end if;
  else
    z <= (others => 'Z');
  end if;
end process;
```

Notice how the combinational part of the behaviour has been kept separate from the tristate driver part, with the multiplexer logic completely contained within the first branch of the outer `if` statement that represents the tristate driver.

12.2 Finite State Machines

The basic form of a finite state machine (FSM) is a sequential circuit in which the next state and the circuit outputs depend on the current state and the inputs. The most common application for FSMs is in control circuits. The basic form of an FSM is shown in Figure 12.4.

An FSM can be modelled in VHDL as a combinational block and a register block, and in that sense is nothing special. However, most synthesisers have the capability of performing state optimisation on FSMs to minimise the circuit area. This optimisation is only available if the FSM model fits one of the templates that allows *FSM inference* to take place.

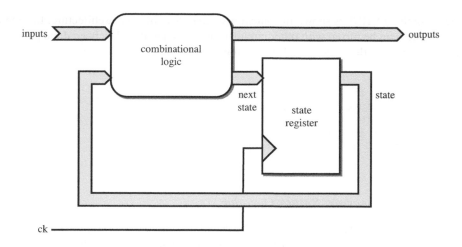

Figure 12.4   Finite state machine.

As usual, the range of different templates vary from one synthesiser to another and some synthesisers support several other variants, but the templates presented here are the common denominator and are the recommended forms.

The key feature of the FSM templates is that the current state and next state are represented by an enumeration type, with one value for each state. The inputs and outputs can be of any type.

The example is a very simple state machine that detects a certain bit sequence (a signature) on a one-bit wide serial input. The state machine is defined by the state-transition diagram in Figure 12.5.

The state-transition diagram shows the states as circles with the state name inside. The states themselves are also annotated with the output value that is required from the state machine when it is in that state. The state transitions are labelled with the input value that causes the transition. This example is a Moore machine, since the output only depends on the current state and is independent of the input.

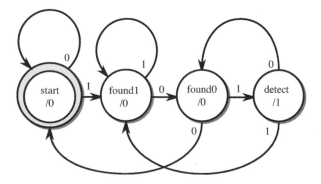

Figure 12.5   Signature detector state-transition diagram.

The example in VHDL is shown in context in a separate entity and architecture. In general, state machines can be mixed with other circuitry and the separation has been done purely for clarity. The entity is the same for all the examples:

```
library ieee;
use ieee.std_logic_1164.all;
entity signature_detector is
  port (d : in std_logic;
        ck : in std_logic;
        found : out std_logic);
end;
```

12.2.1 Two Processes, Single Decoder

This variant of the FSM template puts all of the decoding logic into one combinational process with a second registered process that just generates the state register but has no logic in it.

```
architecture behaviour of signature_detector is

  type state_type is (start, found1, found0, detect);
  signal state, next_state : state_type;

begin

  -- register block
  process
  begin
    wait until rising_edge(ck);
    state <= next_state;
  end process;

  -- combinational logic block
  process (state, d)
  begin
    case state is
      when start =>
        found <= '0';
        if d = '1' then
          next_state <= found1;
        else
          next_state <= start;
        end if;
      when found1 =>
        found <= '0';
        if d = '0' then
          next_state <= found0;
        else
          next_state <= found1;
        end if;
      when found0 =>
        found <= '0';
        if d = '1' then
          next_state <= detect;
```

```
          else
             next_state <= start;
          end if;
       when detect =>
          found <= '1';
          if d = '1' then
             next_state <= found1;
          else
             next_state <= found0;
          end if;
    end case;
  end process;

end;
```

The combinational logic block should be modelled as a process with a `case` statement branching on the current value of the state. The contents of each branch of the `case` statement should contain simple assignments of values to the next state and output signals. Branches in the state-transition diagram are modelled by `if` statements within a branch of the `case` statement, as in this example. To model a Mealy machine, the outputs would also be conditional on the inputs, so the assignments to the outputs would also be inside the `if` statements.

The case statement must be complete, which means that it covers all the states, and it should be purely combinational with no latches, which means that all outputs get a value under all conditions.

This can be rewritten using the combinational process style of assigning a default value before the case statement and then overriding it in the case statement, a style that ensures that the process is combinational:

```
process (state, d)
begin
  next_state <= start;
  found <= '0';
  case state is
    when start =>
      if d = '1' then
        next_state <= found1;
      end if;
    when found1 =>
      if d = '0' then
        next_state <= found0;
      else
        next_state <= found1;
      end if;
    when found0 =>
      if d = '1' then
        next_state <= detect;
      end if;
    when detect =>
      found <= '1';
      if d = '1' then
        next_state <= found1;
      else
        next_state <= found0;
```

```
        end if;
    end case;
end process;
```

12.2.2   Two Processes, Two Decoders

This variant of the FSM template puts all of the output logic into a combinational process and all the state logic in a registered process. Thus, there are two decoders, one decoding the state transitions, the other decoding the state to generate the outputs.

The above example using this alternative style is:

```
architecture behaviour of signature_detector is

    type state_type is (start, found1, found0, detect);
    signal state : state_type;

begin

    -- register block
    process
    begin
      wait until rising_edge(ck);
      case state is
        when start =>
          if d = '1' then
            state <= found1;
          else
            state <= start;
          end if;
        when found1 =>
          if d = '0' then
            state <= found0;
          else
            state <= found1;
          end if;
        when found0 =>
          if d = '1' then
            state <= detect;
          else
            state <= start;
          end if;
        when detect =>
          if d = '1' then
            state <= found1;
          else
            state <= found0;
          end if;
      end case;
    end process;

    -- combinational logic block
    process (state)
```

```
    begin
      case state is
        when start | found1 | found0 =>
          found <= '0';
        when detect =>
          found <= '1';
      end case;
    end process;

  end;
```

This template makes the distinction between Moore and Mealy state machines more obvious. In a Moore machine, the output is generated purely from the state. This means the inputs can change during a cycle without affecting the outputs. In a Mealy machine, the output is a combination of the state and the input. This can cause the outputs to change during a cycle if the inputs change.

In this example, the combinational logic block depends only on the state and does not have any other inputs. This makes it a Moore machine.

Note : this style of FSM also eliminates the `next_state` signal.

12.2.3   Single-Process, Single Decoder

There is a form of FSM that uses just one registered process. This variant puts a register on the FSM outputs as well as the state. This is illustrated by Figure 12.6.

When encoding the outputs for this form, the one-cycle delay introduced by the output register needs to be allowed for in the design. So the signal assignments to the outputs must set the outputs a cycle (state) before they are required. In other words, they are arranged so they

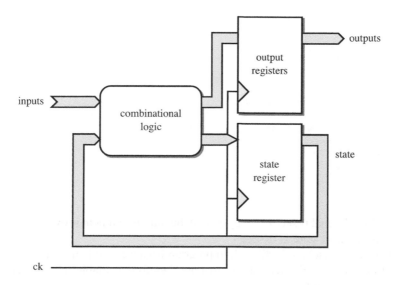

Figure 12.6   Single-process finite state machine.

specify the value that the output *will* have after the state transition rather than the output for the current state.

The previous example using the single-process form of FSM is:

```
architecture behaviour of signature_detector is

    type state_type is (start, found1, found0, detect);
    signal state : state_type;

begin

  -- register block
  process
  begin
    wait until rising_edge(ck);
    found <= '0';
    case state is
      when start =>
        if d = '1' then
          state <= found1;
        else
          state <= start;
        end if;
      when found1 =>
        if d = '0' then
          state <= found0;
        else
          state <= found1;
        end if;
      when found0 =>
        if d = '1' then
          state <= detect;
          found <= '1';
        else
          state <= start;
        end if;
      when detect =>
        if d = '1' then
          state <= found1;
        else
          state <= found0;
        end if;
    end case;
  end process;

end;
```

Note how the found signal is set to ' 0 ' by default, but this assignment is overridden by ' 1 ' in the case where the next state is detect.

This style of state machine is difficult to program because of this need to think one state ahead, but it does have registered outputs that are therefore available immediately after the clock and are guaranteed to be glitch-free.

12.2.4   State Encoding

FSM inference: is automatic and FSM synthesis will choose the most efficient encoding for the state. However, sometimes you want to specify the encoding. Some synthesisers allow the user to specify the binary encoding of the state values. This usually allows a predefined sequence to be specified, such as one-hot encoding or sequential encoding (0, 1, 2. . .). Alternatively, the actual binary encodings can be specified. Some synthesisers will not support user-customisation at all, in which case the encoding will be automatically selected.

Unfortunately, the way you specify the state encoding is completely synthesiser-specific so you have to see the synthesiser's manual to determine how to do it. Typically it is either an attribute in the VHDL file or it is a project setting in a settings file. Here's an example of the use of an attribute:

```
signal state : state_type;
attribute fsm_encoding : string;
attribute fsm_encoding of state : signal is "one_hot";
```

This is a completely made-up attribute – the real name and value will be different in practice and you can only find out what it is from the synthesiser's manual.

In the absence of an explicit encoding, the synthesiser will allocate states automatically. This is usually sufficient and it is rarely necessary to choose an encoding.

12.2.5   Illegal States and Resets

An FSM in VHDL uses an enumeration type that probably will not have exactly a power of two values. When it is mapped onto logic, there will be some state encodings that do not correspond to a state. These are known as *illegal* states.

Even if the enumeration does have a power of two values, the FSM synthesis may choose one-hot encoding that has one register per state and therefore for S states in the RTL model, has 2^S possible states in the hardware, of which all but S are illegal.

If an FSM gets into an illegal state, it may get stuck there since the FSM synthesis does not allow for this possibility – the optimisation of the next-state logic does not account for the illegal states. Indeed, doing so would make one-hot encoding the least effective encoding, whereas it is often (particularly for register-rich FPGA technologies) the best encoding.

Illegal states are not due to erroneous synthesis, they are due to the power-up behaviour of hardware, so it is not the job of FSM synthesis to solve this problem, it is a design issue. The only time an FSM can get into an illegal state is at power-up when the initial state cannot be controlled. Once the FSM is in a legal state, it will always stay in legal states because the FSM synthesis will ensure this. So, in order to avoid illegal states, all that is necessary is to make the FSM resettable. By resetting to a legal state after power-up, the problem of illegal states is solved.

A reset can easily be added to the register part of the template and has no effect on the combinational part. For example, using the two-process, one-decoder FSM model, a synchronous reset to the `start` state could be incorporated by adding a reset input to the entity and rewriting the register part.

```
-- register block
process
begin
  wait until rising_edge(ck);
  if rst = '1' then
    state <= start;
  else
    state <= next_state;
  end if;
end process;
```

This is a synchronous reset. As usual, an asynchronous reset could be used if that is the design requirement. However, the general rule as explained in Section 9.9 is to use synchronous reset unless there is a compelling reason not to.

12.3   RAMs and Register Banks

Conceptually a register bank is not the same as a RAM, since a register bank is a 2-dimensional array of flip-flops and a RAM is a two-dimensional array of memory cells. However, this is an implementation difference and at the level of abstraction used for RTL design, there is no difference. For this reason, the two structures will be discussed together.

The two structures will be referred to collectively as memories, so this term will be used wherever the discussion applies to both structures. The terms RAM and register-bank will then be used to distinguish between the two types of memory.

A memory is generated by using the register model from Chapter 9 on an array signal to create an array of registers. Each element of the array can then be accessed by array indexing. The element type for a register bank can be any synthesisable type, such as any of the synthesis types described in Chapter 6. You could for example have a memory made of an array of `float`. However, if a RAM is required it is usually necessary to only use arrays of `std_logic_vector`.

RAMs are then inferred from the memory model by the synthesiser, so to get a RAM, you must use one of the *RAM inference* templates. This inference decision is based partly on the functionality available in the target technology and partly on the size of the memory: small memories are more efficiently implemented as register banks, whereas larger memories are more efficiently implemented as RAMs.

The problem with RAM inference is that the templates can vary from technology to technology – one technology may implement a particular type of RAM, whilst another does not. If you use a template that's not supported by the technology, the memory will be implemented as a register bank. So if you want a RAM, it is really important to check the documentation to see which templates are available to use with your target synthesiser/ technology combination and to check the synthesis tool's log to ensure that the mapping has taken place.

In the following three sub-sections, three different memory models will be explained. The memory models will be illustrated by examples that will be written as generic components. The data inputs, outputs and the memory itself must be `std_logic_vector` to get a RAM. The address is an unsigned number designating the offset into the array of words, so type `unsigned` is used to represent it.

The three different versions reflect different behaviour on read: synchronous or asynchronous, and different behaviour on a write: whether to read the previous value of a word while writing the new value – conceptually a read-before-write; or whether to read the new value as it is written – conceptually a write-before-read.

In each example, the memory will be parameterised for both word size (`word_size`) and the size of the address bus (`address_size`). The number of words in the register bank will be the full set of addressable words for the address bus size – namely $2^{address\_size}$. The interface is the same for all of the examples except for the entity name so that the differences in the models is more obvious.

```
library ieee;
use ieee.std_logic_1164.all;
use ieee.numeric_std.all;
entity RAM is
  generic (word_size : natural;
           address_size : natural);
  port (d : in std_logic_vector(word_size-1 downto 0);
        ck : in std_logic;
        write : in std_logic;
        address : in unsigned(address_size-1 downto 0);
        q : out std_logic_vector(word_size-1 downto 0));
end;
```

Note how the ports have been sized by the generic parameters.

12.3.1 Asynchronous Read, Synchronous Write

This version of the model implements a memory where the read and write operations are independent, such that the read is combinational (i.e. asynchronous) and the write is synchronous. This means that if you read and write to the same address, the read output will change as the write operation updates the memory contents.

The architecture is:

```
architecture behaviour of RAM is

  type memory_type is array (0 to 2**address_size-1) of
    std_logic_vector(word_size-1 downto 0);
  signal memory : memory_type;

begin

  -- write
  process
  begin
    wait until rising_edge(ck);
    if write = '1' then
      memory(to_integer(address)) <= d;
    end if;
  end process;

  -- read
  q <= memory(to_integer(address));

end;
```

The main complication in writing the architecture is that the `address` bus is represented as an array, in this case a signal of type `unsigned`. This address is supposed to be used to index an array of registers, but it is illegal in VHDL to use an array type to index another array type – a scalar type such as `integer` must be used. The solution is to use a type-conversion function to convert the address bus into an integer value at the point of use; the return value of this function can then be used as an array index. Type conversions do not generate any hardware, so there is no synthesis overhead to this solution although there is a slight simulation overhead. In this case the `to_integer` function from package `numeric_std` will be used to perform the type conversion.

Note how the number of elements in the `memory` has been sized by the generic parameter `address_size` so that the full address range is covered. This has the advantage of making the memory simple – there are no out-of-range addresses for example. However, it is possible to use a third generic parameter to control the address range if an unconventional size is required, for example if the size is not a power of two.

The read assignment uses dynamic indexing of the source of the assignment: one of the elements of `memory` is selected and written to the output q. If a register bank is being generated then the read operation will be implemented as a multiplexer structure selecting one of the register outputs. If a RAM is being generated then it will be implemented as a memory read operation that is functionally equivalent. Note that the read is combinational so if the RAM contents change, the output will change.

The architecture also contains a process that uses the edge function template for a register as explained in Section 9.4. This implements the write behaviour of the memory. Any of the register templates could be used here. The write assignment uses dynamic indexing of the target of the assignment: one of the elements of `memory` is selected to receive the value on input d, provided the control signal `write` is enabled. The other elements preserve their values. The synthesiser implements this functionality by creating a register or memory for each element of signal `memory`, since all the elements of the signal are potentially assigned to in the process. If this is implemented as a register bank, then the dynamic indexing is implemented as a demultiplexer structure that can route the d input to any of the registers in the register bank. If it is implemented as a RAM, it will be implemented as a memory write operation that is functionally equivalent.

This is the simplest form of memory template. However, most real RAMs have synchronous read and write operations. For this reason, this model will nearly always be mapped onto a register bank instead of a RAM.

This template is the recommended form if you want a register bank, but not if you want a RAM. Templates that will be mapped onto RAMs are covered in the following sections.

12.3.2 Synchronous, Read Before Write

This version of the memory model implements a memory where read and write are synchronous and, during a write operation, the old value of a word is presented at the output port whilst the new value is being written to it.

The entity has the same interface as the previous example and only differs in name: RAM_RBW. The architecture is:

```
architecture behaviour of RAM_RBW is
  type memory_type is array (0 to 2**address_size-1) of
    std_logic_vector(word_size-1 downto 0);
  signal memory : memory_type;
begin

  process
  begin
    wait until rising_edge(ck);
    -- read
    q <= memory(to_integer(address));
    -- write
    if write = '1' then
      memory(to_integer(address)) <= d;
    end if;
  end process;

end;
```

The architecture contains a single process that uses the edge function template for a register as before. Both the read and write operations occur in the same process. Any of the register templates could be used here.

As in the previous example, the read assignment uses dynamic indexing of the source of the assignment: one of the elements of memory is selected and written to the output q. The difference is that it is now synchronous because the read takes place in a registered process. If a register bank is being generated then the read operation will be implemented as a multiplexer structure with a registered output. If a RAM is being generated then it will be implemented as a buffered memory read operation, which is functionally equivalent. Note that the read is reading the old value, not the new one, and holding it in an output buffer register. The positioning of the read first in the process is irrelevant and has been positioned there for clarity to the reader – in VHDL terms the reason this reads the old value is that memory is read a delta cycle before it is written.

The write assignment is identical to the previous example.

12.3.3   Synchronous, Write Before Read

This version of the memory model implements a memory where, during a write operation, the new value of a word is presented at the output port whilst the same value is being written to the memory. This is a slightly more complex memory design than the "Read Before Write" model and less likely to be available with every technology, but is still common.

The entity has the same interface as the previous example and only differs in name: RAM_WBR.

The architecture is:

```
architecture behaviour of RAM_WBR is

  type memory_type is array (0 to 2**address_size-1) of
    std_logic_vector(word_size-1 downto 0);
  signal memory : memory_type;
  signal read_address : unsigned(address'range);

begin
```

```
process
begin
  wait until rising_edge(ck);
  -- write
  if write = '1' then
    memory(to_integer(address)) <= d;
  end if;
  read_address <= address;
end process;

-- read
q <= memory(to_integer(read_address));

end;
```

The first part of the process implements the write operation in the same way as the previous example. However, also in this process is an assignment to an internal signal – read_address is a registered version of the address input.

In the concurrent signal assignment at the end of the architecture, the registered version of the address is used to index the memory in a read operation. In other words, the read is reading the value written in the previous clock cycle, by accessing the address specified in the previous clock cycle. This mimics the behaviour of a memory with synchronous output and write-before-read semantics, which is a common alternative form to the read-before-write on programmable devices. The synthesiser restructures this into a functionally equivalent output buffer.

If a register bank is being generated then the read operation will be implemented as a simple multiplexer structure with a registered output. If a RAM is being generated then it will be implemented as a buffered memory read operation, with bypass logic to handle the case where the same address is read and written at the same time, so the input value is fed to the output buffer directly at the same time as it is written to the RAM. It is this bypass logic that makes the write-before-read version larger and slower than the read-before-write.

12.3.4 RAM Read Optimisation

Many designs do not need simultaneous read and write but keep reads and writes as separate operations on different cycles. Some synthesisers allow you to optimise RAM inference by specifying that you do not require simultaneous read and write. This will enable the synthesiser to pick the smallest and probably also the fastest RAM implementation from the target technology regardless of which memory model you used in the design.

Specifying non-overlapping read and write may also enable RAM inference to take place where otherwise the memory would be implemented as a register bank.

Unfortunately, the way you specify this is completely synthesiser-specific so you have to see the synthesiser's manual to determine how to do it. Typically it is either an attribute in the VHDL file or it is a project setting in a settings file. Here's an example of the use of an attribute:

```
signal memory : memory_type;
attribute ram_type : string;
attribute ram_type of memory : signal is "no_rw_overlap";
```

This is a completely made-up attribute – the real name and value will be different in practice.

12.3.5   Getting a Register Bank

Sometimes you do just want a register bank, in which case you need to disable RAM inferencing.

Like all other synthesis options, the way you disable RAM inferencing is completely synthesiser-specific so you have to see the synthesiser's manual to determine how to do it. Typically it is either an attribute in the VHDL file or it is a project setting in a settings file. Here's an example of the use of an attribute:

```
signal memory : memory_type;
attribute ram_type : string;
attribute ram_type of memory : signal is "registers";
```

This is a completely made-up attribute – the real name and value will be different in practice.

12.3.6   Resets

You cannot use either asynchronous nor synchronous resets in memory models where a RAM is required, since RAMs do not generally have this functionality. So only the non-resettable register templates will result in RAM inference taking place.

If you want to be able to reset a RAM, then implement the reset at a higher level in the design as a series of writes of the reset value to all addresses. Do not be tempted to build this functionality into the RAM architecture, this may disable the RAM inference algorithm and result in a register bank instead.

A register bank, by contrast, can have either type of reset because it is just an addressable array of ordinary registers as described in Chapter 9.

12.4   Decoders and ROMs

A ROM is another structure inferred by the synthesiser so you need to use a specific template to get *ROM inference*. Furthermore, if the target technology does not support ROMs, or the synthesiser cannot implement a ROM that meets the design requirements, then it will be implemented as decoding logic instead. In FPGAs, the basic logic element is effectively a ROM already, so very efficient mappings can be achieved by using ROM models.

The difference with ROMs compared with RAMs is that you cannot design a generic ROM and then reuse it wherever you need it, for the simple reason that each ROM will have different data contents and the contents are hard-coded into the design. Thus, for every ROM in the design you need to design a different entity and architecture using the same ROM template.

Note: it is recommended practice to separate the ROM into its own design unit, having a separate entity and architecture, rather than mixing it up with the rest of the design. This is more likely to result in ROM inference taking place and some synthesis tools require it.

12.4.1   Case Statement Decoder

The most common ROM template uses a case statement to decode all possible values of the ROM's address, mapping each value onto a signal assignment where the target is the same signal and the source is a constant value.

For example, here's a decoder from 3-bit Gray code to binary using a ROM template:

```
library ieee;
use ieee.std_logic_1164.all;
use ieee.numeric_std.all;
entity gray_decode is
  port (gray : in std_logic_vector(2 downto 0);
        binary : out unsigned(2 downto 0));
end;
architecture behaviour of gray_decode is
begin
  process(gray)
  begin
    case gray is
      when "000" => binary <= "000";
      when "001" => binary <= "001";
      when "011" => binary <= "010";
      when "010" => binary <= "011";
      when "110" => binary <= "100";
      when "111" => binary <= "101";
      when "101" => binary <= "110";
      when "100" => binary <= "111";
      when others => binary <= "XXX";
    end case;
  end process;
end;
```

Note how the when others clause is used to set the output to unknown during simulation. This is required by simulation because there are many input permutations with metalogical values not covered by the other cases. This value will be ignored in synthesis because the case statement is regarded as complete with all the real input encodings covered.

12.4.2   Table Lookup Decoder

There is also a table lookup style of ROM that can be simpler and more compact, especially suitable where the input encoding is numeric. This can be applied to the reverse encoding from binary to gray code since the binary value is a numeric type:

```
library ieee;
use ieee.std_logic_1164.all;
use ieee.numeric_std.all;
entity gray_encode is
  port (binary : in unsigned(2 downto 0);
        gray : out std_logic_vector(2 downto 0));
end;
architecture behaviour of gray_encode is
  type memory_type is array (0 to 7) of
    std_logic_vector(2 downto 0);
  constant memory : memory_type :=
    ("000", "001", "011", "010", "110", "111", "101", "100");
begin
  gray <= memory(to_integer(binary));
end;
```

The possible output values are stored in a constant array such that the offset in the array of each output value is equal to the input address that selects it, in this case the input port `binary`. This might be clearer if the explicit form of the aggregate is used in the initial values of the constant:

```
constant memory : memory_type :=
   (0 => "000",
    1 => "001",
    2 => "011",
    3 => "010",
    4 => "110",
    5 => "111",
    6 => "101",
    7 => "100");
```

Because of the restriction in VHDL that arrays types such as `unsigned` cannot be used to index arrays, the array type has to be declared with a `natural` range, which is also why the offsets into the aggregate are expressed as integer values rather than as bit-strings. The input signal `binary` is type-converted to `natural` using the `to_integer` function and then the result used to index the array.

13

Test Benches

With most hardware description languages, the circuit description and the test waveforms are described in different ways, with the test waveforms described either using a waveform capture facility in the simulator or using a separate waveform language. Most VHDL simulators do not, however, have any form of waveform-capture facility because VHDL is itself sufficiently expressive to be used as a waveform language. The result is the test-bench, which is simply a naming convention for a VHDL model which generates waveforms with which to test a circuit model.

13.1 Test Benches

Test benches clearly only apply to the use of VHDL in simulation. They are not synthesised. Nevertheless, it is appropriate to have a chapter dedicated to the writing of test benches in a book on synthesis VHDL, since the writing of test benches is an important part of the design process and one where many designers get unnecessarily bogged down.

The necessity for test benches is clear. A synthesisable model should be extensively tested in simulation before synthesis to ensure correctness. A synthesiser works, as far as possible, on the principle that 'what you simulate is what you get' (WYSIWYG), so any errors in the design will be faithfully synthesised as errors in the final circuit. It is up to you as the designer to test carefully. Furthermore, this testing should be carried out on the RTL model prior to synthesis. This is where most errors can and should be found.

Diagnosing errors in synthesiser-generated netlists is almost impossible. Waiting until a gate-level model is obtained from synthesis is, in any case, far too late in the design cycle to start checking a design's integrity. The only checking that should be carried out at this late stage is go/no-go testing of the circuit behaviour and checks to ensure that timing problems have not arisen from the synthesiser's mapping to gates.

Because test benches are not synthesised, the full scope of the VHDL language is available for writing them. The restrictions described in the other chapters for writing synthesisable models do not apply when writing test benches, and a number of forms of VHDL will be used that have not been described before.

VHDL for Logic Synthesis, Third Edition. Andrew Rushton.
© 2011 John Wiley & Sons, Ltd. Published 2011 by John Wiley & Sons, Ltd.

The methods used in writing test benches will be developed gradually over a number of examples, starting with a simple combinational circuit to introduce the main topics and then using gradually more complicated examples to introduce new concepts.

13.2 Combinational Test Bench

The first example will show the basic structure that is recommended for a test bench. The example is simple so that understanding of the example does not block the understanding of the test bench.

The task to be solved is to test a simple multiplexer that has the following interface:

```
library ieee;
use ieee.std_logic_1164.all;
entity mux is
  port (in0, in1, sel : in std_logic; z : out std_logic);
end;
```

There is no need to know the structure of the circuit under test. It is sufficient to say that input sel selects input in0 when sel = 0 and input in1 when sel = 1.

The first stage is to create an entity with no ports and an architecture with a component instance of the circuit under test contained within it. The test bench forms a wrapper that completely encloses the circuit under test, which is why there are no ports on the test bench entity. Also in the architecture will be a signal to connect to every port of the circuit under test. The convention is to use a signal with the same name and type as the component port so that the test bench is readable.

The recommended practice for naming the entity and architecture is to give the entity the same name as the circuit under test but with _test appended. The architecture should have a name that identifies it as a test bench and not a behavioural description – test_bench is a good choice of architecture name.

```
entity mux_test is
end;

library ieee;
use ieee.std_logic_1164.all;
architecture test_bench of mux_test is
  signal in0, in1, sel, z : std_logic;
begin
  CUT: entity work.mux port map (in0, in1, sel, z);
end;
```

This example is of course incomplete – so far it just consists of the circuit under test (CUT) as a component using direct binding and the signals connected to its inputs and outputs.

Note that an entity with no ports has no port specification at all, not just an empty port specification, which would be an error.

The next stage is to build up a test set and a process to generate test stimuli at the required times. Both of these use features of VHDL not previously used.

The recommended way of defining a test set that can be applied to most designs is by declaring a constant array of stimuli. Each stimulus is a record containing the values to be applied to each of the inputs for a single test. The record for the stimulus type contains one field for each input to the CUT. The convention is once again to name each field after the circuit port that it will be driving. The type declaration goes in the architecture declarative part (before the `begin`). The record type is:

```
type sample is record
   in0 : std_logic;
   in1 : std_logic;
   sel : std_logic;
end record;
```

Then there needs to be an array-type definition so that it is possible to create an array of this record type. The array-type declaration will be unconstrained so that an array containing a test set of any size can be created:

```
type sample_array is array (natural range <>) of sample;
```

Now that the type declarations are complete, it is possible to declare a constant array of samples containing the stimulus data for the test bench.

```
constant test_data : sample_array :=
   (
     ('0','0','0'),
     ('0','1','0'),
     ('1','0','0'),
     ('1','1','0'),
     ('0','0','1'),
     ('0','1','1'),
     ('1','0','1'),
     ('1','1','1')
   );
```

This constant declaration needs a bit of explaining. The constant is called `test_data` and is of type `sample_array`. No range has been given for `test_data`, even though the type is an unconstrained array, because in constant declarations the VHDL analyser can work out the range of an unconstrained array from the size of value given to it. It is good practice to allow the analyser to do this, since the test set will probably change often and it could become tedious remembering to re-count how many tests there are just to give the array the correct range when the analyser can do it for you.

The value of the constant has been given as an aggregate of aggregates. The outer level of aggregate corresponds to the array, which in this case contains eight elements. Each element is itself an aggregate because the element type is a record. The inner level of aggregate therefore corresponds to the sample record, which in turn contains three `std_logic` values, represented here as character literals.

The test set in this case carries out an exhaustive test of all the possible input permutations. This method of storing the test data in a constant array makes adding or removing tests extremely easy.

The final stage of building this test bench is to write a process that will apply this test set to the circuit under test. The test driver process for this example is:

```
process
begin
  for i in test_data'range loop
    in0 <= test_data(i).in0;
    in1 <= test_data(i).in1;
    sel <= test_data(i).sel;
    wait for 10 ns;
  end loop;
  wait;
end process;
```

The process contains a `for loop` that steps through the `test_data` array one element at a time. The use of the `range` attribute means that the loop will automatically adjust to the size of the test set if samples are added or removed. Within the loop, the signals that were connected to the ports of the circuit under test are given values from the current sample.

It is worth looking at one of the assignments a little more closely; consider the assignment to `in0`:

```
in0 <= test_data(i).in0;
```

Note that the constant array `test_data` is first indexed by the loop constant i to give one of the sample records, which is then accessed by element selection (`.in0`) to get at the individual `std_logic` value. The other assignments are similar.

Once the circuit inputs have been given their sample values, the process waits for a time delay of 10 ns. This `wait` statement is essential, because signals are not updated until the process pauses on a `wait` statement. When the process pauses at the `wait` statement, the signals are updated and so the sample values will be applied to the circuit under test. The CUT will then respond to the new stimuli, creating a new value on the output ports. This response will also be visible in the waveform display of the simulator. The delay time gives the CUT time to respond before the next sample is applied.

In principle, a combinational RTL model only needs delta time to respond, but it is good practice to use the specified time delay of the final circuit so that the waveform output has the correct timing for the final design. In this way it is easier to understand the waveform and check its correctness.

The final feature of the test process is the final unconditional `wait` statement after the loop has completed. If the `wait` statement was not there, the process would restart and carry out the test again (and again...). The `wait` statement stops the process completely once all the samples have been applied and therefore stops the simulation. A simulator will always stop when there are no more transactions to be processed.

It is good practice to write test benches to be self-stopping in this way so there is no need to specify a time limit to the simulator. Then, if new tests are added or redundant ones removed, it is not necessary to calculate the simulation time required to carry out all the tests. The test bench will effectively control the simulation time for you.

In summary, here's the test bench in its entirety so that the various parts, which have been presented disjointly, can be seen in context:

```
entity mux_test is
end;

library ieee;
use ieee.std_logic_1164.all;
architecture test_bench of mux_test is
  type sample is record
    in0 : std_logic;
    in1 : std_logic;
    sel : std_logic;
  end record;
  type sample_array is array (natural range <>) of sample;
  constant test_data : sample_array :=
    (
      ('0','0','0'),
      ('0','1','0'),
      ('1','0','0'),
      ('1','1','0'),
      ('0','0','1'),
      ('0','1','1'),
      ('1','0','1'),
      ('1','1','1')
    );
  signal in0, in1, sel, z : std_logic;
begin
  process
  begin
    for i in test_data'range loop
      in0 <= test_data(i).in0;
      in1 <= test_data(i).in1;
      sel <= test_data(i).sel;
      wait for 10 ns;
    end loop;
    wait;
  end process;
  CUT: entity work.mux port map (in0, in1, sel, z);
end;
```

This test bench is essentially complete, for a combinational circuit. However, there are a number of enhancements that must be made to cater for clock signals, asynchronous resets and so on. It is also possible to vary the time delay between samples by including the delay time in the sample array. The simplest enhancement though, is to check the responses of the circuit under test. This will be dealt with next.

13.3   Verifying Responses

A test bench that only generates stimuli is useful when designing a circuit and exploring its behaviour, since it is easier to examine the response on a simulator waveform display and confirm that it is right than to hand calculate the correct result. However, once a circuit is complete, it is a good idea to include the response data in the test bench too. This means that the circuit can be tested again at any stage in the future with just a simple go/no-go test. Such a test bench can then become part of a regression test suite. If the test bench finds no response errors, then the circuit is still working correctly. This is particularly useful if design changes have been made to the implementation and it is desired that no changes are made to the behaviour.

The response data is incorporated into the test set by extending the sample record. The record becomes:

```
type sample is record
  in0 : std_logic;
  in1 : std_logic;
  sel : std_logic;
  z : std_logic;
end record;
```

The output signal z has now got an entry in the sample record. Then, the response data must be added to the test set, which becomes:

```
constant test_data : sample_array :=
  (
    ('0','0','0', '0'),
    ('0','1','0', '0'),
    ('1','0','0', '1'),
    ('1','1','0', '1'),
    ('0','0','1', '0'),
    ('0','1','1', '1'),
    ('1','0','1', '0'),
    ('1','1','1', '1')
  );
```

To make it clear, the stimulus and response fields have been separated by an extra space. This is just a convention for readability.

The final change is to include a response check in the test process. The response test is an assertion using the assert statement to check that the actual response is the same as the expected response. If the response does not match, then an assertion error will be reported.

```
process
begin
  for i in test_data'range loop
    in0 <= test_data(i).in0;
    in1 <= test_data(i).in1;
    sel <= test_data(i).sel;
    wait for 10 ns;
    assert z = test_data(i).z
      report "output z is wrong!"
      severity error;
  end loop;
  wait;
end process;
```

If no error reports are printed by the simulation, then all responses are known to be correct. There is therefore no need to examine the waveform display to check the circuit behaviour.

Note that the assertion is placed after the `wait`. The sequence of events for each sample is worth clarifying. First, the signals connected to the inputs of the circuit under test are given the values of a sample stimulus. These signals are updated with their new values when the process pauses at the `wait` statement. In this case the `wait` statement pauses for 10 ns to give the circuit time to respond. Then the response is checked against the expected response.

An assertion is a test for a true condition – so if the condition is true nothing happens. But if the condition is false, the report is printed. This is the opposite logic to an if statement used to check for an error. You sometimes see the `assert` statement used in its abbreviated `report` form combined with an `if` statement:

```
if z /= test_data(i).z then
  report "output z is wrong!" severity error;
end if;
```

Note that the logic of the test is inverted in this case.

This is a matter of personal taste. The assertion is saying "this condition should hold true" and some designers prefer to think that way, using positive logic. The if statement version is saying "this condition should not happen" and other designers prefer to think that way, using negative logic. Sometimes these can be combined where some part of the condition is negative and another part positive. For example:

```
if initialisation_complete then
  assert z = test_data(i).z
    report "output z is wrong!"
    severity error;
end if;
```

This disables the test while a circuit is being initialised, then enables the test once initialisation is complete. This logic is easier to follow than just an assertion:

```
assert (not initialisation_complete) and (z = test_data(i).z)
  report "output z is wrong!"
  severity error;
```

If the circuit under test is a behavioural model, then it will have a zero time delay, so the choice of wait time is completely arbitrary. However, if there is a known time requirement, it is good practice to run the circuit at actual speed. This means that the same test bench can be used to test the synthesised circuit.

13.4 Clocks and Resets

So far, the examples have been combinational circuits. For synchronous RTL designs a clock and optionally a reset control will be needed. So the test bench can be extended to generate these signals.

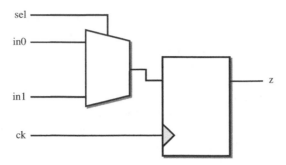

Figure 13.1   Registered multiplexer.

The best way to perform clock generation is within the test process. This has the advantage that the clock will stop when the tests stop. Remember that the test bench should be written so that the simulation is self-stopping, and it is in the generation of clocks that this guideline is usually broken. This is because the most common implementation of a clock generator is as a separate process that simply toggles the clock signal periodically for ever. Such a clock generator will cause the simulator to keep simulating for ever.

It is very easy to implement a single clock within the test process. Multiple clocks can also be accommodated without much extra difficulty. This example, however, will be restricted to a single clock.

The only change in the test bench to test clocked circuits is in the test process. There are no other changes. In particular, there are no changes to the test set.

Consider a simple clocked circuit under test. To keep the example as simple as possible, the example is nothing more than the multiplexer introduced in the previous example, but now with a registered output. The circuit is illustrated in Figure 13.1.

The register in the multiplexer is a rising edge sensitive register. This changed circuit obviously will require the addition of a clock signal to the entity:

```
library ieee;
use ieee.std_logic_1164.all;
entity dmux is
  port (in0, in1, sel, ck : in std_logic; z : out std_logic);
end;
```

There will also be corresponding changes to the component instance and the set of local signals declared in the test-bench architecture:

```
entity dmux_test is
end;

library ieee;
use ieee.std_logic_1164.all;
architecture test_bench of dmux_test is
  signal in0, in1, sel, ck, z : std_logic;
begin
  CUT: entity work.dmux port map (in0, in1, sel, ck, z);
end;
```

The test set will be declared exactly as in the previous example, so that part of the test bench will not be reproduced. The final stage in writing this test bench is the test driver process, which is similar to the previous version of the test process, but now with a clock generator:

```
process
begin
  for i in test_data'range loop
    in0 <= test_data(i).in0;
    in1 <= test_data(i).in1;
    sel <= test_data(i).sel;
    ck <= '0';
    wait for 5 ns;
    ck <= '1';
    wait for 5 ns;
    assert z = test_data(i).z
      report "output z is wrong!"
      severity error;
  end loop;
  wait;
end process;
```

Note that this process sets up the new stimuli on the falling (inactive) edge of the clock. Therefore, no change in the register output will happen at this time. The reason for doing this is to keep the changes in data inputs separate from the active clock edge so that there are no problems caused when data inputs change at exactly the same time as the register is clocked. This can be just as much a problem with behavioural designs as it is with gate-level designs. In behavioural designs, the clock signal will typically propagate through fewer intermediate signal assignments than the datapath, so can arrive at the register process one or more delta cycles before the data values. This means that the correct data values will not be loaded into the register.

The two assignments to the clock, with a wait statement after each one, means that a complete clock cycle is generated for each time round the test loop. The response is only checked 5 ns after the rising (active) edge of the clock. The exact relationships between the two clock edges and the sample time can be adjusted by varying the delay times in the wait statements.

If the circuit needs an asynchronous reset, this can be added before the test loop:

```
process
begin
  rst <= '1';
  wait for 5 ns;
  rst <= '0';
  wait for 5 ns;
  for i in test_data'range loop
```

This only works for an asynchronous reset since the clock is not running during the reset phase. For a synchronous reset, a combination of clock generator and reset generator can be used. For example, to set the reset signal high for 5 clock cycles:

```
process
begin
  rst <= '1';
```

```
for i in 1 to 5 loop
  ck <= '0';
  wait for 5 ns;
  ck <= '1';
  wait for 5 ns;
end loop;
rst <= '0';
for i in test_data'range loop
```

This implements a synchronous reset.

13.5   Other Standard Types

All the examples so far have been based on type std_logic. However, there is no restriction on the type of ports on the circuit under test. This test-bench writing technique can be used with any port type. The key is to simply match the types in the sample record to the types of the ports on the circuit under test. Even array ports can be handled in this way.

As an example, consider the following circuit under test, which is a combinational circuit that counts the number of bits set in the input bus.

```
library ieee;
use ieee.std_logic_1164.all;
use ieee.numeric_std.all;
entity count_ones is
  port (a : in std_logic_vector (15 downto 0);
        count : out unsigned(4 downto 0));
end;
```

The test bench for this circuit is much the same as for the multiplexer example – the count_ones circuit is combinational so there is no clock generation part to it. The main difference is that the test set and the internal signals of the test bench should be of types that match the port types of the circuit under test.

```
entity count_ones_test is
end;
library ieee;
use ieee.std_logic_1164.all;
use ieee.numeric_std.all;
architecture test_bench of count_ones_test is
  type sample is record
    a : std_logic_vector (15 downto 0);
    count : unsigned(4 downto 0);
  end record;
  type sample_array is array (natural range <>) of sample;
  constant test_data : sample_array :=
    (
      ("0000000000000000", "00000"),
      ("0000000000001111", "00100"),
      ("0000111100000000", "00100"),
      ("0000111111110000", "01000"),
      ("0001001000110100", "00101"),
      ("0111011001010100", "01000"),
      ("1111111111111111", "10000")
    );
```

```
    signal a : std_logic_vector (15 downto 0);
    signal count : unsigned(4 downto 0);
begin
  process begin
    for i in test_data'range loop
      a <= test_data(i).a;
      wait for 10 ns;
      assert count = test_data(i).count
        report "output count is wrong!"
        severity error;
    end loop;
    wait;
  end process;
  CUT: entity work.count_ones port map (a, count);
end;
```

Note that the sample data for the input has been written in binary using the string notation. It is also possible to write the samples in hexadecimal, using the bit-string notation, rather than writing out the `std_logic_vector` in its full width using the string notation. The hexadecimal form of the sample array would then be:

```
constant test_data : sample_array :=
  (
    (X"0000", "00000"),
    (X"000F", "00100"),
    (X"0F00", "00100"),
    (X"0FF0", "01000"),
    (X"1234", "00101"),
    (X"7654", "01000"),
    (X"FFFF", "10000")
  );
```

The values for the response are still expressed in binary, since the count output is five bits wide and hexadecimal bit strings can only be used for signals that are a multiple of four bits wide. However, the test bench could be written using integer response data, since the response is meant to represent an integer value, and then the type conversion `to_unsigned` can be used to convert the expected response into the same type as the output of the circuit under test. This approach makes the test bench more readable and therefore less prone to error. The resulting sample record and test data are shown below.

```
type sample is record
  a : std_logic_vector (15 downto 0);
  count : integer;
end record;
constant test_data : sample_array :=
  (
    ("0000000000000000", 0),
    ("0000000000001111", 4),
    ("0000111100000000", 4),
    ("0000111111110000", 8),
    ("0001001000110100", 5),
    ("0111011001010100", 8),
    ("1111111111111111", 16)
  );
```

The assertion that checks the expected response now needs to include the type conversion:

```
assert count = to_unsigned(test_data(i).count,count'length)
  report "output count is wrong!"
  severity error;
```

13.6 Don't Care Outputs

Many circuits do not generate valid data outputs on every clock cycle. For example, a pipeline will generate meaningless outputs for the first few clock cycles whilst the pipeline flushes. Similarly, circuits designed to calculate a result over multiple clock cycles will generate invalid outputs whilst calculating intermediate values. There are plenty of other examples where the output of a circuit is not valid for some reason. In these cases, the test bench should be designed to ignore the invalid outputs and only check the valid responses.

One way of achieving this is to add an extra field to the sample record that indicates whether a valid response is expected as a result of that sample. This valid flag is a boolean, which can be set to true when a correct response is expected and false otherwise.

Consider an example circuit that is simply a 1-bit register pipeline that delays its input by three clock cycles. This means that the pipeline will need to be flushed at the start of the simulation and so the first two responses will be invalid. The pipeline example has the following interface:

The test bench for this circuit is similar to the previous ones, so only the differences due to the introduction of the valid flag will be shown.

The first change is to the sample record:

```
type sample is record
  d : std_logic;
  q : std_logic;
  valid : boolean;
end record;
```

Then, for each sample in the test set, the flag should be set according to whether the response is valid for that sample or not:

```
constant test_data : sample_array :=
  (
    ('1', '0',false),
    ('0', '0',false),
    ('1', '1',true),
    ('1', '0',true),
    ('0', '1',true),
    ('1', '1',true)
  );
```

The final change to the test bench is to switch off testing of the response data when the valid flag is false. The most readable way of doing this is by enclosing the assertion in an if statement:

```
if test_data(i).valid then
  assert test_data(i).q = q report "q is invalid" severity error;
end if;
```

An assertion error will be reported only if the condition on the `if` statement is `true` and the assertion condition evaluates to `false`. If the `valid` flag is `false`, then the assertion will not be tested. In other words, checking of the response is disabled. If, on the other hand, the `valid` flag is `true`, then the assertion will verify whether the `q` output matches the expected response.

An alternative approach can be used when the response is one of the synthesis types introduced in Chapter 6. These packages define a set of functions called `std_match` for each of these types that can be used to test the response against an expected value that contains `'-'` (don't care) values. This can be used instead of the `valid` flag. To switch off response checking, just make the expected response data contain `'-'` values. This can be used with some finesse – individual bits can be ignored by placing `'-'` values in the expected response fields, so only the remaining bits get tested.

The sample record now needs no `valid` flag:

```
type sample is record
   d : std_logic;
   q : std_logic;
end record;
```

Then, for each sample in the test set, the response should be set to an actual value or to `'-'` according to whether the response is valid for that sample or not:

```
constant test_data : sample_array :=
   (
      ('1', '-'),
      ('0', '-'),
      ('1', '1'),
      ('1', '0'),
      ('0', '1'),
      ('1', '1')
   );
```

The `assert` statement then uses the `std_match` function directly:

```
assert std_match(test_data(i).q, q)
   report "q is invalid"
   severity error;
```

To be more specific, the `std_match` functions return false if any of the bits in either of the values being compared contains any of the metalogical values `'U'`, `'X'`, `'Z'` or `'W'`. The array forms of the function also return false if the arguments are of different sizes. The functions returns true if the two arguments match element by element, with the value `'-'` treated as a wildcard that matches any of the remaining values. Note, however, that `'L'` only matches `'L'` and not `'0'` – there is no matching of weak values with their strong equivalents.

The biggest drawback of these functions is that the `'Z'` value is considered a metalogical value that cannot match anything, so the test results in an error even if a `'Z'` is in the expected result – in other words `'Z'` does not match a `'Z'`. This makes the functions difficult to use for testing tristate buses. However, this limitation aside, the functions are very useful for writing test benches.

13.7   Printing Response Values

So far, when an error has been detected, all the examples have simply printed the message
"q is not valid" or some such message, which is not very helpful for diagnosing the
problem. The simulator helps by also printing the simulation time of the error, but it can
be difficult to translate this simulation time into a diagnosis of which test in the test set has
failed. It is far more useful to print out the expected and actual values.

The key to improving the reporting is to recognise that the report part of the assertion can
be any string expression. This means it is possible to build up a report by concatenating a
series of strings together using the "&" operator.

In VHDL-2008, there is a complete set of overloaded functions for each type called
to_string. This provides a way of converting any type to its string representation.

Note: if you are still using VHDL-1993, the to_string functions are to be found in the
additions packages that can be downloaded and installed along with the fixed-point and
floating-point packages as explained in Chapter 6. For example standard_additions
contains the to_string functions for the types defined in package standard: integer,
bit, real, boolean, character and time. Similarly, numeric_std_additions
contains to_string functions for signed and unsigned that in VHDL-2008 are found
in numeric_std. These additions packages can be compiled into ieee_proposed and
made available to the test bench with a use clause.

The to_string function for std_ulogic (and therefore also for std_logic) is found
in std_logic_1164 (or std_logic_1164_additions) and has the following
interface:

```
function to_string (arg : std_ulogic) return string;
```

The assertion can then have a more sophisticated report part:

```
assert test_data(i).q = q
   report "q: expected = " & to_string(test_data(i).q) &
          ", actual = " & to_string(q)
   severity error;
```

The string to be printed in the report is built up out of sub-strings that are concatenated using
the "&" operator.

A further convenience would be to have a way of displaying vector types such as signed
and unsigned as integers. The best way to do this is to use type conversions from the array
types to integer. Then, the to_string function for integer can be used to print it out.
Furthermore, it could be useful sometimes to be able to display integer values as an array of
bits and this can be achieved by using type conversions that work the other way round. The type
conversions are explained in Section 6.7.

For example, a signed value can be converted to integer and then displayed using the
to_string function for integer to give a decimal representation:

```
assert q = data(i).q
   report "q: expected = " & to_string(to_integer(q)) &
          ", actual = " & to_string(to_integer(data(i).q))
   severity error;
```

It is also possible to print values of the synthesisable array types (std_logic_vector, signed, unsigned, sfixed, ufixed and float) as hexadecimal or octal strings, using the to_hstring and to_ostring variants of these functions. For example, to print an unsigned value in hexadecimal:

```
assert q = data(i).q
  report "q: expected = " & to_hstring(q) &
         ", actual = " & to_hstring(data(i).q)
  severity error;
```

For consistency, there is a to_bstring that is identical to the to_string function.

Finally, there are long forms of these functions: to_binary_string, to_hex_string and to_octal_string for those that find long names add to the readability of the model.

13.8   Using TextIO to Read Data Files

All the examples of test benches so far have been completely self-contained. That is, all the stimulus and response data is built into the test bench itself. There are situations where it makes more sense to read the stimulus and response data from a data file. An example of such a situation is where the test data has been generated from an automated tool. Rather than go through the tedious task of editing this test data into the test bench, it is better to read it straight from the data file during simulation.

VHDL has an I/O system that allows this to be done. The built-in procedures for performing text I/O are contained in the appropriately named package textio, which is part of the VHDL standard and so is found in library std.

Package textio provides just enough functionality to read data files, so there isn't much room for ambiguity about how to use the package. However, it should be noted that the text I/O in VHDL is quite different from I/O in other (software) languages.

The textio package is listed in Appendix A.10, so that can be used as a reference. The basic subprograms that will be used in this example are:

```
procedure file_open (file f : text;
                     name : in string;
                     kind : in file_open_kind := read_mode);
procedure file_close (file f : text);
function endfile (file f : text) return boolean;
procedure readline (file f : text; l : inout line);
procedure read (l : inout line; value : out type);
```

There are two special types used in these subprograms: text and line. Type text is a type that represents a text file, whilst type line is a type that represents a line of text from that file.

The basic sequence of events in processing text files is, first to open the file, then as long as there is text to process, read a whole line of text from the file. Text can only be read a whole line at a time using the readline procedure. Once a text line has been read, then a set of read procedures can be used to disassemble the text line into its elements and convert the elements into VHDL types. In the simplified interface above, this is represented by the read procedure's last parameter where type can be replaced by any VHDL type supported by textio.

In order to illustrate the use of textio in practice, the original test bench from the beginning of the chapter, which was used to illustrate the basic test-bench structure, will be rewritten to use it. Initially, the example will be written for ports of type bit and later it will be rewritten to use type std_logic. This two-stage introduction is necessary because std_logic is not directly supported by textio so an example using bit, which is directly supported, is used to cover the basics.

Like all the other test benches, the first stage is to declare the component instance with signals connected to each port. This gives a skeleton test bench:

```
entity mux_test is
end;

use std.textio.all;
architecture test_bench of mux_test is
  signal in0, in1, sel, z : bit;
begin
  CUT: entity work.mux port map (in0, in1, sel, z);
end;
```

Note that the package textio has been made visible by a use clause so that it can be used in writing the test bench.

Unlike the previous examples, there will be no test set or data structures to store the test data. Therefore, the only thing left to do is to write the test process. The test process is:

```
process
  file data : text;
  variable sample : line;
  variable in0_var, in1_var, sel_var, z_var : bit;
begin
  file_open(data, "mux.dat", read_mode);
  while not endfile(data) loop
    readline (data, sample);
    read (sample, in0_var);
    read (sample, in1_var);
    read (sample, sel_var);
    read (sample, z_var);
    in0 <= in0_var;
    in1 <= in1_var;
    sel <= sel_var;
    wait for 10 ns;
    assert z = z_var report "z incorrect" severity error;
  end loop;
  file_close(data);
  wait;
end process;
```

This process needs some explanation.

First, the file declaration at the start of the process declares a file, which is then opened at the start of the process using the file_open procedure. It is opened with the identifier data of type text and kind read_mode (readable only), which is to be associated with a file called "mux.dat".

The process contains a while loop. The while loop has not been covered before since it cannot be synthesised. A while loop is a loop that continues for as long as its condition remains true. In this case, the condition is: not endfile(data). This expression is first of all a function, endfile, testing whether the end of the file identified as data has been reached. The not inverts the sense of this so that the loop continues for as long as the end of the file has not been reached.

Within the loop, the first step is to read a line of data into the variable `sample`, which is of type `line`. This is done by the call to the procedure `readline`. Then, there follows a series of `read` operations, each of which reads a `bit` from the line. Note that a `read` operation reads from the line, not from the file itself. The `read` operations effectively consume the line, so each one starts reading where the previous read left off. Any attempt to read a value beyond the end of a line or to read a value of the wrong type, will result in an assertion error being raised.

The `read` operations need to be passed a variable to fill with the value read. This must be a variable and not a signal – the parameter is an `out` parameter of kind `variable`. This is why the example has separate `read` operations on intermediate variables that are then immediately assigned to signals.

The outputs of the circuit under test can be compared directly with the variable, as the assertion in the example shows.

Finally, at the end of the process, the file is closed using the `file_close` procedure.

The file itself will contain a series of lines of text, each one containing values of the correct type to match the values being read. In this case, the correct values are 0 or 1, since the type being read is a `bit`. Note that the values are the digits 0 and 1, without any quotes. Any whitespace (spaces or tabs) between the values will be ignored.

For example, a typical data file for this test bench would be:

```
000 0
010 0
100 1
110 1
001 0
011 1
101 0
111 1
```

In this example, for clarity, the stimuli have been grouped together and whitespace has been used to separate the response data.

Consider a test bench that reapplies the same test data twice. This is not necessary for this example, but it could be useful in a design where the same test data is to be applied to a circuit with a number of different modes. The same data file could be reused for each mode. However, in this example the same data set will simply be applied twice by enclosing the tests in an outer `for` loop.

```
process
  file data : text;
  ...
begin
  for i in 1 to 2 loop
    file_open(data, "mux.dat", read_mode);
    while not endfile (data) loop
      ...
    end loop;
    file_close(data);
  end loop;
  wait;
end process;
```

13.9   Reading Standard Types

Package `textio` is not limited to use with type `bit`. There are `read` operations defined for reading all of the synthesisable types from package `standard` – specifically:

```
bit
bit_vector
boolean
character
string
integer
```

There is also support for non-synthesisable types:

```
real
time
```

These can be useful within a test bench even though they cannot be used in a synthesisable design.

When reading the character types, `bit` and `character`, the `read` operation simply reads a single character from the line and then tries to match it to the type. For `bit`, any leading whitespace (spaces and tabs) is skipped first. For `character`, no skipping is done, since the whitespace characters are valid members of `character`.

When reading the string types, `bit_vector` and `string`, the `read` operation reads as many characters as necessary to fill the variable passed to the `read` procedure and tries to match them to the element type. In other words, the `read` operation adapts to the size of the variable. Once again, for `bit_vector`, leading whitespace is skipped before the first valid character (but not for subsequent characters in the string value), whereas for `string` it is not because the whitespace characters are valid members of a `string`.

To give an example, the following would allow the reading of an 8-bit `bit_vector`:

```
process
   variable byte : bit_vector (7 downto 0);
   ...
begin
   ...
   read(sample, byte);
```

When reading type `integer`, an optional sign followed by numeric digits are read until a non-numeric digit is found and then the result is converted to `integer`. Any leading whitespace will be skipped first. Generally, the best way to terminate an `integer` in a data file is to use whitespace.

The final type of use in writing test benches, which can be read but that is not synthesisable, is type `time`. This means that even the delays used in the `wait` statement that controls the testing can be read from the data file. The following is a template for reading a `time` and then using it in a `wait` statement:

```
process
   variable delay : time;
   ...
```

```
begin
  ...
  read(sample, delay);
  wait for delay;
```

The file format for times is the same as in VHDL – a number representing the time followed by a space and then a string representing the units. A number on its own is not a valid time and therefore will result in an error. Also, the space between the number and the unit is required. For example, here is a data file with the time value at the end of the line:

```
000 0 10 ns
010 0 10 ns
100 1 10 ns
110 1 10 ns
001 0 10 ns
011 1 10 ns
101 0 10 ns
111 1 10 ns
```

Note that the times are relative times – in other words, each delay is an additional 10 ns in this example. To use absolute times (measured from the start of simulation), simply subtract the time now from the delay in the wait statement:

```
process
  variable delay : time;
  ...
begin
  ...
  read (sample, delay);
  wait for delay - now;
```

The word now refers to a function with no parameters that returns the current simulation time. Note that if the delay is less than now, the result of the subtraction is negative, and this will raise an error because it is not possible to wait for a negative time. It is therefore essential that the samples in the file are sorted into time order:

```
000 0 10 ns
010 0 20 ns
100 1 30 ns
110 1 40 ns
001 0 50 ns
011 1 60 ns
101 0 70 ns
111 1 80 ns
```

13.10   TextIO Error Handling

There are in fact two read operations provided for each type:

```
procedure read(l : inout line; value : out type);
procedure read(l : inout line; value : out type;
               good : out boolean);
```

The basic `read` operation that was used in the previous example was the first of these procedures. If an error occurs when trying to read the type, an assertion of severity error will be raised.

The second version of the `read` procedure returns a `boolean` flag called `good` to signify whether the read was successful or not, instead of raising an error. This can be useful to allow the test bench to skip over, for example, blank lines or comments in the data file. To illustrate this, here's a simple modification that allows the test bench to skip over a line if the first read of a `bit` fails:

```
process
   file data : text;
   variable sample : line;
   variable in0_var, in1_var, sel_var, z_var : bit;
   variable OK : boolean;
begin
   file_open(data, "mux.dat", read_mode);
   while not endfile (data) loop
      readline (data, sample);
      read (sample, in0_var, OK);
      if OK then
         read (sample, in1_var);
         read (sample, sel_var);
         read (sample, z_var);
         in0 <= in0_var;
         in1 <= in1_var;
         sel <= sel_var;
         wait for 10 ns;
         assert z = z_var report "z incorrect" severity error;
      end if;
   end loop;
   file_close(data);
   wait;
end process;
```

This would then allow the data file to be annotated with comments and formatted for readability:

```
# test set for the exhaustive simulation of MUX
# in0 in1 sel z
    0   0   0  0
    0   1   0  0
    1   0   0  1
    1   1   0  1
    0   0   1  0
    0   1   1  1
    1   0   1  0
    1   1   1  1
```

It is also possible to append comments to each line, since the `read` operations only read up to the last value in the sample. You are not obliged to read the whole line before moving on to the next, so any comment strings at the end of the line can simply be ignored.

13.11   TextIO for Synthesis Types

The built-in read operations covered by `textio` only cover the standard built-in types of VHDL. They do not cover some of the other types in common use and, in particular, they don't cover `std_ulogic` or any of the synthesis types. This section describes standard extensions to `textio` for the synthesis types.

In the original standards for the synthesis packages `std_logic_1164` and `numeric_std`, no I/O operations were defined. I/O was provided for the `std_logic_1164` types by many vendors providing a non-standard I/O package called `std_logic_textio`. However, being non-standard this package could not be guaranteed to be available. Package `std_logic_textio` is now obsolete and its use is deprecated.

In VHDL-2008 all of the synthesis packages have I/O operations in the packages themselves.

In VHDL-1993, the compatibility versions of the new packages `fixed_pkg` and `float_pkg` have I/O operations built-in. For packages `std_logic_1164` and `numeric_std`, additions packages have been provided in the same way and from the same source as the fixed-point and floating-point packages described in Chapter 6.

So, the following types have I/O procedures. The VHDL-2008 package and, where appropriate, the VHDL-1993 additions package providing the I/O operations for older systems is listed with the types covered by that package:

```
std_logic_1164 (VHDL-1993: std_logic_1164_additions)
  std_ulogic
  std_logic_vector
  std_ulogic_vector
numeric_std (VHDL-1993: numeric_std_additions)
  unsigned
  signed
fixed_pkg
ufixed
 sfixed
float_pkg
 float
```

These packages provide the following procedures:

```
procedure read (l : inout line; value : out type);
procedure read (l : inout line; value : out type;
                 good : out boolean);

procedure write (l : inout line; value : in type;
                  justified : in side := right;
                  field : in width := 0);
```

Where *type* is one of the synthesis types, depending on the I/O package.

When reading the basic type, `std_ulogic`, the read operation simply reads a single character from the line and then tries to match it to the type. Any leading whitespace (spaces and tabs) is skipped first.

Note that the I/O routines for `std_ulogic` can also be used for the more commonly used `std_logic`, since `std_logic` is simply a subtype of `std_ulogic`.

Furthermore, the range of I/O operations has been extended to provide binary, octal and hexadecimal support for bit-array types. These packages also provide the following procedures for the array types only:

```
procedure oread (l : inout line; value : out type);
procedure oread (l : inout line; value : out type;
                 good : out boolean);

procedure hread (l : inout line; value : out type);
procedure hread (l : inout line; value : out type;
                 good : out boolean);

procedure owrite (l : inout line; value : in type;
                  justified : in side := right;
                  field : in width := 0);

procedure hwrite (l : inout line; value : in type;
                  justified : in side := right;
                  field : in width := 0);
```

When reading the array types the `read` operation reads as many characters as necessary to fill the variable passed to the `read` procedure and tries to match them to the element type. Once again, leading whitespace is skipped before the first valid character but not between characters. In octal (oread) and hexadecimal (hread) forms, the value read from the file is expanded into a bit-string and assigned to the variable. The variable must be a multiple of 3 bits long for octal and four bits for hexadecimal.

13.12   TextIO for User-Defined Types

The recommendation for logic synthesis is to use the synthesis types for most datapaths, so it is rare to need to write I/O procedures. However, sometimes user-defined types are needed and then I/O procedures can be provided for those types as well.

The most difficult `read` procedure to create is a `read` procedure for a named enumeration such as is used for writing FSMs (Section 12.2). A shortcut would be to read an `integer` that represents the position value and then convert the value to the target enumeration type using the `val` attribute. It has the disadvantage that the values of the enumeration type are stored in the data file as integer values, which is not very readable, but it is a quick solution.

However, it can be useful sometimes to allow the enumeration values to be written in the data file as string values, so in this case a means must be found of matching the enumeration values with the string values. Indeed, this is how type `boolean` is read. The file representation of `boolean` uses the strings `"true"` and `"false"` (without the quotes of course). To illustrate the method for reading enumeration values, here's a possible implementation for a `read` procedure for reading the following enumeration type:

```
type light_type is (red, amber, green);
```

The `read` procedure will be written to skip any preceding whitespace, then it will try to match the string representation of the enumeration values with the contents of the line.

```
procedure read (l : inout line; value : out light_type;
                good : out boolean) is
begin
  skipwhite(l);
  if (l.all'length >= 3 and
      l.all(l.all'left to l.all'left + 2) = "red")
  then
    value := red;
    skip (l, 3, good);
  elsif (l.all'length >= 5 and
         l.all(l.all'left to l.all'left + 4) = "amber")
  then
    value := amber;
    skip (l, 5, good);
  elsif (l.all'length >= 5 and
         l.all(l.all'left to l.all'left + 4) = "green")
  then
    value := green;
    skip (l, 5, good);
  else
    value := red;
    good := false;
  end if;
end;
```

This procedure uses the knowledge that the line type used in textio is an access type (a pointer in software terminology) to a string. Access types will not be covered in any detail by this book, but are being used here. It is sufficient to know that the .all selection is used to dereference the pointer and access the string itself. In this case, a slice of the line is being compared with string values for the text representing the values of the light_type type. Note that the uppercase representations could easily be incorporated into this procedure.

The read procedure above uses a procedure for skipping whitespace. This works by counting the number of whitespace characters that need skipping and then using uses the built-in read procedure for string to read a string of that length, effectively consuming the whitespace from the line in one go. This is more efficient (i.e. faster) than skipping one character at a time. The skipwhite procedure is:

```
procedure skip (l : inout line; length : in natural) is
  variable str : string (1 to length);
  variable good : boolean;
begin
  read (l, str, good);
end;
procedure skipwhite (l : inout line) is
  variable length : natural := 0;
begin
  for i in l.all'range loop
    exit when l.all(i) /= ' ' and l.all(i) /= HT;
    length := length + 1;
  end loop;
```

```
      if length > 0 then
        skip (l, length);
      end if;
    end;
```

Note that the `skipwhite` procedure skips both spaces and tabs (a tab is indicated by the character literal HT, which stands for Horizontal Tab).

However, the `read` procedure is clearly clumsy since the same logic is used in every comparison. It would be better implemented as a loop:

```
procedure do_compare(l : inout line;
                     value : in light_type;
                     found : out boolean) is
  constant image : string := value'image;
begin
  if ((l.all'length >= image'length) and
      (l.all(l.all'left to l.all'left + image'length) = image))
  then
    skip (l, image'length);
    found := true;
  else
    found := false;
  end if;
end;
procedure read (l : inout line; value : out light_type;
                good : out boolean) is
  variable found : boolean := false;
begin
  good := true;
  skipwhite(l);
  for possible in light_type'range loop
    do_compare(l, possible, found);
    value := possible;
    exit when found;
  end loop;
  if not found then
    good := false;
  end if;
end;
```

The `read` procedure loops through all the possible values of the enumeration type, calling the `do_compare` procedure with each one. If a match is found, then the result is set to the value that caused the match and the loop exits. Otherwise, it keeps looping until all values have been tried. If a match is not found after all the values have been tried, the `good` flag is set false to indicate that the read failed.

The `do_compare` procedure works by first declaring a local constant string that is initialised with the image of the possible value of the type. This is done in this way because the size of the string varies from one call of the procedure to another, and this is the conventional way of capturing an unknown-length array in a local constant so that it can be referenced many times. Then, the comparison of the string value with the contents of the line of the file is performed by the if statement. If this matches, the requisite number of characters in the line is skipped and the `found` flag set.

13.13   Worked Example

13.13.1   Systolic Processor

This example develops a test bench to test the systolic processor developed in Section 10.7. The test data will run the processor through a single calculation and check the results as they shift out of the processor.

The test set will be:

$$
\begin{pmatrix} 1 & 2 & 3 \\ 4 & 5 & 6 \\ 7 & 8 & 9 \end{pmatrix} \begin{pmatrix} 1 \\ 2 \\ 3 \end{pmatrix} = \begin{pmatrix} 14 \\ 32 \\ 50 \end{pmatrix}
$$

The original entity to be tested was:

```
library ieee;
use ieee.std_logic_1164.all;
use ieee.numeric_std.all;
entity systolic_multiplier is
   port (d : in signed (15 downto 0);
         ck, rst : in std_logic;
         q : out signed (15 downto 0));
end;
```

To make the test bench more readable, values will be represented by integer values in the test set. These integer values will then be type converted to type signed, which is the port type of the systolic processor. The conversion will be performed by using the type conversion to_unsigned from package numeric_std. A valid flag will be used since there are many times during the test when the data output is of no interest and there is no way of representing don't cares in the integer representation.

The test bench for this circuit is:

```
entity systolic_multiplier_test is
end;

library ieee;
use ieee.std_logic_1164.all;
use ieee.numeric_std.all;
architecture test_bench of systolic_multiplier_test is

   type sample is record
     d : integer;
     rst : std_logic;
     q : integer;
     valid : boolean;
   end record;
   type sample_array is array (natural range <>) of sample;
   constant test_data : sample_array :=
     (
       (0, '1', 0, false), -- reset
```

```
             (1, '0',  0, false),  -- ld_a11
             (2, '0',  0, false),  -- ld_a12
             (3, '0',  0, false),  -- ld_a13
             (4, '0',  0, false),  -- ld_a21
             (5, '0',  0, false),  -- ld_a22
             (6, '0',  0, false),  -- ld_a23
             (7, '0',  0, false),  -- ld_a31
             (8, '0',  0, false),  -- ld_a32
             (9, '0',  0, false),  -- ld_a33
             (1, '0',  0, false),  -- ld_b1
             (2, '0',  0, false),  -- ld_b2
             (3, '0',  0, false),  -- ld_b3
             (0, '0',  0, false),  -- calc1
             (0, '0',  0, false),  -- calc2
             (0, '0',  0, false),  -- calc3
             (0, '0',  0, false),  -- calc4
             (0, '0', 14, true),   -- calc5
             (0, '0',  0, false),  -- calc6
             (0, '0', 32, true),   -- calc7
             (0, '0',  0, false),  -- calc8
             (0, '0', 50, true)    -- calc9
         );
      signal d, q : signed (15 downto 0);
      signal ck, rst : std_logic;

begin

   process
   begin
     for i in test_data'range loop
       d <= to_unsigned(test_data(i).d, d'length);
       rst <= test_data(i).rst;
       ck <= '0';
       wait for 5 ns;
       ck <= '1';
       wait for 5 ns;
       if test_data(i).valid then
         assert to_unsigned(test_data(i).q, q'length) = q
           report "q: expected = " &
                   to_string(to_unsigned(test_data(i).q)) &
                 ", actual = " & to_string(q);
           severity error;
       end if;
     end loop;
     wait;
   end process;

   CUT: entity work.systolic_multiplier
     port map (d, ck, rst, q);
end;
```

14

Libraries

Libraries are an important concept in VHDL, and yet they seem to cause a lot of problems and confusion amongst VHDL users. For that reason, this short chapter has been included to give a brief overview of libraries and recommendations on how to organise your work in libraries.

14.1 The Library

A library can be regarded as a container that contains compiled design units. A library can also be regarded as the destination for VHDL compilers.

When a VHDL source file is compiled (or *analysed* seems to be the preferred term in VHDL circles) into a VHDL system, it is split into separate design units. If a source file contains, say, ten design units then the file will be split into ten parts. Each design unit is then analysed in isolation and the compiled result, an intermediate form, is saved into a library. At the end of the analysis, the library will contain ten design units.

Also, when a simulation is run, a design unit is simulated directly in its library, so the original source file is not involved in the simulation process. Instead, the model is built by assembling all of the intermediate forms for all of the design units in the model into a runnable model.

Typically, a simulator will analyse VHDL source into object code – so object code is the intermediate form. Then, the simulation consists of linking the object code into a program that is then run. However, this is not the only approach possible and some simulators use different techniques.

A simulator and a synthesiser will both use the same basic mechanism, the main difference being that a design unit is synthesised from the library rather than simulated. However, the synthesiser will use a very different intermediate form from the simulator. This means that simulation and synthesis must be carried out separately in different libraries. Once again, the original source file has no part to play in the synthesis process.

This demonstrates that the libraries of each VHDL tool must be kept separate, because each tool will use a different intermediate form to store design units in its libraries. So, for example, the simulator's library `ieee` will be separate from the synthesiser's library `ieee`.

The main benefit of libraries to the VHDL user is that it allows a natural way of partitioning a design into its subsystems. Each major subsystem can be designed separately

---

VHDL for Logic Synthesis, Third Edition. Andrew Rushton.
© 2011 John Wiley & Sons, Ltd. Published 2011 by John Wiley & Sons, Ltd.

in its own library. This library will contain the subsystem itself plus all the test benches for unit testing of that subsystem. Then, the whole design can be assembled from another library, using the subsystem libraries as resources. This top-level library will contain only references to the other subsystems, not copies, thus ensuring that there is only one definitive copy of each component. The top-level library will also contain the test benches for the overall system tests.

All VHDL systems use this library system because it is part of the VHDL standard. However, the standard says nothing about how the library system should be implemented, it simply states how it should appear in VHDL terms as a purely abstract concept.

In practice, a library is always implemented as a directory that can be placed anywhere on a file system. The directory contains the files that the VHDL system uses to represent the compiled design units. These are generally hidden from the users of the system since there is no need to know anything about what happens in a VHDL library directory. The format of the files are specific to the VHDL system that created them, so a library is not portable between VHDL systems.

14.2   Library Names

Every VHDL library should be given a name. When you refer to a library in VHDL, you refer to it by this name, not its location on the file system. For example, the library `ieee` has already been referred to in previous chapters. It has appeared many times in the example VHDL source code in the form of a library clause:

```
library ieee;
```

The library clause makes all of the design units within the named library available for use in the current design unit. For example, if the library clause appears before an entity, then it means that the entity can use any definitions in that library for defining the ports. It also means that the associated architecture can use the library definitions, since architectures inherit library clauses from their entities.

The library clause is used in conjunction with the use clause. The following combination has been used in many of the examples:

```
library ieee;
use ieee.std_logic_1164.all;
entity ...
```

This first makes the library `ieee` available for use in the following entity and its architecture, then it makes all the definitions from package `std_logic_1164` found in that library available for use. This makes `std_logic`, `std_logic_vector` and all the associated operators available for use.

This is a typical example: making a resource available for use in a design unit is a two-stage process. First, the library containing the resource is made available, and secondly the resource itself is made available.

Library `ieee` is a container for IEEE standard VHDL components that are not directly part of the VHDL standard but that have been standardised for use in application areas that use VHDL. The exact contents of library `ieee` will be discussed in Section 14.4.

The obvious question is, where is library `ieee`? The VHDL definition says nothing about this, and there is no mechanism in the language itself for specifying where to look for a library, so the question of how the VHDL analyser finds a library is simulator or synthesiser specific. All VHDL systems must have some mechanism for mapping the name of a library onto the directory that contains it. Since this is outside the language definition, the methods used vary enormously, although the most common method involves some sort of mapping file that contains a list of library mappings. Typically there will be one such file for each working directory, so you can have a different set of library mappings for each directory you work in.

When you create a library of your own to work in, you should always give it a name. Most VHDL systems enforce this anyway, so you cannot create a nameless library. It is helpful to use a name that gives some clue as to what its purpose is. For example, a library of utilities could be called `utilities`, whilst a library containing the parts of a FIR filter could be called `fir_filter` or even just `fir`.

There are restrictions on the names that can be used. Since the library name will be referred to from VHDL, the name must obey the rules of VHDL names. This means that it must start with a letter, contain only letters, digits and underscores and not have two consecutive underscores. Finally, it must not be a VHDL keyword or the reserved library name `work`. For reference, the VHDL keywords are listed in Appendix B.1. The reserved library name `work` will be explained in the next section.

14.3   Library Work

The reserved library name `work` is used to refer to the library that you are currently working in. It is an alias or pseudonym for the library, since the library will also have a name of its own. You can think of the name `work` as a pointer or reference to a library rather than a library itself. All compilations, simulations or syntheses are carried out in library `work`.

To illustrate the use of libraries, an artificial design will be used. The design will be a two-way communications device consisting of a separate transmitter and receiver. The top level of the design will be the whole communications system that brings together the transmitter and receiver. Finally, such a design will inevitably have common utilities and subcomponents that can be shared between the components.

In this simple example it can be seen that there is a natural division into four libraries. The transmitter will be developed in library `transmitter`, the receiver will be developed in library `receiver`, the utilities will be collected together into a library called `utilities` and then the whole system will be assembled in library `system`. A quick check with Appendix B.2 confirms that none of these names is a VHDL keyword.

Consider the designer working on the transmitter section. That designer will set their `work` library to point to library `transmitter`. Another designer working on the receiver section of the design will set library `work` to `receiver`.

This organisation allows the two designers to work independently. Both will have access to the contents of the `utilities` library and the placement of components in that library will have to be managed between them, but the design work in the two main system libraries `transmitter` and `receiver` can go on in parallel. When the two subsystems are complete, or at least complete enough to try assembling a prototype, the `system` components can be written, incorporating the two subsystems as components in the design. There is no need to copy any of the source code of the two subsystems because their libraries can be used directly in the completed design by referring to them by their names. Thus, the top-level design unit for the receiver can be used as a component in the overall system by simply referring to that component in library `receiver`.

The only difficulty here is when to refer to a library as `work` and when to refer to it by its name. In general, all subcomponents of a subsystem should stay together in one library. They can refer to one another by referring to each other via the library name `work`. The name of the library should not be used if it is the same library as the design unit itself is in. This strategy means that a library can be renamed, for example to resolve a name conflict, and the only source code changes that will have to be made will be in the system design that refers to the library.

For example, suppose the transmitter has a package called `transmitter_types` and a top-level design unit called simply `transmitter`. It is good practice to give the top level of a design the same name, or at least a similar name, to its library. This acts as a reminder of which design unit is the top level. The top-level entity will look something like this:

```
use work.transmitter_types.all;
entity transmitter is
    port (a : in transmitter_data(15 downto 0);
    ...
```

In this case, the package has been referred to via the library name `work` rather than the library name `transmitter` because it is intended that the package should always be compiled into the same library as the entity.

The other special handling when referring to library `work` is the lack of a `library` clause. Library `work` is always available for use in a design, so it is not necessary to have a clause that says `library work`.

This organisation means that, if a library has to change name for any reason, the components within it will still refer to each other correctly and there is no need to change any of the source code. For example, all components in library `receiver` will refer to each other using the name `work`, not the name `receiver`.

14.4 Standard Libraries

There are two libraries that are standard and therefore will be part of any VHDL system that you are likely to use. These two libraries are called `std` and `ieee`. The set of design units found within each library is standardised. In the case of library `std`, the standard packages have remained much the same since VHDL was standardised in 1987, with only slight revisions to the packages themselves in 1993 and 2008. Library `ieee` by contrast has developed and now contains many packages that are applicable to synthesis. Only the synthesis-related packages will be described.

When using the VHDL-1993 compatibility packages to provide VHDL-2008 features to tools that don't support them yet (see Chapter 6 for an explanation), you may also use a temporary library called `ieee_proposed`. Everything in `ieee_proposed` will eventually be merged into library `ieee`.

14.4.1   Library std

Library std is an integral part of the VHDL standard and is so essential that there is an implicit library clause referring to it for any design unit that you compile. This means that you never need a library clause to library std.

There are two design units in library std.

```
package standard
package textio
```

Package standard defines the basic set of standard types. The package is listed in Appendix A.1.

Package standard is so integral to the language (for example an if statement must have a boolean expression) that there is an implicit use clause for every design unit that you compile making package standard available for that design unit.

To summarise, there are two implicit clauses before every design unit that are equivalent to the following VHDL statements:

```
library std;
use std.standard.all;
```

It is never necessary to state these explicitly and it is recommended practice not to state them.

Package textio is also part of library std. This is a text I/O subsystem that allows files to be read or written and reports to be printed to the screen. It includes overloaded read and write operations on the standard types defined in package standard. It has no direct use for synthesis, since there is no way of synthesising such I/O operations. However, it is extensively used in the development of test benches, as described in Chapter 13.

14.4.2   Library ieee

Library ieee is a library containing IEEE-standard design units that form extensions to the language for certain application-specific areas. Since it is not integral to the language, there is no implicit library clause for it, so you have to use an explicit library clause:

```
library ieee;
```

There are six standard synthesis-related packages in library ieee, all of which have been described in previous chapters but mostly in Chapter 6. These are:

```
package std_logic_1164
package numeric_bit
package numeric_std
package fixed_float_types
package fixed_pkg
package float_pkg
```

The first design unit to be added to library ieee was the package std_logic_1164, so-called because it contains the definitions for the standard logic type std_logic and because the package was originally defined by IEEE standard number 1164, though it is now part of the main VHDL standard. Nevertheless, it is a rather clumsy name and a completely unmemorable number. It would be interesting to know how many other standard logic packages the design committee were anticipating when they decided to include the standard number in the name! Fortunately, you will type this name so often that it will either become second nature or an editor macro.

Package `std_logic_1164` was described in Section 6.3 and is listed in Appendix A.3.

Packages `numeric_bit` and `numeric_std` were described in Section 6.4. They are alternative implementations depending on which base type you use for logic modelling. Package `numeric_bit` is rarely used and is not recommended, whilst package `numeric_std` is listed in Appendix A.5.

Both `fixed_pkg` and `float_pkg` use the supplementary package `fixed_float_types` to define their rounding modes. This package is listed in Appendix A.7.

Package `fixed_pkg` was described in Section 6.5 and is listed in Appendix A.8.

Package `float_pkg` was described in Section 6.5 and is listed in Appendix A.9.

All other synthesisable design units in library `ieee` are non-standard and shouldn't be there now that the numeric packages have been standardised. However, there are some packages that are often found there for historical reasons. It will take some time for these packages to be phased out because of the number of designs already in existence that use them. These are:

```
package std_logic_unsigned
package std_logic_signed
package std_logic_arith
package std_logic_misc
package std_logic_textio
```

Packages `std_logic_unsigned` and `std_logic_signed` are rather obsolete bitwise arithmetic packages. Together they are roughly equivalent to `std_logic_arith` but they have a fundamentally flawed design that makes them virtually unusable in practice. They should never be used and will not be covered in this book. The package `std_logic_arith` is a proprietary package to Synopsys Incorporated, but is available in the majority of VHDL systems since Synopsys made the wise decision to make the package public domain, thus ensuring its widespread use. It is very similar to `numeric_std` and indeed was the model for it, although the design committee probably wouldn't admit this. It was covered in Chapter 7 but its use is now deprecated. New designs should use `numeric_std` in preference. Package `std_logic_misc` contains some assorted utilities that extend the functionality of `std_logic_1164` slightly. Finally, `std_logic_textio` extends I/O to the `std_logic_1164` types. It has also been made obsolete by recent development and its use is also deprecated.

Any other packages found in library `ieee` are non-standard (neither IEEE-standard nor de-facto standard) and shouldn't be there at all. The exceptions are the packages standardised for the IEEE VITAL standard, which is a standard that addresses the issue of gate-level simulation. VITAL defines a standard way of writing gate libraries such that they can be automatically back-annotated with timing data to give accurate timing models during gate-level simulation. They are not relevant to the RTL design stages used up to synthesis and so are outside the scope of this book.

14.4.3   Library ieee_proposed

Library `ieee_proposed` is a library containing proposed IEEE-standard design units that form extensions to the language. In particular, it can contain VHDL-1993 compatibility versions of the new VHDL-2008 synthesis packages described in Chapter 6, plus additions packages containing additions to the existing standard packages. Since it is not integral to the language, there is no implicit library clause for it, so you have to use an explicit library clause:

```
library ieee_proposed;
```

If you are using these VHDL-1993 compatibility packages then the following synthesis-related packages will be compiled into this library:

```
package standard_additions
package std_logic_1164_additions
package numeric_std_additions
package fixed_float_types
package fixed_pkg
package float_pkg
```

The additions packages add the missing I/O and string formatting functions used in test benches as described in Section 13.11. The last two packages are the VHDL-1993 versions of the fixed-point and floating-point packages for systems that don't yet provide them in VHDL-2008 form in library `ieee`.

14.5   Organising Your Files

It is common practice and good sense to have all the source files for all the design units in a library stored in one directory. The library mapping file containing the library dependencies for that set of design units can then go in the same directory. Furthermore, the library itself is usually also stored in that directory (as a subdirectory that is) so that whenever any changes are made to the contents of a library, all the pertinent information can be found in one place.

When using both synthesis and simulation from different vendors (and sometimes even with the same vendor), two libraries will be needed. The internal format of a library is vendor-specific and also usually tool-specific, so a library for the simulation system will not be usable with the synthesis system. Indeed, a library for one simulator cannot be used with another simulator. You will need to compile all source files with the simulator and the same set of files excluding the test benches with the synthesiser. The two different versions of the library and the two library mapping files should still be kept in the same directory, using different subdirectories for the libraries. This organisation ensures that there is just one copy of the VHDL source code.

Just to illustrate this point, consider the filter design used earlier to illustrate the use of libraries to subdivide a task. In that example, there were four libraries: `utilities`, `transmitter`, `receiver` and `system`. Each of these libraries will be stored in a separate directory. A typical directory structure for such a project is illustrated by Figure 14.1.

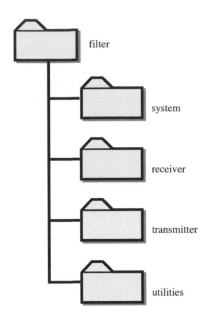

Figure 14.1 Project directory structure.

In this example, all of the library directories are contained in one project directory. This is also good practice, since it keeps the whole project in one place, whilst at the same time allowing different designers to work on each of the libraries.

Consider the `transmitter` section and let's assume that this component has just two source files, `transmitter.vhdl` and `memory.vhdl`. This component uses subcomponents from library `utilities` as well as standard packages from `std` and `ieee`. Thus, there are three other libraries being used as resources in addition to the `work` library for this component. The contents of the design directory are illustrated by Figure 14.2.

Both the simulation and synthesis libraries will have the same VHDL name, even though they are in different subdirectories. The simulation library will be stored in a subdirectory called `simlib`, whilst the synthesis library will be kept in a subdirectory called `synlib`. To avoid any vendor-specific details, I will assume that the simulator uses a library mapping file called `simulator.map` and the synthesiser uses a library mapping file called `synthesiser.map`.

The simulator's library mapping file `simulator.map` will contain a set of library mappings that includes the working library for the transmitter component, plus the three resource libraries used by this component. The resulting library mapping file looks something like this:

```
transmitter = ./simlib
utilities = ../utilities/simlib
ieee = /simulator/libs/ieee
std = /simulator/libs/std

work : transmitter
```

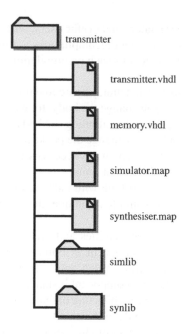

Figure 14.2   Project subdirectory contents.

The synthesiser's library mapping file `synthesiser.map` will map the library name onto the synthesis version of the libraries:

```
transmitter = ./synlib
utilities = ../utilities/synlib
ieee = /synthesiser/libs/ieee
std = /synthesiser/libs/std

work : transmitter
```

This illustration included a mapping for library `std` and a mapping for `ieee`. Most VHDL systems do not require these mappings to be explicitly stated since these libraries are predefined as global libraries. Once again, bear in mind that there will be two versions of each library, one will be part of the simulator's setup, whilst the other will be part of the synthesiser.

14.6   Incremental Compilation

One of the additional advantages of the library system of VHDL, apart from the ability to work independently on different parts of a design, is that it allows incremental compilation.

Incremental compilation means that, if you make a change to a design unit, then it is only necessary to recompile that unit. All other units, which are unchanged, do not need to be recompiled.

This is a simplification, however. All design units are dependent on the packages that they use. If a used package changes, then all design units that *depend* on that package must also be recompiled.

Many VHDL systems, both simulators and synthesisers, will automatically calculate such dependencies and perform these dependent recompilations automatically. However, you as a designer can help to minimise the extent of such recompilations by the way that you manage your VHDL code.

When you add a use clause to a design unit, you are making that unit dependent only on the package header, not on its body. If you change the body, for example to fix a bug in one of the functions provided by the package, then recompiling the body only will avoid any dependent recompilations. On the other hand, if you change the package header, for example to add new declarations or to change an existing one, then it is necessary to recompile the header too. This will force the recompilation of all dependent units.

It is therefore good practice when working with packages to keep the source code for the header and the body separate. By this simple technique, the number and extent of recompilations can be minimised.

A similar dependency exists between entities and architectures. When a component is instantiated in an architecture, the component is bound to its entity by a configuration specification. This creates a dependency on that entity. If the entity is recompiled for any reason, then all architectures containing component instances bound to the entity will have to be recompiled. However, once again the architecture of that entity can be recompiled on its own without causing a recompilation of the dependent units.

Once again, you can minimise the effect of changes by keeping entities and architectures in separate files. Then only the minimum set of files will need to be recompiled if there are any changes.

This is why VHDL has separate primary units and secondary units. Changes to the primary units (entities and package headers) force recompiles of all design units dependent on them. However, primary units change relatively infrequently compared to secondary units. Secondary units (architectures and package bodies) can be changed and recompiled without forcing any recompilations of other design units. By keeping primary and secondary units in separate files, the impact of design changes on the compilation overhead of simulation and synthesis can be minimised.

15

Case Study

This case study brings together all the principles established in the rest of the book. It covers the RTL design of a synchronous system, the design of test benches to test it and the results from those tests.

The example is a low-pass digital filter and the case study shows how the original specification is gradually converted into an RTL design suitable for logic synthesis, with an exploration of the design space to find an optimum data representation and precision for the filter calculations.

The example also shows how test benches are written for this design that explore the frequency response of the filter and verify that the specification has been met.

15.1 Specification

The specification is a Low-Pass digital filter with the following characteristics:

    Maximum Frequency: 80 kHz;
    Cut-off Frequency: 20 kHz;
    Transition Bandwidth: 10 kHz;
    Stop-band Attenuation: 18 dB.

The maximum frequency is the highest input frequency that the filter can handle.

The cut-off frequency is the end of the pass band and the frequency at which the gain should start to drop off.

The transition bandwidth is the frequency gap between the cut-off frequency at the end of the pass band and the beginning of the stop band.

The stop-band attenuation is the reduction in signal power for frequencies in the stop-band.

For this example the cut-off frequency is set at 20 kHz, the transition bandwidth is 10 kHz, so the stop band starts at 30 kHz at which point the specification requires at least 18 dB attenuation.

This is illustrated by the pass-band diagram in Figure 15.1.

VHDL for Logic Synthesis, Third Edition. Andrew Rushton.
© 2011 John Wiley & Sons, Ltd. Published 2011 by John Wiley & Sons, Ltd.

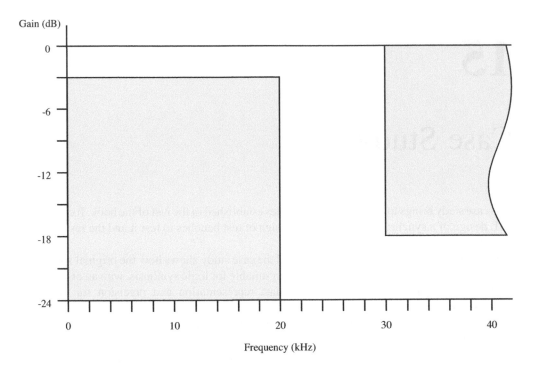

Figure 15.1   Pass-band diagram for the low-pass filter.

The implementation of this filter will use a simple Finite Impulse Response (FIR) digital filter design.

An FIR filter is expressed by a simple formula whereby the current output z_n is calculated from the current input sample x_n and previous samples x_{n-1}, x_{n-2}, etc. combined with the filter coefficients c_0, c_1, c_2, etc.

$$z_n = x_n{}^*c_0 + x_{n-1}{}^*c_1 + \ldots + x_{n-i}{}^*c_i + \ldots$$

This formula can be simplified by replacing the samples with the contents of a delayed sample store such that s_i is the stored input sample from i samples earlier. Thus, s_0 is the current sample, s_1 is the previous sample and so on. In other words, s_i stores the value x_{n-i}. At each new sample, the stored samples move so that the new value becomes s_0, the previous s_0 becomes s_1 and so on. Using this simplification the formula becomes:

$$z_n = s_0{}^*c_0 + s_1{}^*c_1 + \ldots \mid s_i{}^*c_i + \ldots$$

This is illustrated by the block diagram in Figure 15.2.

15.2   System-Level Design

Digital filter theory is beyond the scope of this book. The theory behind this system-level design is explained further in an online tutorial (Robin, 2005).

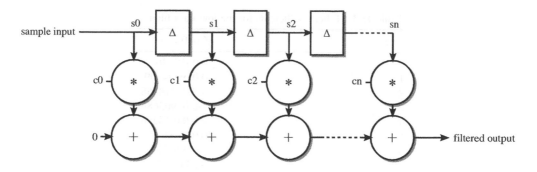

Figure 15.2   Block diagram of the FIR filter.

The first system-level design decision to be made is which *window* function to use when choosing filter coefficients. The window function does two things. First, it imposes a limit on the number of coefficients, otherwise the theoretical number of coefficients is infinite. Secondly, it optionally applies a taper so that older samples contribute less to the output than recent ones. This tapering gives the filter a flatter pass band and a higher attenuation in the stop band than a design without the taper. There are a number of window functions with different tapers that result in different filter characteristics. However, the price of tapered window functions is that they add more stages to the filter. For this example, to keep the design simple, a rectangular window function will be used. This is the simplest window function of all that limits the number of stages but does not apply a taper. It has the disadvantage that the pass band is not completely flat: there can be about 1–2 dB variation in the frequency response across the pass band. Also, a rectangular window filter is limited to a maximum of 21 dB attenuation in the stop band, but that is sufficient for this design.

The next stage is to determine the number of stages in the filter, known as the *order* of the filter. The order N is directly related to the window function and the two key frequencies: the maximum frequency F_{max} and the transition bandwidth F_T.

$$F_T = k.F_{max}/N$$

The value k is a scale factor that depends on the window function.
So, rearranging this to get a formula for N:

$$N = k.F_{max}/F_T$$

For the rectangular window function, $k = 1.84$. Substituting the design objectives for this filter we get:

$$N = 1.84*80000/10000$$

$$N = 14.7$$

The filter order must be an even number, so to meet this specification we need N to be at least 16. This results in 17 coefficients, since the calculation uses the current sample plus 16 previous ones.

Table 15.1 Filter coefficients for the low-pass filter

Stage	Coefficient
0	0.0368626
1	0.019574609
2	−0.023181587
3	−0.05903686
4	−0.04914286
5	0.02388607
6	0.13902785
7	0.2445255
8	0.28715014
9	0.2445255
10	0.13902785
11	0.02388607
12	−0.04914286
13	−0.05903686
14	−0.023181587
15	0.019574609
16	0.0368626

The final stage of the system-level design is to calculate the coefficients. This is a complex calculation involving a Fast-Fourier Transform (FFT). So, for this example an online filter-design program (Robin, 2005) was used that did this step automatically. The program gave the coefficients listed in Table 15.1.

Note that the coefficients have a mirror symmetry around coefficient 8. For example, coefficient 0 is the same as coefficient 16. This is a consequence of using a rectangular window function.

15.3   RTL Design

The next stage is to decide how this theoretical system-level design is to be implemented as an RTL design. The basic calculation for the filter is a 17-stage multiply-accumulate calculation. In designing the hardware, a key decision is how many multipliers to use because we know that multipliers are large circuits. For a high-performance design, it might be necessary to have several, even as many as 17, multipliers. For a low-performance design it is almost certainly worth using only one.

From the maximum frequency we can immediately deduce one of the design characteristics:

   Minimum Sample Frequency: 160 kHz.

This is because, for DSP applications, the minimum sample frequency is always twice the maximum frequency of the system. Since the maximum frequency in the specification is 80 kHz, it follows that the minimum sample frequency is 160 kHz. This is a low sample frequency for even an FPGA-based design and suggests that the design can easily be implemented by one multiplier.

15.3.1 Block Diagram

The calculation of the output sample requires the storing of 17 samples. Each calculation is in three phases.

1. shifting in a new sample;
2. a 17-stage multiply-accumulate loop;
3. output of the result.

The samples will be stored in a shift register. When a new sample comes into the filter, all the values are shifted along the register. All the registered values will be available for the accumulation cycle, not just the last element.

The multiply-accumulate loop can be carried out over 17 clock cycles using a single multiplier and an accumulator register. The inputs to this multiply-accumulator will be multiplexed from the 17 stored samples and the 17 coefficients.

The output needs to be held steady during calculation of the next output, so needs to have a separate output register.

Figure 15.3 shows the basic block diagram of the filter hardware.

The whole will need a controller to co-ordinate the various components.

From these design parameters we can see that the whole process of generating an output from a sample will take at least 19 clock cycles, which gives us an initial estimate of the minimum clock frequency.

In order to meet the design requirements, the filter must process samples at the rate of 160 k samples/s.

$$F_s = 160000$$

The basic filter algorithm requires 19 clock cycles to process a sample. The filter goes into a waiting state between samples, so can have any number of clock cycles between samples as

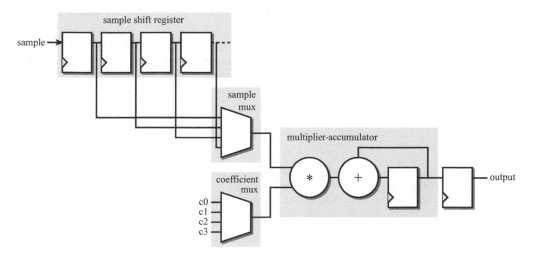

Figure 15.3 Block diagram of the filter hardware.

long as its at least 19. So the filter can be adapted to other clock frequencies. The minimum clock frequency is:

$$F_{ck} \geq 19*160000 \, \text{Hz}$$

$$F_{ck} \geq 3.04 \, \text{MHz}$$

This is a very slow clock for a modern device and confirms the one-multiplier decision made earlier.

15.3.2   Interface

The next stage is to design the interface to the filter circuit. This means choosing what the inputs and outputs will be and also choosing what types to use to represent the data.

This design is a good candidate for using the fixed-point synthesis types described in Section 6.5. All calculations will be performed using fixed-point arithmetic.

Both the input sample and the filtered output are required to be in the range -1 to $+1$. This means that a signed type is needed and that the integer part is 2 bits wide. Note that this range excludes the actual value $+1$ but includes the value -1. If the actual value $+1$ is needed, then the integer part would need to be 3 bits. To allow this to be changed if necessary, the integer part will be made generic. Also, at this stage the size of the fraction part of the datapaths will not be decided but will also be made generic. The circuit will be designed so that different precisions can be tried and compared so that a suitable trade-off between precision and noise can be made. That is, the lower the precision, the greater the error introduced by rounding and this error will appear as output noise. To be able to adjust the precision of the samples means the design's noise characteristics can be measured for different precisions and the best trade-off chosen as a result of that analysis.

The design will use handshaking. That is, the circuit driving the filter will indicate using a single-bit input when a sample is ready. The filter will read the sample in response to that signal. When the next output value has been calculated, the filter will indicate that it is ready by raising a single-bit output signal for one cycle.

So, in the interface, there needs to be a clock input, a reset input, a sample input and a filtered output. In addition, two handshake signals are needed, one input telling the filter when a sample is ready and one output to indicate when a result is ready.

This results in the following entity:

```
entity low_pass_filter is
   generic (integer_bits : integer; fraction_bits : integer);
   port (clock : in std_logic;
         reset : in std_logic;
         sample : in sfixed(integer_bits-1 downto -fraction_bits);
         sample_ready : in std_logic;
         output : out sfixed(integer_bits-1 downto -fraction_bits);
         output_ready : out std_logic);
end;
```

Note that this has been called `low_pass_filter` because that is what it *does*. This is preferable to calling it `fir_filter` which is how it is *implemented*. Entity names and ports are best chosen using conceptual or black-box names with no indication of the implementation.

The usual context clauses for using the fixed-point package will be needed:

```
library ieee;
use ieee.std_logic_1164.all;
use ieee.numeric_std.all;
use ieee.fixed_float_types.all;
use ieee.fixed_pkg.all;
```

This needs to be placed before the entity.

15.3.3   Outline Architecture

At this stage an outline architecture can also be written:

```
architecture behaviour of low_pass_filter is

  constant order : natural := 16;
  constant counter_bits : natural := 5;

begin
end;
```

The first constant, `order`, is the order of the filter, which is used to give the size of the sample store. Because of the slightly odd conventions for denoting the filter's order, the sample store is one element larger than the order. That is, a filter with order 16 uses 17 samples in the output calculation and therefore needs a 17-element sample store indexed from 0 to 16. The second constant, `counter_bits`, is the number of bits needed in a counter to be able to index the sample store and therefore must be able to count from 0 to `order`.

The datapath of the design is the same width throughout, so a subtype can be declared in the architecture that can then be used to define all variables and signals used in the datapath:

```
subtype datapath_type is
  sfixed(integer_bits-1 downto -fraction_bits);
```

This subtype has the same range as the sample and output ports of the entity.

15.3.4   Coefficient Store

The floating-point filter coefficients calculated in Section 15.2 need to be converted into fixed-point before incorporating them into the design. Since the precision decision is being deferred, the coefficients need to be converted at the highest precision that they are likely to be needed. They can then be rounded down to the required size at the point of use. A 2-bit integer part and 32-bit fractional part was chosen as the highest precision, giving a signed 2.32 bit (i.e. 34-bit) fixed-point representation.

This conversion from floating-point to fixed-point cannot be done in the synthesisable model, because floating-point values are not synthesisable. So the coefficients must be converted separately and the resulting fixed-point values used in the design. The conversion from floating-point to fixed-point is extremely difficult to do by hand, so it is easier to use a VHDL model to perform the calculation for us. Because the model is only used to do this calculation and is therefore not required to be synthesisable, it can use the floating-point to fixed-point (i.e. `real` to `sfixed`) conversion provided in `fixed_pkg`. The following architecture was used to perform this calculation:

```
library ieee;
use ieee.std_logic_1164.all;
use ieee.numeric_std.all;
use ieee.fixed_float_types.all;
use ieee.fixed_pkg.all;
use std.textio.all;
entity low_pass_filter_calculator is
end;
architecture info of low_pass_filter_calculator is

  type coeff_array_type is array (natural range <>) of real;
  constant coefficients : coeff_array_type :=
    (
      0.0368626,
      0.019574609,
      -0.023181587,
      -0.05903686,
      -0.04914286,
      0.02388607,
      0.13902785,
      0.2445255,
      0.28715014,
      0.2445255,
      0.13902785,
      0.02388607,
      -0.04914286,
      -0.05903686,
      -0.023181587,
      0.019574609,
      0.0368626
    );

begin

  process
    variable l : line;
  begin
    for i in coefficients'range loop
      write(l, i);
      write(l, string'(":"));
      write(l, coefficients(i));
      write(l, string'(":"));
      write(l, to_sfixed(coefficients(i), 1, -32));
      writeline(output, l);
    end loop;
    wait;
  end process;

end;
```

Table 15.2 Conversion of real coefficients to fixed-point

Index	Real	Fixed-point
0	3.686260e-002	00.00001001011011111101001111001101
1	1.957461e-002	00.00000101000000101101011101110001
2	−2.318159e-002	11.11111010000100001100010110000010
3	−5.903686e-002	11.11110000111000101111010111011001
4	−4.914286e-002	11.11110011011010110101111110100000
5	2.388607e-002	00.00000110000111010110010111000001
6	1.390279e-001	00.00100011100101110101010001000101
7	2.445255e-001	00.00111110100110010011100100100010
8	2.871501e-001	00.01001001100000101010101111101100
9	2.445255e-001	00.00111110100110010011100100100010
10	1.390279e-001	00.00100011100101110101010001000101
11	2.388607e-002	00.00000110000111010110010111000001
12	−4.914286e-002	11.11110011011010110101111110100000
13	−5.903686e-002	11.11110000111000101111010111011001
14	−2.318159e-002	11.11111010000100001100010110000010
15	1.957461e-002	00.00000101000000101101011101110001
16	3.686260e-002	00.00001001011011111101001111001101

This uses TextIO to print the converted filter coefficients to the standard output of the simulator. Running this model in a simulator gave the output in Table 15.2, which has been reformatted as a table for readability.

So, it is now possible to paste these values into the design of the coefficient store that can then be added to the declaration part of the architecture:

```
subtype coeff_type is sfixed(1 downto -32);
type coeff_array_type is
  array (natural range 0 to order) of coeff_type;
constant coefficients : coeff_array_type :=
  (
    B"00_00001001011011111101001111001101",
    B"00_00000101000000101101011101110001",
    B"11_11111010000100001100010110000010",
    B"11_11110000111000101111010111011001",
    B"11_11110011011010110101111110100000",
    B"00_00000110000111010110010111000001",
    B"00_00100011100101110101010001000101",
    B"00_00111110100110010011100100100010",
    B"00_01001001100000101010101111101100",
    B"00_00111110100110010011100100100010",
    B"00_00100011100101110101010001000101",
    B"00_00000110000111010110010111000001",
    B"11_11110011011010110101111110100000",
    B"11_11110000111000101111010111011001",
    B"11_11111010000100001100010110000010",
    B"00_00000101000000101101011101110001",
    B"00_00001001011011111101001111001101"
  );
```

The fixed-point coefficients have been converted into bit-strings and an underscore is used to show where the binary point is.

15.3.5   Sample Store

The sample store is a shift-register with 17 registers. It uses a similar array type to the coefficient store. However, the width of the sample store registers is the datapath width rather than the coefficient width:

```
type sample_array_type is
  array (natural range 0 to order) of datapath_type;
signal samples : sample_array_type;
```

The register bank uses the template for register banks described in Section 12.3. The following process is used for the register bank:

```
process
begin
  wait until rising_edge(clock);
  if samples_shift = '1' then
    samples(0) <= sample;
    for i in 1 to order loop
      samples(i) <= samples(i-1);
    end loop;
  end if;
end process;
```

So, the standard registered process template is used. Within that, the shift operation is controlled by the input control signal `samples_shift`:

```
signal samples_shift : std_logic;
```

This follows a common convention that the control signals for a structure have the structure name as the first part of the signal name. This makes it easier to understand what signals are doing in the design.

When this signal is low, nothing happens and the register bank simply preserves its contents. When the control signal goes high, the shift register operates, capturing the current sample in register 0 and moving all the existing samples along one register. The control signal will be generated by the controller. The design of the register bank requires this control to go high for just one clock cycle, since if it stays high for longer, the shift register will shift again. This requirement of the shifter becomes a design constraint on the controller.

15.3.6   Calculation and Accumulator

The next stage is to design the accumulator circuit. This is based on an accumulator register that is declared as a signal with the same width as the rest of the datapath:

```
signal accumulator : datapath_type;
```

The calculation for each cycle of the multiply-accumulator sequence is done within a registered process using intermediate variable assignments:

```
sample := samples(to_integer(address));
coefficient :=
  resize(coefficients(to_integer(address)), coefficient);
product := resize(sample * coefficient, product);
```

To simplify and clarify the calculation, intermediate variables have been used.

The first variable assignment simply gets the addressed sample from the sample store. No resizing takes place because it is already the correct width for the datapath. This temporary variable was used just to clarify the address indexing. Note how the unsigned address is first converted to integer using the to_integer function at the point of use, as recommended in Section 6.1, then used to index the sample store array.

The second variable assignment gets the coefficient from the coefficient store using the same array indexing technique as for the sample. The coefficient is resized to the datapath width from its maximal size. This is a reduction in the fractional part only, so overflow cannot happen but underflow is possible. The default underflow behaviour has been used for the resize, meaning that the value will be rounded.

The third variable assignment performs the multiplication. It forms the product of the sample and the coefficient using the * operator, which will generate a double-sized result. This is wrapped in a call to the resize function to reduce it down to the datapath size. The default overflow and underflow parameters have been used and, since both are possible, this means that the product will overflow by saturating and underflow by rounding.

This calculation now needs to be wrapped in a registered process with control signals that determine when to reset the accumulator and when to perform the accumulation. Once again, the design will assume that the controller will generate the appropriate control signals in the right order. There are two controls needed: a signal to reset the accumulator to zero and a signal to enable the calculation. In addition, the address generator will also generate the address used to access the sample and coefficient stores. It is the controller's responsibility to synchronise the address counter with the accumulator control signals.

So, the accumulator is a registered process with a synchronous reset:

```
process
  variable coefficient : datapath_type;
  variable sample : datapath_type;
  variable product : datapath_type;
begin
  wait until rising_edge(clock);
  if accumulator_clear = '1' then
    accumulator <= (others => '0');
  elsif accumulator_calculating = '1' then
    sample := samples(to_integer(address));
    coefficient :=
      resize(coefficients(to_integer(address)),
             coefficient);
    product :=
      resize(sample * coefficient, product);
```

```
      accumulator <=
        resize(accumulator + product, accumulator);
    end if;
  end process;
```

The reset branch is simple, setting the accumulator to zero using an aggregate with an others clause. It would not be possible to use a string literal here because the datapath is of unspecified width controlled by the generic parameter. The aggregate will adjust to the width of the accumulator.

In the calculation branch of the main if statement, a single accumulate step is calculated. The calculation of the product was explained earlier. The full process also includes the accumulation step, which is also wrapped in a resize call.

15.3.7   Address Generator

The address generator is required to count up from 0 to 16 (the filter's order) on demand from the controller. It is used by the multiply-accumulator block to access the coefficient and sample stores.

The address signal is an unsigned value and can count over the range of the samples store and coefficients store. Its size is controlled by the generic parameter counter_bits:

```
  signal address : unsigned(counter_bits-1 downto 0);
```

The address generator is a simple counter register that holds its value when disabled and counts upwards when enabled. It has a synchronous reset to set the address to zero:

```
  process
  begin
    wait until rising_edge(clock);
    if address_clear = '1' then
      address <= (others => '0');
    elsif address_counting = '1' then
      address <= address + 1;
    end if;
  end process;
```

The counter is controlled by two control signals generated by the controller. There is no need for the address counter to stop counting when it gets to the top of the range; it is the controller's responsibility to stop the count on completion of the multiply-accumulate calculation by disabling address_counting.

15.3.8   Output Register

The output register stores the result from the accumulator at the end of the calculation so that it can be held steady for the whole sample period. It also generates an output handshake signal to indicate when the output has changed.

The convention is to use internal signals for registers and to have a combinational assignment to the out port separate from the rest of the design. So, an internal signal is created to form the register that will hold the output value and the handshake signal:

```
signal result : datapath_type;
signal result_ready : std_logic;
```

The output ports are then connected to these internal signals:

```
output <= result;
output_ready <= result_ready;
```

The output register is then defined entirely in terms of the internal signals:

```
process
begin
  wait until rising_edge(clock);
  result_ready <= '0';
  if result_save = '1' then
    result <= accumulator;
    result_ready <= '1';
  end if;
end process;
```

The behaviour of the output register is quite simple although complicated slightly by the extra `result_ready` signal. When the controller sets the `result_save` signal, the register stores the output of the accumulator in `result`. The rest of the time it just preserves the value. The process also includes the logic for the handshake signal. When the new result is saved to the output register, an extra one-bit register `result_ready` is set to high for just one clock cycle. Actually, it is set high for as long as the `result_save` signal is high, but the controller ensures that this only happens for one cycle. By registering the `result_ready` signal rather than decoding it combinationally, it is synchronised with the output register.

15.3.9   Controller

The final stage in the design is to design the controller. The filter goes through a sequence of states, so it makes sense to design the controller using a finite-state machine (FSM) as described in Section 12.2. The FSM will use the two-process model: a registered process to update the state and a combinational process to decode the logic for both the next state and the control signals.

There are six states:

waiting
    the default state, waiting for a sample. When a sample arrives, move to sampling state.
sampling
    stays in this state for one cycle only while the samples shift-register is shifted, capturing the new sample. Then moves immediately to the calculating_first state.
calculating_first
    starts the calculation by clearing the accumulator and address register and allowing one cycle for the first product to be calculated. Moves immediately to the calculating state.

calculating
> performs the accumulation and enables the address counter to step through the samples. Stays in this state until the penultimate step of the calculation, then moves to calculating_last state.

calculating_last
> disables the address counter but keeps accumulating for one last cycle to capture the last product in the result. Moves immediately to the outputting state.

outputting
> stays in this state for one cycle only while the output register is updated. Then moves back to the waiting state.

State transitions are made on clock edges because the state signal is registered.

Notice that there is a `sampling` state when the input sample gets captured. It takes a single clock cycle for the FSM to move from the waiting state to the sampling state. This means that the sample is captured one clock cycle after the `sample_ready` signal goes high. In order to capture the sample in the same cycle as the `sample_ready` signal, the shift-register enable would have to be decoded combinationally from a primary input and this would be a potential glitch hazard.

The three states used in the calculation reflect the different timings for the address register and accumulator. Both need to be reset to zero before the calculation starts. Then, the calculation takes place, enabling the address counter but allowing a clock cycle to elapse between setting the address and accumulating the calculation of that sample. This one-cycle delay means that the address counter is stopped when it reaches the last sample, but the accumulator needs to continue for one more cycle to accumulate the last product.

The state is stored in a registered signal using an enumerated state type for readability and to enable FSM inference:

```
type state_type is
  (waiting,
   sampling,
   calculating_first, calculating, calculating_last,
   outputting);
signal state, next_state : state_type;
```

The following process describes the registered process of the FSM:

```
process
begin
  wait until rising_edge(clock);
  if reset = '1' then
    state <= waiting;
  else
    state <= next_state;
  end if;
end process;
```

This process includes a synchronous reset. This design follows the convention in RTL design that the system is reset by resetting the controller. It also follows the convention that resets should be synchronous unless there is a really compelling reason to use asynchronous reset.

The second process in the controller is the combinational decoding of these states into the control signals for the other filter components:

```
process(state, address, sample_ready)
begin
  samples_shift <= '0';
  address_counting <= '0';
  address_clear <= '0';
  accumulator_calculating <= '0';
  accumulator_clear <= '0';
  result_save <= '0';
  case state is
    when waiting =>
      if sample_ready = '1' then
        next_state <= sampling;
      end if;
    when sampling =>
      samples_shift <= '1';
      next_state <= calculating_first;
    when calculating_first =>
      accumulator_clear <= '1';
      address_clear <= '1';
      next_state <= calculating;
    when calculating =>
      address_counting <= '1';
      accumulator_calculating <= '1';
      if address = order-1 then
        next_state <= calculating_last;
      end if;
    when calculating_last =>
      accumulator_calculating <= '1';
      next_state <= outputting;
    when outputting =>
      next_state <= waiting;
      result_save <= '1';
  end case;
end process;
```

Note that only state changes are conditional on the inputs and all control outputs are dependent only on the state itself. This makes it a Moore machine (see Section 12.2) and ensures that all internal control signals are synchronised to the clock. The style of the decoder follows a common convention in which all the control outputs are initialised low before the case statement, but then overridden with a high value under the appropriate conditions.

The state-transition logic implements the state transitions described earlier. The reset puts the FSM into its waiting state. When a sample is input, the controller detects the sample_ready signal going high and triggers a state transition into sampling state. It stays in that state for just one clock cycle while the sample store shifts in the new value. It then moves unconditionally to the calculating_first state, where it resets the accumulator and the address counter. It then moves to the calculating state, starting the address generator and enabling the accumulator to start the calculation of the next output. It stays in the calculating state until the address generator gets to one short of the end of its count, when it moves into the `calculating_last` state. This disables the address counter but continues the accumulation of the last product. This then transitions unconditionally `outputting` state. It stays in that state for just one cycle, enabling the output register so that the accumulator output is captured, unconditionally making a state transition to the `waiting` state again.

Note that the filter only responds to the `sample_ready` signal when its in the `waiting` state. The rest of the time the input is ignored and therefore cannot interrupt a calculation.

This controller takes 22 clock cycles to process a sample. This is different from the original estimate of 19 since extra cycles have been added to enable the controller to be a simple, synchronous design. This leads to a revision of the minimum clock frequency of the system to meet the specification:

$$F_{ck} \geq 22*160000\,\mathrm{Hz}$$

$$F_{ck} \geq 3.52\,\mathrm{MHz}$$

In principle, if the design had to be as fast as possible, the controller could be optimised to eliminate these extra cycles by overlapping the accumulator reset with sampling and the start of the accumulator calculation. The result register could also be overlapped with the next sample. This would reduce the controller to 19 cycles. However, in this case the design does not need to be as fast as possible and the controller design is adequate for the task

15.4   Trial Synthesis

At this point the design was tested by compiling into both the simulator and synthesis systems to see if any errors were detected by either system before moving on to writing simulation test benches.

Bear in mind that the design of the filter is generic. So to synthesise the design, a non-generic top level was needed. Here's a 16-bit filter using a single instance of the generic filter:

```
library ieee;
use ieee.std_logic_1164.all;
use ieee.numeric_std.all;
use ieee.fixed_float_types.all;
use ieee.fixed_pkg.all;
entity low_pass_filter_16 is
  port (clock : in std_logic;
        reset : in std_logic;
        sample : in sfixed(1 downto -14);
        sample_ready : in std_logic;
        output : out sfixed(1 downto -14);
        output_ready : out std_logic);
end;

architecture behaviour of low_pass_filter_16 is
begin
  LPF16 : entity work.low_pass_filter
    generic map(2,14)
```

```
        port map (clock, reset,
                    sample, sample_ready,
                    output, output_ready);
    end;
```

This design was then synthesised.

The design compiled successfully, but the synthesis phase generated warnings that the controller required latches. This was not expected since the intention was to design a purely synchronous decoder. Specifically, the `next_state` signal needed latches, indicating that it was not being assigned to under all circumstances in the combinational controller process. This was causing latch inference to be invoked as explained in Section 8.6. Latches can only be inferred from combinational processes, so this knowledge helps to narrow down the search for the error to the FSM's combinational process.

On examination of that process it was clear that some branches of the decoding logic for the state machine make the state transition conditional without an else clause or a default value. This implies storage between process executions. This was an error in the design.

The intended behaviour is that the state should remain the same unless a state transition is required, default behaviour that can be added by having an assignment to the `next_state` signal just before the case statement:

```
    next_state <= state;
    case state is ...
```

Then, within the case statement, this default behaviour is overridden under those circumstances when a state transition is required.

This modified design was re-synthesised and this time the result was purely synchronous as intended. Of course, all the test benches were run again to ensure this change didn't introduce another error.

This is an illustration of the use of synthesis as well as simulation to detect design errors. Simulation would not have detected this error because signals in VHDL do preserve their previous value unless assigned. However, the synthesised circuit would not have been as intended. The existence of latches in the synthesised circuit indicated that the design had an error.

It is good practice to determine in advance whether latches are desired and if so how many are expected in the design. Then, the synthesis reports can be checked to make sure that the expected number were actually generated. It is also good practice to perform trial synthesis runs as early as possible in the design process to catch this kind of synthesis-specific error.

15.5   Testing the Design

The design will be tested using a series of test benches. The first will test the basic functioning of the circuit, iron out any bugs in the design and verify that the circuit is behaving as intended. This will be followed by a more complex test bench to measure the noise characteristics of the filter so that an optimum design can be selected.

Each test bench is implemented as a different architecture of the same empty entity.

The basic framework is:

```
library ieee;
use ieee.std_logic_1164.all;
use ieee.numeric_std.all;
use ieee.fixed_float_types.all;
use ieee.fixed_pkg.all;
use ieee.math_real.all;

entity low_pass_filter_test is
end;
```

Then, each architecture adds a test to the test set. Note that the package `math_real` has been added to the `use` clauses. This has not been introduced before because it is completely unsynthesisable, but it will prove very useful for writing these test benches, as will become clear later. It is listed in Appendix A.13.

One common denominator is the clocking scheme.

In order to meet the design requirements, the filter must process samples at the rate of 160 k samples/s.

$$F_s = 160000$$

This gives us a sample period of:

$$T_s = 1/F_s$$

$$T_s = 6.25\,\mu s$$

The basic filter algorithm requires 22 clock cycles to process a sample. The filter goes into a waiting state between samples, so can have any number of clock cycles between samples as long as its at least 22. So the filter can be adapted to higher clock frequencies. The minimum clock frequency is:

$$F_{ck} \geq 22*160000\,Hz$$

$$F_{ck} \geq 3.52\,MHz$$

This gives us a maximum clock period of:

$$T_{ck} \leq 1/F_{ck}$$

$$T_{ck} \leq 284\,ns$$

For the tests, a clock generator will be used that clocks at least this fast, but that is independent of the sample generator that will work at exactly 160 k samples/s.

Just to get an idea of how the frequency response of the filter relates to the sample frequency, the following calculations can be useful for reference:

- Cut-off frequency of 20 kHz is equivalent to a full sine wave in 8 samples.
- Stop-Band frequency of 30 kHz is equivalent to a full sine wave in 5.3 samples.
- Maximum frequency of 80 kHz is equivalent to a full sine wave in 2 samples.

The first stage is to develop a test bench that will test the basic functionality of this design.

15.5.1 Basic Test

This test bench just checks the basic functionality of the filter. It simply sends samples to the filter so that the internal signals can be monitored and checked against the expected behaviour.

The first stage is to create the basic architecture with the circuit under test in place along with the signals to connect to it:

```
architecture basic_test of low_pass_filter_test is
  constant integer_bits : natural := 2;
  constant fraction_bits : natural := 32;
  subtype datapath_type is
    sfixed(integer_bits-1 downto -fraction_bits);
  signal clock : std_logic;
  signal reset : std_logic;
  signal sample : datapath_type;
  signal sample_ready : std_logic;
  signal result : datapath_type;
  signal result_ready : std_logic;
begin
  CUT : entity work.low_pass_filter
    generic map (integer_bits,fraction_bits)
    port map (clock, reset,
              sample, sample_ready,
              result, result_ready);
end;
```

The intention in this design is to drive the filter clock independently of the sample generation so that they can run at different and independent speeds. So there will be two separate generator processes.

The clock generator will run until stopped by a stop signal from the sample generator, so that when all samples have been produced, the whole simulation stops.

```
process
  constant clock_period : time := 250 ns;
  procedure generate_clock_cycle is
    constant high_time : time := clock_period / 2;
    constant low_time : time := clock_period - high_time;
  begin
    clock <= '0';
    wait for low_time;
    clock <= '1';
    wait for high_time;
  end;
begin
  reset <= '1';
  generate_clock_cycle;
  reset <= '0';
  while clock_running loop
    generate_clock_cycle;
  end loop;
  wait;
end process;
```

The process generates a reset signal for one clock cycle, then continuously generates clocks until the control signal `clock_running` goes false. The calculation of the high and low times ensures that the clock period doesn't get changed due to rounding errors if the period is an odd multiple of the resolution limit.

The control signal is boolean:

```
signal clock_running : boolean := true;
```

The sample generator for this test produces sine waves at different frequencies across the whole frequency range of the filter.

This process is quite complex so it will be broken down into parts.

The basic outline of the process is:

```
process
  constant sample_period : time := 6.25 us;
begin
  ... perform tests
  clock_running <= false;
  wait;
end process;
```

The sample period results in samples at the rate of 160 k samples/s. In this case we first need a procedure that manages the sending of a sample, setting the `sample_ready` flag high for a fixed time period and then setting it low for the rest of the sample period. This is similar to the clock generator procedure:

```
procedure generate_sample(value : real) is
  constant ready_time : time := 500 ns;
  constant wait_time : time := sample_period - ready_time;
begin
  sample_real <= value;
  sample_ready <= '1';
  wait for ready_time;
  sample_ready <= '0';
  wait for wait_time;
end;
```

This declaration is placed in the process declarative part so that it has access to the `sample_period` value.

The procedure generates a sample on a signal of type real:

```
signal sample_real : real;
```

This is easier to generate than the fixed-point value and is also easier to view since most simulators have a way of displaying real values as waveforms, whereas the fixed-point types used in the filter design cannot be viewed yet by most simulators.

The real sample is then assigned to the fixed-point input to the filter by a concurrent signal assignment in the architecture body:

```
sample <= to_sfixed(sample_real, sample);
```

Similarly, to make it easier to view the filter output as a waveform, the output signal will also be converted from fixed-point to real by another concurrent signal assignment:

```
result_real <= to_real(result);
```

The next stage is to generate samples at a range of frequencies. To do this a pair of nested loops is used. The outer loop steps through the test frequencies, then the inner loop generates a series of samples at that frequency.

```
for f in 1 to 40 loop
  step := real(f) * math_pi / 80.0;
  for i in 1 to 160 loop
    generate_sample(sin(step * real(i)));
  end loop;
end loop;
```

The outer loop frequencies are measured in kHz from 1 kHz to 40 kHz. There are 160 samples generated per frequency, representing a millisecond of real time. Each sample is generated by the `sin` function from `math_real`. This calculates the sine of its input in radians.

Package `math_real` provides common mathematical operations such as sine and cosine on real numbers, it is totally unsynthesisable but sometimes of use in writing test benches as in this case and is listed in Appendix A.13.

To work out the step between samples the frequency is converted into a rotation angle per sample. The first step is to convert the frequency f into an angular velocity R measured in radians per second, where:

$$R = 2\pi f$$

So, for a sample frequency of f_s, this equates to an angular distance between samples ΔR measured in radians per sample of:

$$\Delta R = 2\pi f / f_s$$

This calculation for the step size can be seen in the outer loop. The inner loop then generates the samples by multiplying the step size by the sample number.

In order to test the frequency response, it would be possible to examine the waveform on the screen of a simulator and try to find the maximum output for each frequency. However, it is easier to get the test bench to calculate it. The idea is to keep a running tally of the largest magnitude output for each frequency. This is done by adding the following code around the inner sample loop:

```
minmax := 0.0;
for i in 1 to 160 loop
  generate_sample(sin(step * real(i)));
  if i > 50 then
    minmax :=
      realmax(minmax, sign(result_real) * result_real);
  end if;
end loop;
```

```
write(l, f);
write(l, string'(" kHz = "));
write(l, minmax);
writeline(output,l);
```

The minmax value is a variable of type real. So, before the inner loop, the minmax value is initialised to zero. Then, within the loop it is updated if the current result from the filter has a larger magnitude than the current value of minmax. The formula:

```
sign(result_real) * result_real
```

generates the absolute value of a real number – the sign function from math_real returns -1.0 if negative and 1.0 if positive (and 0.0 if zero). Multiplying this by the sample gives a positive number equal to the magnitude.

The test for maximum magnitude is delayed for 50 samples to allow the filter to settle down after a frequency change. When the frequency changes there is a sharp change in the input waveform that can generate an impulse on the output that takes a few cycles to settle down.

Finally, after the inner loop has finished running, TextIO is used to output the result for that frequency.

The result is a series of text outputs from the simulator like this:

```
1 kHz = 9.502971E-01
2 kHz = 9.449322E-01
3 kHz = 9.369100E-01...
```

The output magnitudes were converted to a power gain in dB by the formula:

```
gain = 20*log10(magnitude)
```

The gain values were plotted as a graph, overlaid on the pass-band diagram as shown in Figure 15.4.

This completes the basic functionality test bench.

15.5.2   Noise Calculations

In this section a second test bench will be written to measure the effect on varying the precision of the fixed-point representation on the noise characteristics of the circuit. In this case, the noise we are interested in is that introduced by rounding during the calculation of the output.

So, the main objective of the test bench is to measure the output noise of the filter for different precisions. In order to perform a comparison, two circuits under test will be used, one at maximum precision and one under reduced precision. This will allow the outputs to be compared and a *relative* noise figure produced.

The test bench uses the same clock and sample generator as the basic test above. However, it has two circuits under test so that they can be compared:

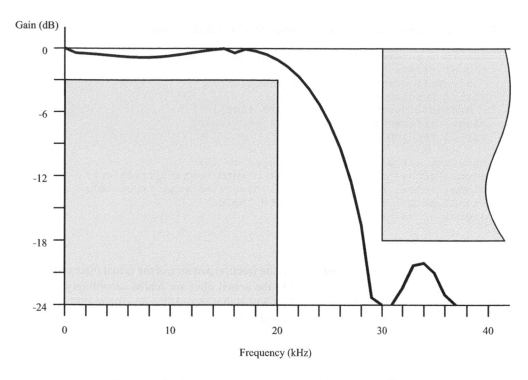

Figure 15.4 Actual frequency response of the low-pass filter.

```
reference_sample <= to_sfixed(sample_real, reference_sample);
reference : entity work.low_pass_filter
   generic map (2,reference_fraction)
   port map (clock, reset, reference_sample, sample_ready,
            reference_result, reference_result_ready);
reference_real <= to_real(reference_result);

actual_sample <= to_sfixed(sample_real, actual_sample);
actual : entity work.low_pass_filter
   generic map (2,actual_fraction)
   port map (clock, reset, actual_sample, sample_ready,
            actual_result, actual_result_ready);
actual_real <= to_real(actual_result);

difference <= actual_real - reference_real;
```

So, the sample is converted to fixed-point by a concurrent signal assignment. The reference sample is 2.32 bits, the maximum allowed for the design. The actual sample can have its fraction bits varied for each run. The last line calculates a difference between the outputs of the two filters, which is assigned to the difference signal so that it can be viewed in a simulation waveform viewer.

There are, of course, corresponding duplicate signal declarations:

```
constant reference_fraction : integer := 32;
signal reference_sample :
  sfixed(1 downto -reference_fraction);
signal reference_result :
  sfixed(1 downto -reference_fraction);
signal reference_result_ready : std_logic;
signal reference_real : real;

constant actual_fraction : integer := 16;
signal actual_sample : sfixed(1 downto -actual_fraction);
signal actual_result : sfixed(1 downto -actual_fraction);
signal actual_result_ready : std_logic;
signal actual_real : real;

signal difference : real;
```

The `actual_fraction` constant controls the fraction part size of the actual filter and can be varied between runs. All signals attached to the actual filter are scaled accordingly.

In the main sample generator loop, code is added that records the maximum difference for the whole run. The sample generation loop therefore becomes:

```
minmax := 0.0;
for f in 1 to 40 loop
  step := real(f) * math_pi / 80.0;
  for i in 1 to 160 loop
    generate_sample(sin(step * real(i)));
    minmax := realmax(minmax, sign(difference) * difference);
  end loop;
end loop;
write(l, actual_fraction);
write(l, string'(" = "));
write(l, minmax);
writeline(output,l);
```

The final part prints the size of the actual fraction part and the maximum error detected.

This can then be compiled as a table, one result per simulation run and with different fraction sizes for each run. Each value has been calculated to 2 significant digits as shown in Table 15.3.

So, even a 12-bit fraction part (i.e. using 2.12 bit signed fixed-point format) results in an error of 0.0013, representing 0.13% of the range or a signal-to-noise ratio of 58 dB. Adding four bits

Table 15.3 Maximum error for different fixed-point sizes

Size	Error	Signal/noise (dB)
2.16 (18)	0.000077	82
2.14 (16)	0.00025	72
2.12 (14)	0.0013	58
2.8 (10)	0.020	34
2.6 (8)	0.080	22

to give a 16-bit fraction (i.e. using 2.16 bit signed fixed-point format) increases the signal-to-noise to 82 dB.

15.6   Floating-Point Version

It was stated at the start of the chapter that the filter should be a fixed-point design. However, to explore whether this statement was a good design decision, the design was converted to use the synthesisable floating-point package described in Section 6.6.

The conversion was simple: first the entity was converted to use different generic parameters and to have `float` ports sized by those parameters:

```
library ieee;
use ieee.std_logic_1164.all;
use ieee.numeric_std.all;
use ieee.fixed_float_types.all;
use ieee.fixed_pkg.all;
use ieee.float_pkg.all;
entity low_pass_filter_float is
  generic(
    exponent_bits : natural;
    fraction_bits : natural);
  port(
    clock : in std_logic;
    reset : in std_logic;
    sample : in float(exponent_bits downto -fraction_bits);
    sample_ready : in std_logic;
    output : out float(exponent_bits downto -fraction_bits);
    output_ready : out std_logic);
end;
```

Then, within the architecture, the type used to define datapath signals was changed:

```
subtype datapath_type is
  float(exponent_bits downto -fraction_bits);
```

Finally, the coefficients were converted into bit-string values in 8:23 float format using a modified version of the coefficient calculator model used to generate the fixed-point coefficients:

```
subtype coeff_type is float32;
type coeff_array_type is array (natural range 0 to order)
  of coeff_type;
constant coefficients : coeff_array_type :=
  (
    0 => B"0_01111010_00101101111110100111101",
    1 => B"0_01111001_01000000101101011101110",
    2 => B"1_01111001_01111011110011101010000",
    3 => B"1_01111010_11100011101000010100010",
    4 => B"1_01111010_10010010100101000000110",
    5 => B"0_01111001_10001110101100010111000",
    6 => B"0_01111100_00011100101110101010001",
    7 => B"0_01111100_11110100110010011100101",
    8 => B"0_01111101_00100110000010101011000",
```

```
 9 => B"0_01111100_11110100110010011100101",
10 => B"0_01111100_00011100101110101010001",
11 => B"0_01111001_10000111010110010111000",
12 => B"1_01111010_10010010100101000000110",
13 => B"1_01111010_11100011101000010100010",
14 => B"1_01111001_01111011110011101010000",
15 => B"0_01111001_01000000101101011101110",
16 => B"0_01111010_00101101111110100111101"
);
```

Table 15.4   Maximum error for different floating-point sizes

Size	Error	Signal/noise (dB)
6 : 9 (16)	0.0027	51
4 : 11 (16)	0.00088	61
6 : 7 (14)	0.012	38
4 : 9 (14)	0.0029	51

The only other change required was to the multiplier-accumulator that now uses operators from the floating-point package. The difference with floating-point compared with fixed-point is that the result of an operation is the same size as the arguments, so no resizing is needed to give a fixed datapath width. The modified calculation is:

```
sample := samples(to_integer(address));
coefficient :=
  resize(coefficients(to_integer(address)), coefficient);
product := sample * coefficient;
accumulator <= accumulator + product;
```

Both of the test benches were also converted to floating-point in the same way.

The frequency response of the floating-point version was identical to the fixed-point version. However, the noise characteristics were different. Table 15.4 shows the noise measurements for different datapath sizes:

Different trade-offs between exponent size and fraction size have been used and show that a shorter exponent with a longer fraction tends to give a lower noise value for the same word size. Interestingly, the fixed-point version has a lower noise factor for the same word-length in every case. For example, a 2:14 (16-bit) fixed-point filter had a signal-to-noise ratio of 72 dB, whereas the two 16-bit floating-point versions had noise factors of 51 dB and 61 dB, respectively.

15.7   Final Synthesis

Since this is a case study, it was decided to implement several versions of the filter. Four variants were chosen using all the permutations of fixed-point or floating-point with either a 16-bit or a 32-bit sample. All four were synthesised to the same FPGA technology so that they could be compared.

This means that four top-level designs were created to instantiate the chosen generic filter circuits with the required generic parameters. These top-level designs also reduced the interface to the simple types std_logic and std_logic_vector according to the convention for synthesisable top-level circuits explained in Section 6.10 to ensure that the post-synthesis netlist has the same interface.

As an example, here's the 4:11 (16-bit) floating-point variant:

```
library ieee;
use ieee.std_logic_1164.all;
entity low_pass_filter_float_16 is
  port (clock : in std_logic;
        reset : in std_logic;
        sample : in std_logic_vector(15 downto 0);
        sample_ready : in std_logic;
        output : out std_logic_vector(15 downto 0);
        output_ready : out std_logic);
end;

use ieee.numeric_std.all;
use ieee.fixed_float_types.all;
use ieee.fixed_pkg.all;
use ieee.float_pkg.all;
architecture behaviour of low_pass_filter_float_16 is
begin
  LPF16 : entity work.low_pass_filter_float
    generic map(4, 11)
    port map (clock => clock,
              reset => reset,
              sample => to_float(sample, 4, 11),
              sample_ready => sample_ready,
              to_slv(output) => output,
              output_ready => output_ready);
end;
```

Note the type conversions on the array ports.

All four designs were generated in the same way and then synthesised. Table 15.5 gives the statistics for the four designs.

As can be seen from these statistics, the floating-point variants were about half the speed and 3–4 times the number of logic cells of the fixed-point variants. Yet the floating-point variants have inferior noise performance for a particular word size. This confirms that the choice of fixed-point arithmetic was the right choice. However, it also confirms that it is realistic to synthesise floating-point arithmetic operations.

The synthesiser reported a maximum clock frequency for this design of 67 MHz for the targeted technology, if the 16-bit fixed-point version is used, and this is well within the design requirement of 3.52 MHz. Even allowing for a pessimistic view of the synthesiser's

Table 15.5 Synthesis results for the low-pass filter

Datapath type	Size (bits)	Maximum clock (MHz)	Logic cells	Register cells	I/O pins
sfixed	2.14 (16)	67	281	315	36
sfixed	2.30 (32)	55	478	619	68
float	4:11 (16)	31	1027	315	36
float	8:23 (32)	24	2083	626	68

timing estimator, this would suggest the filter could be run at over 10 times the target frequency if desired.

15.8   Generic Version

This design implements a specific filter with a specific set of coefficients. However, the core multiply-accumulator circuit and its controller is general-purpose and can be converted into a generic filter design. To do this, the coefficient store must be made external to the design and extra ports provided for accessing this external coefficients store. This reorganisation of the design allows different sets of coefficients to be used with different designs but using the same filter core.

In fact, this means the core can be used to implement low-pass, high-pass and band-pass filters given the right coefficients to do so. For this reason a more general name will be chosen so the entity will be called filter_core.

The first stage is to define the interface for the generic core of the filter:

```
library ieee;
use ieee.std_logic_1164.all;
use ieee.numeric_std.all;
use ieee.fixed_float_types.all;
use ieee.fixed_pkg.all;
entity filter_core is
  generic(
    integer_bits : natural;
    fraction_bits : natural;
    counter_bits : natural;
    order : natural);
  port(
    clock : in std_logic;
    reset : in std_logic;
    sample : in
      sfixed(integer_bits-1 downto -fraction_bits);
    sample_ready : in std_logic;
    coefficient_address : out
      unsigned(counter_bits-1 downto 0);
    coefficient : in
      sfixed(integer_bits-1 downto -fraction_bits);
    output : out
      sfixed(integer_bits-1 downto -fraction_bits);
    output_ready : out std_logic);
end;
```

This entity has two more generic parameters to control the order of the filter and the size of the counter for the coefficient store address.

This entity also has two extra ports, the address output for the coefficient store and the coefficient input from the coefficient store.

To create the architecture, the existing architecture is copied and also renamed:

```
architecture behaviour of filter_core is
```

Then, the existing design for the low-pass filter is copied into this architecture as the starting point for the design. The coefficient store is removed and the two new ports incorporated into the design. So, the calculation becomes:

```
sample := samples(to_integer(address));
product := resize(sample * coefficient, product);
accumulator <= resize(accumulator + product, accumulator);
```

Note how the calculation of the product now uses the coefficient input port rather than the local coefficient variable that is removed. Also, the address counter must be connected to the coefficient address output using a simple concurrent assignment at the end of the architecture:

```
coefficient_address <= address;
```

Now, a new version of the low-pass filter can be created that uses an instance of this generic filter core. The coefficient store is incorporated into this level and connected to the core via the two new ports. The architecture now looks like:

```
architecture behaviour of low_pass_filter is

  constant order : natural := 16;
  constant counter_bits : natural := 5;

  -- coefficient store
  subtype coeff_type is sfixed(1 downto -32);
  type coeff_array_type is array (natural range 0 to order)
    of coeff_type;
  constant coefficients : coeff_array_type :=
    (
      B"00_00001001011011111101001111001101",
      B"00_00000101000000101101011101110001",
      B"11_11111010000100001100010110000010",
      B"11_11110000111000101111010111011001",
      B"11_11110011011010110101111110100000",
      B"00_00000110000111010110010111000001",
      B"00_00100011100101110101010001000101",
      B"00_00111110100110010011100100100010",
      B"00_01001001100000101010101111101100",
      B"00_00111110100110010011100100100010",
      B"00_00100011100101110101010001000101",
      B"00_00000110000111010110010111000001",
      B"11_11110011011010110101111110100000",
      B"11_11110000111000101111010111011001",
      B"11_11111010000100001100010110000010",
      B"00_00000101000000101101011101110001",
      B"00_00001001011011111101001111001101"
    );

  -- internal signals
  signal address :
    unsigned(counter_bits-1 downto 0);
  signal coefficient :
    sfixed(integer_bits-1 downto -fraction_bits);
```

```
begin

    filter : entity work.filter_core
        generic map(integer_bits, fraction_bits,
                    counter_bits, order)
        port map (clock, reset,
                    sample, sample_ready,
                    address, coefficient,
                    output, output_ready);

    coefficient <=
        resize(coefficients(to_integer(address)), coefficient);

end;
```

Note how the coefficient is selected from the coefficient store using the address output from the filter.

This design was re-simulated with all the test benches and re-synthesised to confirm that no errors were introduced in the re-organisation.

15.9 Conclusions

The design process has been stopped after the first stage of synthesis. The fitting of the circuit to a device, specifying pin-outs, etc. is very tool and technology specific and is a matter for the user manual for the synthesis tool.

This design is a simple work-through of a digital filter, which is itself a simple circuit. Only four possible implementations of the design were tried varying only in the type and size of the datapath operators. There are many other permutations of this design that could be tried:

- For higher frequency of operation, more than one multiplier could be used.
- The symmetry of the coefficients could be used to halve the number of multiplications and therefore the number of clock cycles required to perform the accumulation.
- The multiplier-accumulator could be pipelined to increase the throughput, with a pipeline register between the multiplier and the accumulator.
- It was assumed that overflow should use Saturate Mode. Saturation mode prevents an overflow from being turned into an abrupt change from very positive to very negative due to wrapping. Nevertheless, in a performance-critical design it might be an acceptable trade-off to use Wrap Mode.
- It was assumed that underflow should use Rounding Mode. This minimises the noise generated by rounding errors. However, no experiments were performed to calculate the noise factor in Truncate Mode. It might yield a higher-performance design to truncate the result and then compensate for the noisier output by adding 1–2 more fraction bits.
- Some of the controller states could be overlapped to reduce the number of cycles per calculation.

Digital filters scale proportionately with sample rate. For example, given the same set of coefficients, this filter could be used with the following specification:

Sample Frequency: 1.6 M samples/s;
Maximum Frequency: 800 kHz;
Cut-off Frequency: 200 kHz;
Stop Band: 300 kHz;
Stop Band Attenuation: 18 dB;
Minimum Clock Frequency: 35.2 MHz.

The design implemented in this case study is capable of running at this speed.

This also illustrates why the sample rate must be controlled accurately. If the input is sampled at any speed other than 160 k samples/s, the pass band moves in proportion to the sample frequency.

Appendix A

Package Listings

This Appendix lists the package headers of all the standard packages in the `std` and `ieee` libraries that are relevant to synthesis.

A.1 Package Standard

Package `standard` defines the basic types provided by VHDL in all circumstances. This is the VHDL-1993 version and is supplemented by package `standard_additions` (see Appendix A.2). In VHDL-2008, the additions are merged into package `standard`.

```
package standard is

    type boolean is (false, true);
    function "and"(l, r: boolean) return boolean;
    function "or"(l, r: boolean) return boolean;
    function "nand"(l, r: boolean) return boolean;
    function "nor"(l, r: boolean) return boolean;
    function "xor"(l, r: boolean) return boolean;
    function "xnor"(l, r: boolean) return boolean;
    function "not"(l: boolean) return boolean;
    function "="(l, r: boolean) return boolean;
    function "/="(l, r: boolean) return boolean;
    function "<"(l, r: boolean) return boolean;
    function "<="(l, r: boolean) return boolean;
    function ">"(l, r: boolean) return boolean;
    function ">="(l, r: boolean) return boolean;

    type bit is ('0', '1');
    function "and"(l, r: bit) return bit;
    function "or"(l, r: bit) return bit;
    function "nand"(l, r: bit) return bit;
    function "nor"(l, r: bit) return bit;
    function "xor"(l, r: bit) return bit;
    function "xnor"(l, r: bit) return bit;
    function "not"(l: bit) return bit;
    function "="(l, r: bit) return boolean;
    function "/="(l, r: bit) return boolean;
```

VHDL for Logic Synthesis, Third Edition. Andrew Rushton.
© 2011 John Wiley & Sons, Ltd. Published 2011 by John Wiley & Sons, Ltd.

```
function "<"(l, r: bit) return boolean;
function "<="(l, r: bit) return boolean;
function ">"(l, r: bit) return boolean;
function ">="(l, r: bit) return boolean;

type character is (...); -- ascii 8-bit values
function "="(l, r: character) return boolean;
function "/="(l, r: character) return boolean;
function "<"(l, r: character) return boolean;
function "<="(l, r: character) return boolean;
function ">"(l, r: character) return boolean;
function ">="(l, r: character) return boolean;

type severity_level is (note, warning, error, failure);
function "="(l, r: severity_level) return boolean;
function "/="(l, r: severity_level) return boolean;
function "<"(l, r: severity_level) return boolean;
function "<="(l, r: severity_level) return boolean;
function ">"(l, r: severity_level) return boolean;
function ">="(l, r: severity_level) return boolean;

-- universal types cannot be used but are implicit
type _uni_int is range implementation_defined;
function "="(l, r: _uni_int) return boolean;
function "/="(l, r: _uni_int) return boolean;
function "<"(l, r: _uni_int) return boolean;
function "<="(l, r: _uni_int) return boolean;
function ">"(l, r: _uni_int) return boolean;
function ">="(l, r: _uni_int) return boolean;
function "+"(l: _uni_int) return _uni_int;
function "-"(l: _uni_int) return _uni_int;
function "abs"(l: _uni_int) return _uni_int;
function "+"(l, r: _uni_int) return _uni_int;
function "-"(l, r: _uni_int) return _uni_int;
function "*"(l, r: _uni_int) return _uni_int;
function "/"(l, r: _uni_int) return _uni_int;
function "mod"(l, r: _uni_int) return _uni_int;
function "rem"(l, r: _uni_int) return _uni_int;

type _uni_real is range implementation_defined;
function "="(l, r: _uni_real) return boolean;
function "/="(l, r: _uni_real) return boolean;
function "<"(l, r: _uni_real) return boolean;
function "<="(l, r: _uni_real) return boolean;
function ">"(l, r: _uni_real) return boolean;
function ">="(l, r: _uni_real) return boolean;
function "+"(l: _uni_real) return _uni_real;
function "-"(l: _uni_real) return _uni_real;
function "abs"(l: _uni_real) return _uni_real;
function "+"(l, r: _uni_real) return _uni_real;
function "-"(l, r: _uni_real) return _uni_real;
function "*"(l, r: _uni_real) return _uni_real;
function "/"(l, r: _uni_real) return _uni_real;
function "*"(l: _uni_real; anonymous: _uni_int)return _uni_real;
function "*"(l: _uni_int; anonymous: _uni_real)return _uni_real;
```

```
function "/"(l: _uni_real; anonymous: _uni_int)return _uni_real;
type integer is range implementation_defined;
function "**"(l: _uni_int; anonymous: integer)return _uni_int;
function "**"(l: _uni_real; anonymous: integer)return _uni_real;
function "="(l, r: integer) return boolean;
function "/="(l, r: integer) return boolean;
function "<"(l, r: integer) return boolean;
function "<="(l, r: integer) return boolean;
function ">"(l, r: integer) return boolean;
function ">="(l, r: integer) return boolean;
function "+"(l: integer) return integer;
function "-"(l: integer) return integer;
function "abs"(l: integer) return integer;
function "+"(l, r: integer) return integer;
function "-"(l, r: integer) return integer;
function "*"(l, r: integer) return integer;
function "/"(l, r: integer) return integer;
function "mod"(l, r: integer) return integer;
function "rem"(l, r: integer) return integer;
function "**"(l: integer; anonymous: integer) return integer;

type real is range implementation_defined;
function "="(l, r: real) return boolean;
function "/="(l, r: real) return boolean;
function "<"(l, r: real) return boolean;
function "<="(l, r: real) return boolean;
function ">"(l, r: real) return boolean;
function ">="(l, r: real) return boolean;
function "+"(l: real) return real;
function "-"(l: real) return real;
function "abs"(l: real) return real;
function "+"(l, r: real) return real;
function "-"(l, r: real) return real;
function "*"(l, r: real) return real;
function "/"(l, r: real) return real;
function "**"(l: real; r: integer) return real;

type time is range implementation_defined
  units
    fs;
    ps=1000 fs;
    ns=1000 ps;
    us=1000 ns;
    ms=1000 us;
    sec=1000 ms;
    min=60 sec;
    hr=60 min;
  end units;
function "="(l, r: time) return boolean;
function "/="(l, r: time) return boolean;
function "<"(l, r: time) return boolean;
function "<="(l, r: time) return boolean;
function ">"(l, r: time) return boolean;
function ">="(l, r: time) return boolean;
function "+"(l: time) return time;
```

```
function "-"(l: time) return time;
function "abs"(l: time) return time;
function "+"(l, r: time) return time;
function "-"(l, r: time) return time;
function "*"(l: time; r: integer) return time;
function "*"(l: time; r: real) return time;
function "*"(l: integer; r: time) return time;
function "*"(l: real; r: time) return time;
function "/"(l: time; r: integer) return time;
function "/"(l: time; r: real) return time;
function "/"(l, r: time) return _uni_int;

subtype delay_length is time range 0 fs to time'high;
pure function now return delay_length;

subtype natural is integer range 0 to integer'high;
subtype positive is integer range 1 to integer'high;

type string is array (positive range <>) of character;
function "="(l, r: string) return boolean;
function "/="(l, r: string) return boolean;
function "<"(l, r: string) return boolean;
function "<="(l, r: string) return boolean;
function ">"(l, r: string) return boolean;
function ">="(l, r: string) return boolean;
function "&"(l: string; r: string)  return string;
function "&"(l: string; r: character) return string;
function "&"(l: character; r: string) return string;
function "&"(l: character; r: character)return string;

type bit_vector is array (natural range <>) of bit;
function "and"(l, r: bit_vector) return bit_vector;
function "or"(l, r: bit_vector) return bit_vector;
function "nand"(l, r: bit_vector) return bit_vector;
function "nor"(l, r: bit_vector) return bit_vector;
function "xor"(l, r: bit_vector) return bit_vector;
function "xnor"(l, r: bit_vector) return bit_vector;
function "not"(l: bit_vector) return bit_vector;
function "sll"(l: bit_vector; r: integer)return bit_vector;
function "srl"(l: bit_vector; r: integer)return bit_vector;
function "sla"(l: bit_vector; r: integer)return bit_vector;
function "sra"(l: bit_vector; r: integer)return bit_vector;
function "rol"(l: bit_vector; r: integer)return bit_vector;
function "ror"(l: bit_vector; r: integer)return bit_vector;
function "="(l, r: bit_vector) return boolean;
function "/="(l, r: bit_vector) return boolean;
function "<"(l, r: bit_vector) return boolean;
function "<="(l, r: bit_vector) return boolean;
function ">"(l, r: bit_vector) return boolean;
function ">="(l, r: bit_vector) return boolean;
function "&"(l: bit_vector; r: bit_vector)return bit_vector;
function "&"(l: bit_vector; r: bit) return bit_vector;
function "&"(l: bit; r: bit_vector) return bit_vector;
function "&"(l: bit; r: bit)  return bit_vector;
```

```
    type file_open_kind is (read_mode, write_mode, append_mode);
    function "="(l, r: file_open_kind) return boolean;
    function "/="(l, r: file_open_kind) return boolean;
    function "<"(l, r: file_open_kind) return boolean;
    function "<="(l, r: file_open_kind) return boolean;
    function ">"(l, r: file_open_kind) return boolean;
    function ">="(l, r: file_open_kind) return boolean;

    type file_open_status is
      (open_ok, status_error, name_error, mode_error);
    function "="(l, r: file_open_status) return boolean;
    function "/="(l, r: file_open_status) return boolean;
    function "<"(l, r: file_open_status) return boolean;
    function "<="(l, r: file_open_status) return boolean;
    function ">"(l, r: file_open_status) return boolean;
    function ">="(l, r: file_open_status) return boolean;

  attribute foreign: string;

end;
```

A.2   Package Standard_Additions

Package `standard_additions` is the VHDL-1993 supplement to package `standard` (see Appendix A.1). In VHDL-2008, the additions are merged into package `standard` and so this package is empty.

```
package standard_additions is

  -- std_match operators
  -- implemented as extended-named functions in VHDL-1993
  -- implemented as operators in VHDL-2008
  function \?=\ (l, r  : boolean)
    return boolean;
  function \?/=\ (l, r : boolean)
    return boolean;
  function \?<\ (l, r  : boolean)
    return boolean;
  function \?<=\ (l, r : boolean)
    return boolean;
  function \?>\ (l, r  : boolean)
    return boolean;
  function \?>=\ (l, r : boolean)
    return boolean;
  function \?=\ (l, r  : bit)
    return bit;
  function \?/=\ (l, r : bit)
    return bit;
  function \?<\ (l, r  : bit)
    return bit;
  function \?<=\ (l, r : bit)
    return bit;
```

```
function \?>\ (l, r  : bit)
  return bit;
function \?>=\ (l, r : bit)
  return bit;
function \??\ (l : bit)
  return boolean;
function \?=\ (l, r  : bit_vector)
  return bit;
function \?/=\ (l, r : bit_vector)
  return bit;

-- minimum/maximum functions
function minimum (l, r : boolean)
  return boolean;
function maximum (l, r : boolean)
  return boolean;
function minimum (l, r : bit)
  return bit;
function maximum (l, r : bit)
  return bit;
function minimum (l, r : character)
  return character;
function maximum (l, r : character)
  return character;
function minimum (l, r : severity_level)
  return severity_level;
function maximum (l, r : severity_level)
  return severity_level;
function minimum (l, r : integer)
  return integer;
function maximum (l, r : integer)
  return integer;
function minimum (l, r : real)
  return real;
function maximum (l, r : real)
  return real;
function minimum (l, r : time)
  return time;
function maximum (l, r : time)
  return time;
function minimum (l, r : string)
  return string;
function maximum (l, r : string)
  return string;
function minimum (l : string)
  return character;
function maximum (l : string)
  return character;
function minimum (l, r : bit_vector)
  return bit_vector;
function maximum (l, r : bit_vector)
  return bit_vector;
function minimum (l : bit_vector)
  return bit;
function maximum (l : bit_vector)
```

```
    return bit;
function minimum (l, r : file_open_kind)
  return file_open_kind;
function maximum (l, r : file_open_kind)
  return file_open_kind;
function minimum (l, r : file_open_status)
  return file_open_status;
function maximum (l, r : file_open_status)
  return file_open_status;

-- edge detection for possible clock types
function rising_edge (signal s  : boolean)
  return boolean;
function falling_edge (signal s : boolean)
  return boolean;
function rising_edge (signal s  : bit)
  return boolean;
function falling_edge (signal s : bit)
  return boolean;

-- selecting boolean operators
function "and" (l  : bit_vector; r : bit)
  return bit_vector;
function "and" (l  : bit; r : bit_vector)
  return bit_vector;
function "or" (l   : bit_vector; r : bit)
  return bit_vector;
function "or" (l   : bit; r : bit_vector)
  return bit_vector;
function "nand" (l : bit_vector; r : bit)
  return bit_vector;
function "nand" (l : bit; r : bit_vector)
  return bit_vector;
function "nor" (l  : bit_vector; r : bit)
  return bit_vector;
function "nor" (l  : bit; r : bit_vector)
  return bit_vector;
function "xor" (l  : bit_vector; r : bit)
  return bit_vector;
function "xor" (l  : bit; r : bit_vector)
  return bit_vector;
function "xnor" (l : bit_vector; r : bit)
  return bit_vector;
function "xnor" (l : bit; r : bit_vector)
  return bit_vector;

-- reducing boolean operators
-- implemented as functions in VHDL-1993
-- implemented as boolean operators in VHDL-2008
function and_reduce (l  : bit_vector)
  return bit;
function or_reduce (l   : bit_vector)
  return bit;
function nand_reduce (l  : bit_vector)
  return bit;
```

```
function nor_reduce (l  : bit_vector)
  return bit;
function xor_reduce (l  : bit_vector)
  return bit;
function xnor_reduce (l  : bit_vector)
  return bit;

-- arithmetic operations
function "mod" (l, r : time)
  return time;
function "rem" (l, r : time)
  return time;

-- String formatting functions
function to_string (value : bit_vector)
  return string;
alias to_bstring is
  to_string [ bit_vector return string];
alias to_binary_string is
  to_string [ bit_vector return string];
function to_ostring (value : bit_vector)
  return string;
alias to_octal_string is
  to_ostring [ bit_vector return string];
function to_hstring (value : bit_vector)
  return string;
alias to_hex_string is
  to_hstring [ bit_vector return string];
function to_string (value : boolean)
  return string;
function to_string (value : bit)
  return string;
function to_string (value : character)
  return string;
function to_string (value : severity_level)
  return string;
function to_string (value : integer)
  return string;
function to_string (value : real)
  return string;
function to_string (value : time)
  return string;
function to_string (value : file_open_kind)
  return string;
function to_string (value : file_open_status)
  return string;
function to_string (value : real; digits : natural)
  return string;
function to_string (value : real; format : string)
  return string;
function to_string (value : time; unit : time)
  return string;

-- new type boolean_vector
type boolean_vector is array (natural range <>) of boolean;
```

```
function "and" (l, r  : boolean_vector)
  return boolean_vector;
function "or" (l, r   : boolean_vector)
  return boolean_vector;
function "nand" (l, r : boolean_vector)
  return boolean_vector;
function "nor" (l, r  : boolean_vector)
  return boolean_vector;
function "xor" (l, r  : boolean_vector)
  return boolean_vector;
function "xnor" (l, r : boolean_vector)
  return boolean_vector;
function "not" (l : boolean_vector)
  return boolean_vector;
function "and" (l : boolean_vector; r : boolean)
  return boolean_vector;
function "and" (l : boolean; r : boolean_vector)
  return boolean_vector;
function "or" (l : boolean_vector; r : boolean)
  return boolean_vector;
function "or" (l : boolean; r : boolean_vector)
  return boolean_vector;
function "nand" (l : boolean_vector; r : boolean)
  return boolean_vector;
function "nand" (l : boolean; r : boolean_vector)
  return boolean_vector;
function "nor" (l : boolean_vector; r : boolean)
  return boolean_vector;
function "nor" (l : boolean; r : boolean_vector)
  return boolean_vector;
function "xor" (l : boolean_vector; r : boolean)
  return boolean_vector;
function "xor" (l : boolean; r : boolean_vector)
  return boolean_vector;
function "xnor" (l : boolean_vector; r : boolean)
  return boolean_vector;
function "xnor" (l : boolean; r : boolean_vector)
  return boolean_vector;
function and_reduce (l  : boolean_vector)
  return boolean;
function or_reduce (l  : boolean_vector)
  return boolean;
function nand_reduce (l  : boolean_vector)
  return boolean;
function nor_reduce (l  : boolean_vector)
  return boolean;
function xor_reduce (l  : boolean_vector)
  return boolean;
function xnor_reduce (l  : boolean_vector)
  return boolean;
function "sll" (l : boolean_vector; r : integer)
  return boolean_vector;
function "srl" (l : boolean_vector; r : integer)
  return boolean_vector;
function "sla" (l : boolean_vector; r : integer)
```

```
   return boolean_vector;
function "sra" (l : boolean_vector; r : integer)
   return boolean_vector;
function "rol" (l : boolean_vector; r : integer)
   return boolean_vector;
function "ror" (l : boolean_vector; r : integer)
   return boolean_vector;
function "=" (l, r  : boolean_vector)
   return boolean;
function "/=" (l, r : boolean_vector)
   return boolean;
function "<" (l, r  : boolean_vector)
   return boolean;
function "<=" (l, r : boolean_vector)
   return boolean;
function ">" (l, r  : boolean_vector)
   return boolean;
function ">=" (l, r : boolean_vector)
   return boolean;
function \?=\ (l, r  : boolean_vector)
   return boolean;
function \?/=\ (l, r : boolean_vector)
   return boolean;
function "&" (l : boolean_vector; r : boolean_vector)
   return boolean_vector;
function "&" (l : boolean_vector; r : boolean)
   return boolean_vector;
function "&" (l : boolean; r : boolean_vector)
   return boolean_vector;
function "&" (l : boolean; r : boolean)
   return boolean_vector;
function minimum (l, r : boolean_vector)
   return boolean_vector;
function maximum (l, r : boolean_vector)
   return boolean_vector;
function minimum (l : boolean_vector)
   return boolean;
function maximum (l : boolean_vector)
   return boolean;

-- New type integer_vector
type integer_vector is array (natural range <>) of integer;
function "=" (l, r  : integer_vector)
   return boolean;
function "/=" (l, r : integer vector)
   return boolean;
function "<" (l, r  : integer_vector)
   return boolean;
function "<=" (l, r  : integer_vector)
   return boolean;
function ">" (l, r  : integer_vector)
   return boolean;
function ">=" (l, r  : integer_vector)
   return boolean;
function "&" (l : integer_vector; r : integer_vector)
```

```
      return integer_vector;
   function "&" (l : integer_vector; r : integer)
      return integer_vector;
   function "&" (l : integer; r : integer_vector)
      return integer_vector;
   function "&" (l : integer; r : integer)
      return integer_vector;
   function minimum (l, r : integer_vector)
      return integer_vector;
   function maximum (l, r : integer_vector)
      return integer_vector;
   function minimum (l : integer_vector)
      return integer;
   function maximum (l : integer_vector)
      return integer;

   -- New type real_vector
   type real_vector is array (natural range <>) of real;
   function "=" (l, r  : real_vector)
      return boolean;
   function "/=" (l, r  : real_vector)
      return boolean;
   function "<" (l, r  : real_vector)
      return boolean;
   function "<=" (l, r  : real_vector)
      return boolean;
   function ">" (l, r  : real_vector)
      return boolean;
   function ">=" (l, r  : real_vector)
      return boolean;
   function "&" (l : real_vector; r : real_vector)
      return real_vector;
   function "&" (l : real_vector; r : real)
      return real_vector;
   function "&" (l : real; r : real_vector)
      return real_vector;
   function "&" (l : real; r : real)
      return real_vector;
   function minimum (l, r : real_vector)
      return real_vector;
   function maximum (l, r : real_vector)
      return real_vector;
   function minimum (l : real_vector)
      return real;
   function maximum (l : real_vector)
      return real;

   -- New type time_vector
   type time_vector is array (natural range <>) of time;
   function "=" (l, r  : time_vector)
      return boolean;
   function "/=" (l, r  : time_vector)
      return boolean;
   function "<" (l, r  : time_vector)
      return boolean;
```

```
      function "<=" (l, r  : time_vector)
        return boolean;
      function ">" (l, r  : time_vector)
        return boolean;
      function ">=" (l, r  : time_vector)
        return boolean;
      function "&" (l : time_vector; r : time_vector)
        return time_vector;
      function "&" (l : time_vector; r : time)
        return time_vector;
      function "&" (l : time; r : time_vector)
        return time_vector;
      function "&" (l : time; r : time)
        return time_vector;
      function minimum (l, r : time_vector)
        return time_vector;
      function maximum (l, r : time_vector)
        return time_vector;
      function minimum (l : time_vector)
        return time;
      function maximum (l : time_vector)
        return time;

    end;
```

A.3   Package Std_Logic_1164

Package `std_logic_1164` defines the 9-value logic type `std_ulogic` and its arrays.
This is the VHDL-1993 version and is supplemented by package `std_logic_1164_`
`additions` (see Appendix A.4). In VHDL-2008, the additions are merged into package
`std_logic_1164`.

```
    package std_logic_1164 is

      -- logic state system
      type std_ulogic is ('U',  -- Uninitialised
                          'X',  -- Forcing  Unknown
                          '0',  -- Forcing  0
                          '1',  -- Forcing  1
                          'Z',  -- High Impedance
                          'W',  -- Weak     Unknown
                          'L',  -- Weak     0
                          'H',  -- Weak     1
                          '-'   -- Don't care
                          );
      function resolved (s : std_ulogic_vector) return std_ulogic;
      subtype std_logic is resolved std_ulogic;

      type std_ulogic_vector is array (natural range <>) of std_ulogic;
      type std_logic_vector is array (natural range <>) of std_logic;
```

```
-- common subtypes

subtype X01 is resolved std_ulogic range 'X' TO '1';
subtype X01Z is resolved std_ulogic range 'X' TO 'Z';
subtype UX01 is resolved std_ulogic range 'U' TO '1';
subtype UX01Z is resolved std_ulogic range 'U' TO 'Z';

-- overloaded logical operators

function "and"  (l : std_ulogic; r : std_ulogic) return UX01;
function "nand" (l : std_ulogic; r : std_ulogic) return UX01;
function "or"   (l : std_ulogic; r : std_ulogic) return UX01;
function "nor"  (l : std_ulogic; r : std_ulogic) return UX01;
function "xor"  (l : std_ulogic; r : std_ulogic) return UX01;
function "xnor" (l : std_ulogic; r : std_ulogic) return UX01;
function "not"  (l : std_ulogic) return UX01;

-- vectorized overloaded logical operators

function "and"  (l, r : std_logic_vector)
  return std_logic_vector;
function "and"  (l, r : std_ulogic_vector)
  return std_ulogic_vector;

function "nand" (l, r : std_logic_vector)
  return std_logic_vector;
function "nand" (l, r : std_ulogic_vector)
  return std_ulogic_vector;

function "or"   (l, r : std_logic_vector)
  return std_logic_vector;
function "or"   (l, r : std_ulogic_vector)
  return std_ulogic_vector;

function "nor"  (l, r : std_logic_vector)
  return std_logic_vector;
function "nor"  (l, r : std_ulogic_vector)
  return std_ulogic_vector;

function "xor"  (l, r : std_logic_vector)
  return std_logic_vector;
function "xor"  (l, r : std_ulogic_vector)
  return std_ulogic_vector;

function "xnor" (l, r : std_logic_vector)
  return std_logic_vector;
function "xnor" (l, r : std_ulogic_vector)
  return std_ulogic_vector;

function "not"  (l : std_logic_vector)
  return std_logic_vector;
function "not"  (l : std_ulogic_vector)
  return std_ulogic_vector;

-- conversion functions
```

```
function To_bit (s : std_ulogic; xmap : bit := '0')
  return bit;

function To_bitvector (s : std_logic_vector; xmap : bit := '0')
  return bit_vector;

function To_bitvector (s : std_ulogic_vector; xmap : bit := '0')
  return bit_vector;

function To_StdULogic (b : bit)
  return std_ulogic;
function To_StdLogicVector (b : bit_vector)
  return std_logic_vector;
function To_StdLogicVector (s : std_ulogic_vector)
  return std_logic_vector;
function To_StdULogicVector (b : bit_vector)
  return std_ulogic_vector;
function To_StdULogicVector (s : std_logic_vector)
  return std_ulogic_vector;

-- strength strippers and type convertors

function To_X01  (s : std_logic_vector)
  return  std_logic_vector;
function To_X01  (s : std_ulogic_vector)
  return  std_ulogic_vector;
function To_X01  (s : std_ulogic)
  return  X01;
function To_X01  (b : bit_vector)
  return  std_logic_vector;
function To_X01  (b : bit_vector)
  return  std_ulogic_vector;
function To_X01  (b : bit)
  return  X01;

function To_X01Z (s : std_logic_vector)
  return  std_logic_vector;
function To_X01Z (s : std_ulogic_vector)
  return  std_ulogic_vector;
function To_X01Z (s : std_ulogic)
  return  X01Z;
function To_X01Z (b : bit_vector)
  return  std_logic_vector;
function To_X01Z (b : bit_vector)
  return  std_ulogic_vector;
function To_X01Z (b : bit)
  return  X01Z;

function To_UX01  (s : std_logic_vector)
  return  std_logic_vector;
function To_UX01  (s : std_ulogic_vector)
  return  std_ulogic_vector;
function To_UX01  (s : std_ulogic)
  return  UX01;
function To_UX01  (b : bit_vector)
```

```
    return  std_logic_vector;
  function To_UX01  (b : bit_vector)
    return  std_ulogic_vector;
  function To_UX01  (b : bit)
    return  UX01;

  -- edge detection

  function rising_edge  (signal s : std_ulogic) return boolean;
  function falling_edge (signal s : std_ulogic) return boolean;

  -- object contains an unknown

  function Is_X (s : std_ulogic_vector) return  boolean;
  function Is_X (s : std_logic_vector) return  boolean;
  function Is_X (s : std_ulogic) return  boolean;

end;
```

A.4 Package Std_Logic_1164_Additions

Package std_logic_1164_additions is the VHDL-1993 supplement to package std_logic_1164 (see Appendix A.3). In VHDL-2008, the additions are merged into package std_logic_1164 and so this package is empty.

```
package std_logic_1164_additions is

  -- std_match operators
  -- implemented as extended-named functions in VHDL-1993
  -- implemented as operators in VHDL-2008
  function \?=\ (l, r : std_ulogic) return std_ulogic;
  function \?=\ (l, r : std_logic_vector) return std_ulogic;
  function \?=\ (l, r : std_ulogic_vector) return std_ulogic;
  function \?/=\ (l, r : std_ulogic) return std_ulogic;
  function \?/=\ (l, r : std_logic_vector) return std_ulogic;
  function \?/=\ (l, r : std_ulogic_vector) return std_ulogic;
  function \?>\ (l, r : std_ulogic) return std_ulogic;
  function \?>=\ (l, r : std_ulogic) return std_ulogic;
  function \?<\ (l, r : std_ulogic) return std_ulogic;
  function \?<=\ (l, r : std_ulogic) return std_ulogic;
  function \??\ (s : std_ulogic) return boolean;

  -- minimum/maximum functions
  function maximum (l, r : std_ulogic_vector)
    return std_ulogic_vector;
  function maximum (l, r : std_logic_vector)
    return std_logic_vector;
  function maximum (l, r : std_ulogic)
    return std_ulogic;
  function minimum (l, r : std_ulogic_vector)
    return std_ulogic_vector;
  function minimum (l, r : std_logic_vector)
```

```
    return std_logic_vector;
function minimum (l, r : std_ulogic)
  return std_ulogic;

-- selecting boolean operators
function "and" (l : std_logic_vector; r : std_ulogic)
  return std_logic_vector;
function "and" (l : std_ulogic_vector; r : std_ulogic)
  return std_ulogic_vector;
function "and" (l : std_ulogic; r : std_logic_vector)
  return std_logic_vector;
function "and" (l : std_ulogic; r : std_ulogic_vector)
  return std_ulogic_vector;
function "nand" (l : std_logic_vector; r : std_ulogic)
  return std_logic_vector;
function "nand" (l : std_ulogic_vector; r : std_ulogic)
  return std_ulogic_vector;
function "nand" (l : std_ulogic; r : std_logic_vector)
  return std_logic_vector;
function "nand" (l : std_ulogic; r : std_ulogic_vector)
  return std_ulogic_vector;
function "or" (l : std_logic_vector; r : std_ulogic)
  return std_logic_vector;
function "or" (l : std_ulogic_vector; r : std_ulogic)
  return std_ulogic_vector;
function "or" (l : std_ulogic; r : std_logic_vector)
  return std_logic_vector;
function "or" (l : std_ulogic; r : std_ulogic_vector)
  return std_ulogic_vector;
function "nor" (l : std_logic_vector; r : std_ulogic)
  return std_logic_vector;
function "nor" (l : std_ulogic_vector; r : std_ulogic)
  return std_ulogic_vector;
function "nor" (l : std_ulogic; r : std_logic_vector)
  return std_logic_vector;
function "nor" (l : std_ulogic; r : std_ulogic_vector)
  return std_ulogic_vector;
function "xor" (l : std_logic_vector; r : std_ulogic)
  return std_logic_vector;
function "xor" (l : std_ulogic_vector; r : std_ulogic)
  return std_ulogic_vector;
function "xor" (l : std_ulogic; r : std_logic_vector)
  return std_logic_vector;
function "xor" (l : std_ulogic; r : std_ulogic_vector)
  return std_ulogic_vector;
function "xnor" (l : std_logic_vector; r : std_ulogic)
  return std_logic_vector;
function "xnor" (l : std_ulogic_vector; r : std_ulogic)
  return std_ulogic_vector;
function "xnor" (l : std_ulogic; r : std_logic_vector)
  return std_logic_vector;
function "xnor" (l : std_ulogic; r : std_ulogic_vector)
  return std_ulogic_vector;

-- reducing boolean operators
```

```
-- implemented as functions in VHDL-1993
-- implemented as boolean operators in VHDL-2008
function and_reduce (l : std_logic_vector)
  return std_ulogic;
function and_reduce (l : std_ulogic_vector)
  return std_ulogic;
function nand_reduce (l : std_logic_vector)
  return std_ulogic;
function nand_reduce (l : std_ulogic_vector)
  return std_ulogic;
function or_reduce (l : std_logic_vector)
  return std_ulogic;
function or_reduce (l : std_ulogic_vector)
  return std_ulogic;
function nor_reduce (l : std_logic_vector)
  return std_ulogic;
function nor_reduce (l : std_ulogic_vector)
  return std_ulogic;
function xor_reduce (l : std_logic_vector)
  return std_ulogic;
function xor_reduce (l : std_ulogic_vector)
  return std_ulogic;
function xnor_reduce (l : std_logic_vector)
  return std_ulogic;
function xnor_reduce (l : std_ulogic_vector)
  return std_ulogic;

-- shift operators
function "sll" (l : std_logic_vector; r : integer)
  return std_logic_vector;
function "sll" (l : std_ulogic_vector; r : integer)
  return std_ulogic_vector;
function "srl" (l : std_logic_vector; r : integer)
  return std_logic_vector;
function "srl" (l : std_ulogic_vector; r : integer)
  return std_ulogic_vector;
function "rol" (l : std_logic_vector; r : integer)
  return std_logic_vector;
function "rol" (l : std_ulogic_vector; r : integer)
  return std_ulogic_vector;
function "ror" (l : std_logic_vector; r : integer)
  return std_logic_vector;
function "ror" (l : std_ulogic_vector; r : integer)
  return std_ulogic_vector;

-- type conversions
alias to_bv is
  ieee.std_logic_1164.to_bitvector
  [ std_logic_vector, bit return bit_vector];
alias to_bv is
  ieee.std_logic_1164.to_bitvector
  [ std_ulogic_vector, bit return bit_vector];
alias to_bit_vector is
  ieee.std_logic_1164.to_bitvector
  [ std_logic_vector, bit return bit_vector];
```

```
alias to_bit_vector is
  ieee.std_logic_1164.to_bitvector
  [ std_ulogic_vector, bit return bit_vector] ;
alias to_slv is
  ieee.std_logic_1164.to_stdlogicvector
  [ bit_vector return std_logic_vector] ;
alias to_slv is
  ieee.std_logic_1164.to_stdlogicvector
  [ std_ulogic_vector return std_logic_vector] ;
alias to_std_logic_vector is
  ieee.std_logic_1164.to_stdlogicvector
  [ bit_vector return std_logic_vector] ;
alias to_std_logic_vector is
  ieee.std_logic_1164.to_stdlogicvector
  [ std_ulogic_vector return std_logic_vector] ;
alias to_suv is
  ieee.std_logic_1164.to_stdulogicvector
  [ bit_vector return std_ulogic_vector] ;
alias to_suv is
  ieee.std_logic_1164.to_stdulogicvector
  [ std_logic_vector return std_ulogic_vector] ;
alias to_std_ulogic_vector is
  ieee.std_logic_1164.to_stdulogicvector
  [ bit_vector return std_ulogic_vector] ;
alias to_std_ulogic_vector is
  ieee.std_logic_1164.to_stdulogicvector
  [ std_logic_vector return std_ulogic_vector] ;

-- String formatting functions
function to_string (value : std_ulogic)
  return string;
function to_string (value : std_ulogic_vector)
  return string;
function to_string (value : std_logic_vector)
  return string;
alias to_bstring is
  to_string [ std_ulogic_vector return string] ;
alias to_bstring is
  to_string [ std_logic_vector return string] ;
alias to_binary_string is
  to_string [ std_ulogic_vector return string] ;
alias to_binary_string is
  to_string [ std_logic_vector return string] ;
function to_ostring (value : std_ulogic_vector)
  return string;
function to_ostring (value : std_logic_vector)
  return string;
alias to_octal_string is
  to_ostring [ std_ulogic_vector return string] ;
alias to_octal_string is
  to_ostring [ std_logic_vector return string] ;
function to_hstring (value : std_ulogic_vector)
  return string;
function to_hstring (value : std_logic_vector)
```

```
      return string;
alias to_hex_string is
  to_hstring [ std_ulogic_vector return string] ;
alias to_hex_string is
  to_hstring [ std_logic_vector return string] ;

-- Text I/O procedures
procedure read (l : inout line;
                value : out std_ulogic;
                good : out boolean);
procedure read (l : inout line;
                value : out std_ulogic);
procedure read (l : inout line;
                value : out std_ulogic_vector;
                good : out boolean);
procedure read (l : inout line;
                value : out std_ulogic_vector);
procedure read (l : inout line;
                value : out std_logic_vector;
                good : out boolean);
procedure read (l : inout line;
                value : out std_logic_vector);
alias bread is
  read [ line, std_ulogic_vector, boolean] ;
alias bread is
  read [ line, std_ulogic_vector] ;
alias bread is
  read [ line, std_logic_vector, boolean] ;
alias bread is
  read [ line, std_logic_vector] ;
alias binary_read is
  read [ line, std_ulogic_vector, boolean] ;
alias binary_read is
  read [ line, std_ulogic_vector] ;
alias binary_read is
  read [ line, std_logic_vector, boolean] ;
alias binary_read is
  read [ line, std_logic_vector] ;

procedure oread (l : inout line;
                 value : out std_ulogic_vector;
                 good : out boolean);
procedure oread (l : inout line;
                 value : out std_ulogic_vector);
procedure oread (l : inout line;
                 value : out std_logic_vector;
                 good : out boolean);
procedure oread (l : inout line;
                 value : out std_logic_vector);
alias octal_read is
  oread [ line, std_ulogic_vector, boolean] ;
alias octal_read is
  oread [ line, std_ulogic_vector] ;
alias octal_read is
```

```
      oread [ line, std_logic_vector, boolean] ;
  alias octal_read is
    oread [ line, std_logic_vector] ;

  procedure hread (l : inout line;
                     value : out std_ulogic_vector;
                     good : out boolean);
  procedure hread (l : inout line;
                     value : out std_ulogic_vector);
  procedure hread (l : inout line;
                     value : out std_logic_vector;
                     good : out boolean);
  procedure hread (l : inout line;
                     value : out std_logic_vector);
  alias hex_read is
    hread [ line, std_ulogic_vector, boolean] ;
  alias hex_read is
    hread [ line, std_ulogic_vector] ;
  alias hex_read is
    hread [ line, std_logic_vector, boolean] ;
  alias hex_read is
    hread [ line, std_logic_vector] ;

  procedure write (l : inout line;
                     value : in std_ulogic;
                     justified : in    side := right;
                     field : in width := 0);
  procedure write (l : inout line;
                     value : in std_ulogic_vector;
                     justified : in    side := right;
                     field : in width := 0);
  procedure write (l : inout line;
                     value : in std_logic_vector;
                     justified : in    side := right;
                     field : in width := 0);
  alias bwrite is
    write [ line, std_ulogic_vector, side, width] ;
  alias bwrite is
    write [ line, std_logic_vector, side, width] ;
  alias binary_write is
    write [ line, std_ulogic_vector, side, width] ;
  alias binary_write is
    write [ line, std_logic_vector, side, width] ;
  procedure owrite (l : inout line;
                     value : in std_ulogic_vector;
                     justified : in    side := right;
                     field : in width := 0);
  procedure owrite (l : inout line;
                     value : in std_logic_vector;
                     justified : in    side := right;
                     field : in width := 0);
  alias octal_write is
    owrite [ line, std_ulogic_vector, side, width] ;
  alias octal_write is
```

```
       owrite [ line, std_logic_vector, side, width] ;

   procedure hwrite (l : inout line;
                     value : in std_ulogic_vector;
                     justified : in    side := right;
                     field : in width := 0);
   procedure hwrite (l : inout line;
                     value : in std_logic_vector;
                     justified : in    side := right;
                     field : in width := 0);
   alias hex_write is
     hwrite [ line, std_ulogic_vector, side, width] ;
   alias hex_write is
     hwrite [ line, std_logic_vector, side, width] ;

end;
```

A.5   Package Numeric_Std

Package numeric_std provided the synthesisable arbitrary-precision integer arithmetic types signed and unsigned. This is the VHDL-1993 version and is supplemented by package numeric_std_additions (see Appendix A.6). In VHDL-2008, the additions are merged into package numeric_std.

```
package numeric_std is

   ------------------------------------------------------
   -- numeric array type definitions
   ------------------------------------------------------

   type unsigned is array (natural range <>) of std_logic;
   type signed   is array (natural range <>) of std_logic;


   ------------------------------------------------------
   -- arithmetic operators:
   ------------------------------------------------------

   function "abs"(l: signed) return signed;
   function "-"(l: signed) return signed;


   ------------------------------------------------------

   function "+"(l, r: unsigned) return unsigned;
   function "+"(l, r: signed) return signed;
   function "+"(l: unsigned; r: natural) return unsigned;
   function "+"(l: natural; r: unsigned) return unsigned;
   function "+"(l: integer; r: signed) return signed;
   function "+"(l: signed; r: integer) return signed;


   ------------------------------------------------------

   function "-"(l, r: unsigned) return unsigned;
```

```
function "-"(l, r: signed) return signed;
function "-"(l: unsigned;r: natural) return unsigned;
function "-"(l: natural; r: unsigned) return unsigned;
function "-"(l: signed; r: integer) return signed;
function "-"(l: integer; r: signed) return signed;


-----------------------------------------------------------


function "*"(l, r: unsigned) return unsigned;
function "*"(l, r: signed) return signed;
function "*"(l: unsigned; r: natural) return unsigned;
function "*"(l: natural; r: unsigned) return unsigned;
function "*"(l: signed; r: integer) return signed;
function "*"(l: integer; r: signed) return signed;


-----------------------------------------------------------


function "/"(l, r: unsigned) return unsigned;
function "/"(l, r: signed) return signed;
function "/"(l: unsigned; r: natural) return unsigned;
function "/"(l: natural; r: unsigned) return unsigned;
function "/"(l: signed; r: integer) return signed;
function "/"(l: integer; r: signed) return signed;


-----------------------------------------------------------


function "rem"(l, r: unsigned) return unsigned;
function "rem"(l, r: signed) return signed;
function "rem"(l: unsigned; r: natural) return unsigned;
function "rem"(l: natural; r: unsigned) return unsigned;
function "rem"(l: signed; r: integer) return signed;
function "rem"(l: integer; r: signed) return signed;


-----------------------------------------------------------


function "mod"(l, r: unsigned) return unsigned;
function "mod"(l, r: signed) return signed;
function "mod"(l: unsigned; r: natural) return unsigned;
function "mod"(l: natural; r: unsigned) return unsigned;
function "mod"(l: signed; r: integer) return signed;
function "mod"(l: integer; r: signed) return signed;


-----------------------------------------------------------
-- comparison operators
-----------------------------------------------------------


function ">"(l, r: unsigned) return boolean;
function ">"(l, r: signed) return boolean;
function ">"(l: natural; r: unsigned) return boolean;
function ">"(l: integer; r: signed) return boolean;
function ">"(l: unsigned; r: natural) return boolean;
function ">"(l: signed; r: integer) return boolean;


-----------------------------------------------------------
```

```
function "<"(l, r: unsigned) return boolean;
function "<"(l, r: signed) return boolean;
function "<"(l: natural; r: unsigned) return boolean;
function "<"(l: integer; r: signed) return boolean;
function "<"(l: unsigned; r: natural) return boolean;
function "<"(l: signed; r: integer) return boolean;


--------------------------------------------------------


function "<="(l, r: unsigned) return boolean;
function "<="(l, r: signed) return boolean;
function "<="(l: natural; r: unsigned) return boolean;
function "<="(l: integer; r: signed) return boolean;
function "<="(l: unsigned; r: natural) return boolean;
function "<="(l: signed; r: integer) return boolean;


--------------------------------------------------------


function ">="(l, r: unsigned) return boolean;
function ">="(l, r: signed) return boolean;
function ">="(l: natural; r: unsigned) return boolean;
function ">="(l: integer; r: signed) return boolean;
function ">="(l: unsigned; r: natural) return boolean;
function ">="(l: signed; r: integer) return boolean;


--------------------------------------------------------
function "="(l, r: unsigned) return boolean;
function "="(l, r: signed) return boolean;
function "="(l: natural; r: unsigned) return boolean;
function "="(l: integer; r: signed) return boolean;
function "="(l: unsigned; r: natural) return boolean;
function "="(l: signed; r: integer) return boolean;


--------------------------------------------------------


function "/="(l, r: unsigned) return boolean;
function "/="(l, r: signed) return boolean;
function "/="(l: natural; r: unsigned) return boolean;
function "/="(l: integer; r: signed) return boolean;
function "/="(l: unsigned; r: natural) return boolean;
function "/="(l: signed; r: integer) return boolean;


--------------------------------------------------------
-- shift and rotate functions
--------------------------------------------------------


function shift_left(l: unsigned; n: natural) return unsigned;
function shift_right(l: unsigned; n: natural) return unsigned;
function shift_left(l: signed; n: natural) return signed;
function shift_right(l: signed; n: natural) return signed;
function rotate_left(l: unsigned; n: natural) return unsigned;
function rotate_right(l: unsigned; n: natural) return unsigned;
function rotate_left(l: signed; n: natural) return signed;
function rotate_right(l: signed; n: natural) return signed;
```

```
function "sll"(l: unsigned; n: integer) return unsigned;
function "sll"(l: signed; n: integer) return signed;
function "srl"(l: unsigned; n: integer) return unsigned;
function "srl"(l: signed; n: integer) return signed;
function "rol"(l: unsigned; n: integer) return unsigned;
function "rol"(l: signed; n: integer) return signed;
function "ror"(l: unsigned; n: integer) return unsigned;
function "ror"(l: signed; n: integer) return signed;


-------------------------------------------------------
-- resize functions
-------------------------------------------------------

function resize(l: signed; s: natural) return signed;
function resize(l: unsigned; s: natural) return unsigned;


-------------------------------------------------------
-- conversion functions
-------------------------------------------------------

function to_integer(l: unsigned) return natural;
function to_integer(l: signed) return integer;
function to_unsigned(l: natural; s: natural) return unsigned;
function to_signed(l: integer; s: natural) return signed;


-------------------------------------------------------
-- logical operators
-------------------------------------------------------

function "not"(l: unsigned) return unsigned;
function "and"(l, r: unsigned) return unsigned;
function "or"(l, r: unsigned) return unsigned;
function "nand"(l, r: unsigned) return unsigned;
function "nor"(l, r: unsigned) return unsigned;
function "xor"(l, r: unsigned) return unsigned;
function "xnor"(l, r: unsigned) return unsigned;
function "not"(l: signed) return signed;
function "and"(l, r: signed) return signed;
function "or"(l, r: signed) return signed;
function "nand"(l, r: signed) return signed;
function "nor"(l, r: signed) return signed;
function "xor"(l, r: signed) return signed;
function "xnor"(l, r: signed) return signed;


-------------------------------------------------------
-- match functions
-------------------------------------------------------

function std_match(l, r: std_ulogic) return boolean;
function std_match(l, r: unsigned) return boolean;
function std_match(l, r: signed) return boolean;
function std_match(l, r: std_logic_vector) return boolean;
function std_match(l, r: std_ulogic_vector) return boolean;
```

```
----------------------------------------------------
-- translation functions
----------------------------------------------------

function to_01(s: unsigned; xmap: std_logic := '0')
  return unsigned;
function to_01(s: signed; xmap: std_logic := '0')
  return signed;

end;
```

A.6   Package Numeric_Std_Additions

Package numeric_std_additions is the VHDL-1993 supplement to package numeric_std (see Appendix A.5). In VHDL-2008, the additions are merged into package numeric_std and so this package is empty.

```
library ieee;
use ieee.std_logic_1164.all;
use ieee.numeric_std.all;
use std.textio.all;

package numeric_std_additions is

  -- unresolved variant of signed/unsigned
  -- these are only unresolved in VHDL-2008
  -- in VHDL-1993 they are subtypes of the resolved type
  subtype unresolved_unsigned is unsigned;
  subtype unresolved_signed is signed;
  subtype u_unsigned is unsigned;
  subtype u_signed is signed;

  -- add and subtract with a one-bit value
  function "+"(l : unsigned; r : std_ulogic)
    return unsigned;
  function "+"(l : std_ulogic; r : unsigned)
    return unsigned;
  function "+"(l : signed; r : std_ulogic)
    return signed;
  function "+"(l : std_ulogic; r : signed)
    return signed;
  function "-"(l : unsigned; r : std_ulogic)
    return unsigned;
  function "-"(l : std_ulogic; r : unsigned)
    return unsigned;
  function "-"(l : signed; r : std_ulogic)
    return signed;
  function "-"(l : std_ulogic; r : signed)
    return signed;

  -- std_match operators
```

```
-- implemented as extended-named functions in VHDL-1993
-- implemented as operators in VHDL-2008
function \?=\ (l, r : unsigned)
  return std_ulogic;
function \?/=\ (l, r : unsigned)
  return std_ulogic;
function \?>\ (l, r : unsigned)
  return std_ulogic;
function \?>=\ (l, r : unsigned)
  return std_ulogic;
function \?<\ (l, r : unsigned)
  return std_ulogic;
function \?<=\ (l, r : unsigned)
  return std_ulogic;
function \?=\ (l : unsigned; r : natural)
  return std_ulogic;
function \?/=\ (l : unsigned; r : natural)
  return std_ulogic;
function \?>\ (l : unsigned; r : natural)
  return std_ulogic;
function \?>=\ (l : unsigned; r : natural)
  return std_ulogic;
function \?<\ (l : unsigned; r : natural)
  return std_ulogic;
function \?<=\ (l : unsigned; r : natural)
  return std_ulogic;
function \?=\ (l : natural; r : unsigned)
  return std_ulogic;
function \?/=\ (l : natural; r : unsigned)
  return std_ulogic;
function \?>\ (l : natural; r : unsigned)
  return std_ulogic;
function \?>=\ (l : natural; r : unsigned)
  return std_ulogic;
function \?<\ (l : natural; r : unsigned)
  return std_ulogic;
function \?<=\ (l : natural; r : unsigned)
  return std_ulogic;

function \?=\ (l, r : signed)
  return std_ulogic;
function \?/=\ (l, r : signed)
  return std_ulogic;
function \?>\ (l, r : signed)
  return std_ulogic;
function \?>=\ (l, r : signed)
  return std_ulogic;
function \?<\ (l, r : signed)
  return std_ulogic;
function \?<=\ (l, r : signed)
  return std_ulogic;
function \?=\ (l : signed; r : integer)
  return std_ulogic;
function \?/=\ (l : signed; r : integer)
  return std_ulogic;
```

```
function \?>\ (l : signed; r : integer)
  return std_ulogic;
function \?>=\ (l : signed; r : integer)
  return std_ulogic;
function \?<\ (l : signed; r : integer)
  return std_ulogic;
function \?<=\ (l : signed; r : integer)
  return std_ulogic;
function \?=\ (l : integer; r : signed)
  return std_ulogic;
function \?/=\ (l : integer; r : signed)
  return std_ulogic;
function \?>\ (l : integer; r : signed)
  return std_ulogic;
function \?>=\ (l : integer; r : signed)
  return std_ulogic;
function \?<\ (l : integer; r : signed)
  return std_ulogic;
function \?<=\ (l : integer; r : signed)
  return std_ulogic;

-- selecting boolean operators
function "and" (l : std_ulogic; r : unsigned)
  return unsigned;
function "and" (l : unsigned; r : std_ulogic)
  return unsigned;
function "or" (l : std_ulogic; r : unsigned)
  return unsigned;
function "or" (l : unsigned; r : std_ulogic)
  return unsigned;
function "nand" (l : std_ulogic; r : unsigned)
  return unsigned;
function "nand" (l : unsigned; r : std_ulogic)
  return unsigned;
function "nor" (l : std_ulogic; r : unsigned)
  return unsigned;
function "nor" (l : unsigned; r : std_ulogic)
  return unsigned;
function "xor" (l : std_ulogic; r : unsigned)
  return unsigned;
function "xor" (l : unsigned; r : std_ulogic)
  return unsigned;
function "xnor" (l : std_ulogic; r : unsigned)
  return unsigned;
function "xnor" (l : unsigned; r : std_ulogic)
  return unsigned;
function "and" (l : std_ulogic; r : signed)
  return signed;
function "and" (l : signed; r : std_ulogic)
  return signed;
function "or" (l : std_ulogic; r : signed)
  return signed;
function "or" (l : signed; r : std_ulogic)
  return signed;
function "nand" (l : std_ulogic; r : signed)
```

```
    return signed;
function "nand" (l : signed; r : std_ulogic)
    return signed;
function "nor" (l : std_ulogic; r : signed)
    return signed;
function "nor" (l : signed; r : std_ulogic)
    return signed;
function "xor" (l : std_ulogic; r : signed)
    return signed;
function "xor" (l : signed; r : std_ulogic)
    return signed;
function "xnor" (l : std_ulogic; r : signed)
    return signed;
function "xnor" (l : signed; r : std_ulogic)
    return signed;

-- reducing boolean operators
-- implemented as reduction functions in VHDL-1993
-- proper operators in VHDL-2008
function and_reduce(l : signed)
    return std_ulogic;
function nand_reduce(l : signed)
    return std_ulogic;
function or_reduce(l : signed)
    return std_ulogic;
function nor_reduce(l : signed)
    return std_ulogic;
function xor_reduce(l : signed)
    return std_ulogic;
function xnor_reduce(l : signed)
    return std_ulogic;
function and_reduce(l : unsigned)
    return std_ulogic;
function nand_reduce(l : unsigned)
    return std_ulogic;
function or_reduce(l : unsigned)
    return std_ulogic;
function nor_reduce(l : unsigned)
    return std_ulogic;
function xor_reduce(l : unsigned)
    return std_ulogic;
function xnor_reduce(l : unsigned)
    return std_ulogic;

-- arithmetic shift operators
function "sla" (arg : signed; count : integer)
    return signed;
function "sla" (arg : unsigned; count : integer)
    return unsigned;
function "sra" (arg : signed; count : integer)
    return signed;
function "sra" (arg : unsigned; count : integer)
    return unsigned;
```

```
-- maximum-minimum functions
function maximum (l, r : unsigned)
  return unsigned;
function maximum (l, r : signed)
  return signed;
function minimum (l, r : unsigned)
  return unsigned;
function minimum (l, r : signed)
  return signed;
function maximum (l : unsigned; r : natural)
  return unsigned;
function maximum (l : signed; r : integer)
  return signed;
function minimum (l : unsigned; r : natural)
  return unsigned;
function minimum (l : signed; r : integer)
  return signed;
function maximum (l : natural; r : unsigned)
  return unsigned;
function maximum (l : integer; r : signed)
  return signed;
function minimum (l : natural; r : unsigned)
  return unsigned;
function minimum (l : integer; r : signed)
  return signed;

function find_rightmost (arg : unsigned; y : std_ulogic)
  return integer;
function find_rightmost (arg : signed; y : std_ulogic)
  return integer;
function find_leftmost (arg : unsigned; y : std_ulogic)
  return integer;
function find_leftmost (arg : signed; y : std_ulogic)
  return integer;

-- type conversions
function to_unresolved_unsigned (arg, size : natural)
  return unresolved_unsigned;
alias to_u_unsigned is
  to_unresolved_unsigned
  [natural, natural return unresolved_unsigned];
function to_unresolved_signed (arg : integer; size : natural)
  return unresolved_signed;
alias to_u_signed is
  to_unresolved_signed
  [natural, natural return unresolved_signed];

-- strength changers - not synthesisable
function to_x01 (s : unsigned)
  return unsigned;
function to_x01 (s : signed)
  return signed;
```

```
function to_x01z (s : unsigned)
  return unsigned;
function to_x01z (s : signed)
  return signed;
function to_ux01 (s : unsigned)
  return unsigned;
function to_ux01 (s : signed)
  return signed;
function is_x (s : unsigned)
  return boolean;
function is_x (s : signed)
  return boolean;

-- printable string values
function to_string (value : unsigned)
  return string;
function to_string (value : signed)
  return string;
alias to_bstring is
  to_string [ unsigned return string] ;
alias to_bstring is
  to_string [ signed return string] ;
alias to_binary_string is
  to_string [ unsigned return string] ;
alias to_binary_string is
  to_string [ signed return string] ;

function to_ostring (value : unsigned)
  return string;
function to_ostring (value : signed)
  return string;
alias to_octal_string is
  to_ostring [ unsigned return string] ;
alias to_octal_string is
  to_ostring [ signed return string] ;

function to_hstring (value : unsigned)
  return string;
function to_hstring (value : signed)
  return string;
alias to_hex_string is
  to_hstring [ unsigned return string] ;
alias to_hex_string is
  to_hstring [ signed return string] ;

-- Text I/O extensions
procedure read(l : inout line;
               value : out unsigned;
               good : out boolean);
procedure read(l : inout line;
               value : out unsigned);
procedure read(l : inout line;
               value : out signed;
               good : out boolean);
```

```
procedure read(l : inout line;
               value : out signed);
alias binary_read is
  read [ line, unsigned, boolean] ;
alias binary_read is
  read [ line, signed, boolean] ;
alias binary_read is
  read [ line, unsigned] ;
alias binary_read is
  read [ line, signed] ;

procedure oread (l : inout line;
               value : out unsigned;
               good : out boolean);
procedure oread (l : inout line;
               value : out signed;
               good : out boolean);
procedure oread (l : inout line;
               value : out unsigned);
procedure oread (l : inout line;
               value : out signed);
alias octal_read is
  oread [ line, unsigned, boolean] ;
alias octal_read is
  oread [ line, signed, boolean] ;
alias octal_read is
  oread [ line, unsigned] ;
alias octal_read is
  oread [ line, signed] ;

procedure hread (l : inout line;
               value : out unsigned;
               good : out boolean);
procedure hread (l : inout line;
               value : out signed;
               good : out boolean);
procedure hread (l : inout line;
               value : out unsigned);
procedure hread (l : inout line;
               value : out signed);
alias hex_read is
  hread [ line, unsigned, boolean] ;
alias hex_read is
  hread [ line, signed, boolean] ;
alias hex_read is
  hread [ line, unsigned] ;
alias hex_read is
  hread [ line, signed] ;

procedure write (l : inout line;
               value : in unsigned;
               justified : in side := right;
               field : in width := 0);
procedure write (l : inout line;
```

```
                            value : in signed;
                            justified : in side := right;
                            field : in width := 0);
    alias binary_write is
      write [ line, unsigned, side, width] ;
    alias binary_write is
      write [ line, signed, side, width] ;

    procedure owrite (l : inout line;
                            value : in unsigned;
                            justified : in side := right;
                            field : in width := 0);
    procedure owrite (l : inout line;
                            value : in signed;
                            justified : in side := right;
                            field : in width := 0);
    alias octal_write is
      owrite [ line, unsigned, side, width] ;
    alias octal_write is
      owrite [ line, signed, side, width] ;

    procedure hwrite (l : inout line;
                            value : in unsigned;
                            justified : in side := right;
                            field : in width := 0);
    procedure hwrite (l : inout line;
                            value : in signed;
                            justified : in side := right;
                            field : in width := 0);
    alias hex_write is
      hwrite [ line, unsigned, side, width] ;
    alias hex_write is
      hwrite [ line, signed, side, width] ;

  end;
```

A.7   Package Fixed_Float_Types

Package `fixed_float_types` is used to define enumeration types used as generic parameters to `fixed_generic_pkg` and `float_generic_pkg`. For consistency, these enumerations are still defined here for the VHDL-1993 compatibility versions of the packages even though they do not have generic parameters.

```
  package fixed_float_types is

    -- used in fixed_pkg
    type fixed_round_style_type is
      (fixed_round, fixed_truncate);
    type fixed_overflow_style_type is
      (fixed_saturate, fixed_wrap);
```

```
    -- used in float_pkg
    type round_type is
      (round_nearest, round_inf, round_neginf, round_zero);

  end;
```

A.8   Package Fixed_Pkg

Package `fixed_pkg` listed here is the VHDL-1993 compatibility version of the VHDL-2008 instantiation of `fixed_generic_pkg` with the default generic parameters. It is therefore usable in both VHDL-1993 and VHDL-2008 systems without change.

```
package fixed_pkg is

  -- generics converted into constants for VHDL-1993
  constant fixed_round_style : fixed_round_style_type :=
    fixed_round;
  constant fixed_overflow_style : fixed_overflow_style_type :=
    fixed_saturate;
  constant fixed_guard_bits : natural := 3;
  constant no_warning : boolean := false;

  type unresolved_ufixed is array (integer range <>) of std_ulogic;
  type unresolved_sfixed is array (integer range <>) of std_ulogic;

  subtype u_ufixed is unresolved_ufixed;
  subtype u_sfixed is unresolved_sfixed;

  subtype ufixed is u_ufixed;
  subtype sfixed is u_sfixed;

  ---------------------------------------------------
  -- arithmetic operators

  -- ufixed
  function "+" (l, r : ufixed) return ufixed;
  function "-" (l, r : ufixed) return ufixed;
  function "*" (l, r : ufixed) return ufixed;
  function "/" (l, r : ufixed) return ufixed;
  function "rem" (l, r : ufixed) return ufixed;
  function "mod" (l, r : ufixed) return ufixed;

  procedure add_carry (
    l, r   : in  ufixed; c_in : in  std_ulogic;
    result : out ufixed; c_out : out std_ulogic);

  function divide (l, r : ufixed;
    round_style : fixed_round_style_type :=
      fixed_round_style;
    guard_bits  : natural := fixed_guard_bits)
    return ufixed;
```

```
function reciprocal (arg : ufixed;
  round_style : fixed_round_style_type :=
    fixed_round_style;
  guard_bits  : natural := fixed_guard_bits)
  return ufixed;

function remainder (l, r : ufixed;
  round_style : fixed_round_style_type :=
    fixed_round_style;
  guard_bits  : natural := fixed_guard_bits)
  return ufixed;

function modulo (l, r : ufixed;
  round_style : fixed_round_style_type :=
    fixed_round_style;
  guard_bits  : natural := fixed_guard_bits)
  return ufixed;

function scalb (y : ufixed; n : integer) return ufixed;
function scalb (y : ufixed; n : signed) return ufixed;

-- sfixed
function is_negative (arg : sfixed) return boolean;

function "abs" (arg : sfixed) return sfixed;
function "-" (arg : sfixed)return sfixed;
function "+" (l, r : sfixed) return sfixed;
function "-" (l, r : sfixed) return sfixed;
function "*" (l, r : sfixed) return sfixed;
function "/" (l, r : sfixed) return sfixed;
function "rem" (l, r : sfixed) return sfixed;
function "mod" (l, r : sfixed) return sfixed;

procedure add_carry (
  l, r   : in  sfixed; c_in  : in  std_ulogic;
  result : out sfixed; c_out : out std_ulogic);

function divide (l, r : sfixed;
  round_style : fixed_round_style_type :=
    fixed_round_style;
  guard_bits  : natural := fixed_guard_bits)
  return sfixed;

function reciprocal (arg : sfixed;
  round_style : fixed_round_style_type :=
    fixed_round_style;
  guard_bits  : natural := fixed_guard_bits)
  return sfixed;

function remainder (l, r : sfixed;
  round_style : fixed_round_style_type :=
    fixed_round_style;
  guard_bits  : natural := fixed_guard_bits)
  return sfixed;
```

```
function modulo (l, r : sfixed;
  overflow_style : fixed_overflow_style_type :=
    fixed_overflow_style;
  round_style    : fixed_round_style_type :=
    fixed_round_style;
  guard_bits     : natural := fixed_guard_bits)
  return sfixed;

function scalb (y : sfixed; n : integer) return sfixed;
function scalb (y : sfixed; n : signed) return sfixed;


-----------------------------------------------------
-- comparison operators

-- ufixed
function ">"  (l, r : ufixed) return boolean;
function "<"  (l, r : ufixed) return boolean;
function "<=" (l, r : ufixed) return boolean;
function ">=" (l, r : ufixed) return boolean;
function "="  (l, r : ufixed) return boolean;
function "/=" (l, r : ufixed) return boolean;

function \?=\   (l, r : ufixed) return std_ulogic;
function \?/=\  (l, r : ufixed) return std_ulogic;
function \?>\   (l, r : ufixed) return std_ulogic;
function \?>=\  (l, r : ufixed) return std_ulogic;
function \?<\   (l, r : ufixed) return std_ulogic;
function \?<=\  (l, r : ufixed) return std_ulogic;
function std_match (l, r : ufixed) return boolean;

function maximum (l, r : ufixed) return ufixed;
function minimum (l, r : ufixed) return ufixed;

-- sfixed
function ">"  (l, r : sfixed) return boolean;
function "<"  (l, r : sfixed) return boolean;
function "<=" (l, r : sfixed) return boolean;
function ">=" (l, r : sfixed) return boolean;
function "="  (l, r : sfixed) return boolean;
function "/=" (l, r : sfixed) return boolean;

function \?=\   (l, r : sfixed) return std_ulogic;
function \?/=\  (l, r : sfixed) return std_ulogic;
function \?>\   (l, r : sfixed) return std_ulogic;
function \?>=\  (l, r : sfixed) return std_ulogic;
function \?<\   (l, r : sfixed) return std_ulogic;
function \?<=\  (l, r : sfixed) return std_ulogic;
function std_match (l, r : sfixed) return boolean;

function maximum (l, r : sfixed) return sfixed;
function minimum (l, r : sfixed) return sfixed;


-----------------------------------------------------
-- shift and rotate functions
```

```
-- ufixed
function "sll" (l : ufixed; n : integer) return ufixed;
function "srl" (l : ufixed; n : integer) return ufixed;
function "rol" (l : ufixed; n : integer) return ufixed;
function "ror" (l : ufixed; n : integer) return ufixed;
function "sla" (l : ufixed; n : integer) return ufixed;
function "sra" (l : ufixed; n : integer) return ufixed;

function shift_left  (l : ufixed; n : natural) return ufixed;
function shift_right (l : ufixed; n : natural) return ufixed;

-- sfixed
function "sll" (l : sfixed; n : integer) return sfixed;
function "srl" (l : sfixed; n : integer) return sfixed;
function "rol" (l : sfixed; n : integer) return sfixed;
function "ror" (l : sfixed; n : integer) return sfixed;
function "sla" (l : sfixed; n : integer) return sfixed;
function "sra" (l : sfixed; n : integer) return sfixed;

function shift_left  (l : sfixed; n : natural) return sfixed;
function shift_right (l : sfixed; n : natural) return sfixed;


--------------------------------------------------
-- logical functions

-- ufixed
function "not"  (l    : ufixed) return ufixed;
function "and"  (l, r : ufixed) return ufixed;
function "or"   (l, r : ufixed) return ufixed;
function "nand" (l, r : ufixed) return ufixed;
function "nor"  (l, r : ufixed) return ufixed;
function "xor"  (l, r : ufixed) return ufixed;
function "xnor" (l, r : ufixed) return ufixed;

function and_reduce (l : ufixed) return std_ulogic;
function nand_reduce (l : ufixed) return std_ulogic;
function or_reduce (l : ufixed) return std_ulogic;
function nor_reduce (l : ufixed) return std_ulogic;
function xor_reduce (l : ufixed) return std_ulogic;
function xnor_reduce (l : ufixed) return std_ulogic;

function find_leftmost (arg : ufixed; y : std_ulogic)
  return integer;
function find_rightmost (arg : ufixed; y : std_ulogic)
  return integer;

-- sfixed
function "not"  (l    : sfixed) return sfixed;
function "and"  (l, r : sfixed) return sfixed;
function "or"   (l, r : sfixed) return sfixed;
function "nand" (l, r : sfixed) return sfixed;
function "nor"  (l, r : sfixed) return sfixed;
function "xor"  (l, r : sfixed) return sfixed;
function "xnor" (l, r : sfixed) return sfixed;
```

```
function and_reduce (l : sfixed) return std_ulogic;
function nand_reduce (l : sfixed) return std_ulogic;
function or_reduce (l : sfixed) return std_ulogic;
function nor_reduce (l : sfixed) return std_ulogic;
function xor_reduce (l : sfixed) return std_ulogic;
function xnor_reduce (l : sfixed) return std_ulogic;

function find_leftmost (arg : sfixed; y : std_ulogic)
  return integer;
function find_rightmost (arg : sfixed; y : std_ulogic)
  return integer;

-----------------------------------------------------
--   resize functions

-- ufixed
function resize (arg : ufixed;
  left_index    : integer;
  right_index   : integer;
  overflow_style : fixed_overflow_style_type :=
    fixed_overflow_style;
  round_style   : fixed_round_style_type :=
    fixed_round_style)
  return ufixed;

function resize (arg : ufixed;
  size_res      : ufixed;
  overflow_style : fixed_overflow_style_type :=
    fixed_overflow_style;
  round_style   : fixed_round_style_type :=
    fixed_round_style)
  return ufixed;

-- sfixed
function resize (arg : sfixed;
  left_index    : integer;
  right_index   : integer;
  overflow_style : fixed_overflow_style_type :=
    fixed_overflow_style;
  round_style   : fixed_round_style_type :=
    fixed_round_style)
  return sfixed;

function resize (arg : sfixed;
  size_res      : sfixed;
  overflow_style : fixed_overflow_style_type :=
    fixed_overflow_style;
  round_style   : fixed_round_style_type :=
    fixed_round_style)
  return sfixed;

-----------------------------------------------------
-- conversion functions

-- ufixed
```

```
function to_ufixed (arg : natural;
  left_index      : integer;
  right_index     : integer := 0;
  overflow_style : fixed_overflow_style_type :=
    fixed_overflow_style;
  round_style     : fixed_round_style_type :=
    fixed_round_style)
  return ufixed;
function to_ufixed (arg : natural;
  size_res        : ufixed;
  overflow_style : fixed_overflow_style_type :=
    fixed_overflow_style;
  round_style   ·   : fixed_round_style_type :=
    fixed_round_style)
  return ufixed;

function to_ufixed (arg : unsigned) return ufixed;

function to_unsigned (arg : ufixed;
  size            : natural;
  overflow_style : fixed_overflow_style_type :=
    fixed_overflow_style;
  round_style     : fixed_round_style_type :=
    fixed_round_style)
  return unsigned;

function to_unsigned (arg : ufixed;
  size_res        : unsigned;
  overflow_style : fixed_overflow_style_type :=
    fixed_overflow_style;
  round_style     : fixed_round_style_type :=
    fixed_round_style)
  return unsigned;

function to_real (arg : ufixed) return real;

function to_integer (arg : ufixed;
  overflow_style : fixed_overflow_style_type :=
    fixed_overflow_style;
  round_style     : fixed_round_style_type :=
    fixed_round_style)
  return natural;

function ufixed_high (left_index, right_index : integer;
  operation      : character := 'X';
  left_index2   : integer    := 0;
  right_index2 : integer    := 0)
  return integer;

function ufixed_high (size_res   : ufixed;
  operation : character := 'X';
  size_res2 : ufixed)
  return integer;
function ufixed_low (left_index, right_index : integer;
```

```
        operation    : character := 'X';
        left_index2  : integer    := 0;
        right_index2 : integer    := 0)
        return integer;

    function ufixed_low (size_res  : ufixed;
        operation : character := 'X';
        size_res2 : ufixed)
        return integer;

    function saturate (left_index, right_index : integer)
        return ufixed;
    function saturate (size_res : ufixed) return ufixed;

    -- sfixed
    function to_sfixed (arg : integer;
        left_index      : integer;
        right_index     : integer := 0;
        overflow_style : fixed_overflow_style_type :=
          fixed_overflow_style;
        round_style     : fixed_round_style_type :=
          fixed_round_style)
        return sfixed;

    function to_sfixed (arg : integer;
        size_res        : sfixed;
        overflow_style : fixed_overflow_style_type :=
          fixed_overflow_style;
        round_style     : fixed_round_style_type :=
          fixed_round_style)
        return sfixed;

    function to_sfixed (arg : signed) return sfixed;
    function to_sfixed (arg : ufixed) return sfixed;

    function to_signed (arg : sfixed;
        size            : natural;
        overflow_style : fixed_overflow_style_type :=
          fixed_overflow_style;
        round_style     : fixed_round_style_type :=
          fixed_round_style)
        return signed;

    function to_signed (arg : sfixed;
        size_res        : signed;
        overflow_style : fixed_overflow_style_type :=
          fixed_overflow_style;
        round_style     : fixed_round_style_type :=
          fixed_round_style)
        return signed;

    function to_real (arg : sfixed) return real;

    function to_integer (arg : sfixed;
```

```
    overflow_style : fixed_overflow_style_type :=
      fixed_overflow_style;
    round_style    : fixed_round_style_type :=
      fixed_round_style)
    return integer;

  function sfixed_high (left_index, right_index : integer;
    operation    : character := 'X';
    left_index2  : integer   := 0;
    right_index2 : integer   := 0)
    return integer;

  function sfixed_low (left_index, right_index : integer;
    operation    : character := 'X';
    left_index2  : integer   := 0;
    right_index2 : integer   := 0)
    return integer;

  function sfixed_high (size_res : sfixed;
    operation : character := 'X';
    size_res2 : sfixed)
    return integer;

  function sfixed_low (size_res : sfixed;
    operation : character := 'X';
    size_res2 : sfixed)
    return integer;

  function saturate (left_index, right_index : integer)
    return sfixed;
  function saturate (size_res : sfixed)
    return sfixed;


  ----------------------------------------------------
  -- translation functions

  -- ufixed
  function to_01 (s : ufixed; xmap : std_ulogic := '0')
    return ufixed;
  function is_x     (arg : ufixed) return boolean;
  function to_x01   (arg : ufixed) return ufixed;
  function to_x01z  (arg : ufixed) return ufixed;
  function to_ux01  (arg : ufixed) return ufixed;

  function to_slv  (arg : ufixed)
    return std_logic_vector;
  alias to_stdlogicvector is
    to_slv [ ufixed return std_logic_vector] ;
  alias to_std_logic_vector is
    to_slv [ ufixed return std_logic_vector] ;

  function to_sulv (arg : ufixed)
    return std_ulogic_vector;
  alias to_stdulogicvector is
```

```
    to_sulv [ ufixed return std_ulogic_vector] ;
alias to_std_ulogic_vector is
    to_sulv [ ufixed return std_ulogic_vector] ;

function to_ufixed (arg : std_ulogic_vector;
   left_index  : integer;
   right_index : integer)
   return ufixed;

function to_ufixed (arg : std_ulogic_vector;
   size_res : ufixed)
   return ufixed;

function to_ufixed (arg : std_logic_vector;
   left_index  : integer;
   right_index : integer)
   return ufixed;

function to_ufixed (arg : std_logic_vector;
   size_res : ufixed)
   return ufixed;

function to_ufix (arg : std_ulogic_vector;
   width    : natural;
   fraction : natural)
   return ufixed;

function to_ufix (arg : std_logic_vector;
   width    : natural;
   fraction : natural)
   return ufixed;

function ufix_high (width, fraction : natural;
   operation : character := 'X';
   width2    : natural   := 0;
   fraction2 : natural   := 0)
   return integer;

function ufix_low (width, fraction : natural;
   operation : character := 'X';
   width2    : natural   := 0;
   fraction2 : natural   := 0)
   return integer;

-- sfixed
function to_01 (s : sfixed; xmap : std_ulogic := '0')
   return sfixed;
function is_x    (arg : sfixed) return boolean;
function to_x01  (arg : sfixed) return sfixed;
function to_x01z (arg : sfixed) return sfixed;
function to_ux01 (arg : sfixed) return sfixed;
function to_slv (arg : sfixed) return std_logic_vector;
alias to_stdlogicvector is
```

```
    to_slv [ sfixed return std_logic_vector] ;
alias to_std_logic_vector is
    to_slv [ sfixed return std_logic_vector] ;

function to_sulv (arg : sfixed) return std_ulogic_vector;
alias to_stdulogicvector is
    to_sulv [ sfixed return std_ulogic_vector] ;
alias to_std_ulogic_vector is
    to_sulv [ sfixed return std_ulogic_vector] ;

function to_sfixed (arg : std_ulogic_vector;
    left_index  : integer;
    right_index : integer)
    return sfixed;

function to_sfixed (arg : std_ulogic_vector;
    size_res : sfixed)
    return sfixed;

function to_sfixed (arg : std_logic_vector;
    left_index  : integer;
    right_index : integer)
    return sfixed;

function to_sfixed (arg : std_logic_vector;
    size_res : sfixed)
    return sfixed;

function to_sfix (arg : std_ulogic_vector;
    width    : natural;
    fraction : natural)
    return sfixed;

function to_sfix (arg : std_logic_vector;
    width    : natural;
    fraction : natural)
    return sfixed;

function sfix_high (width, fraction : natural;
    operation : character := 'X';
    width2    : natural   := 0;
    fraction2 : natural   := 0)
    return integer;

function sfix_low (width, fraction : natural;
    operation : character := 'X';
    width2    : natural   := 0;
    fraction2 : natural   := 0)
    return integer;

--------------------------------------------------
-- textio functions

-- ufixed
```

```
    procedure read(l     : inout line;
                value : out ufixed);
alias bread is read [ line, ufixed] ;
alias binary_read is read [ line, ufixed] ;

    procedure read(l     : inout line;
                value : out    ufixed;
                good  : out    boolean);
alias bread is read [ line, ufixed, boolean] ;
alias binary_read is read [ line, ufixed, boolean] ;

    procedure write (l          : inout line;
                    value    : in    ufixed;
                    justified : in    side  := right;
                    field    : in    width := 0);
alias bwrite is write [ line, ufixed, side, width] ;
alias binary_write is write [ line, ufixed, side, width] ;

    procedure oread(l     : inout line;
                value : out    sfixed);
alias octal_read is oread [ line, ufixed] ;

    procedure oread(l     : inout line;
                value : out    sfixed;
                good  : out    boolean);
alias octal_read is oread [ line, ufixed, boolean] ;

    procedure owrite (l          : inout line;
                    value    : in    ufixed;
                    justified : in    side  := right;
                    field    : in    width := 0);
alias octal_write is owrite [ line, ufixed, side, width] ;

    procedure hread(l     : inout line;
                value : out    ufixed);
alias hex_read is hread [ line, ufixed] ;

    procedure hread(l     : inout line;
                value : out    ufixed;
                good  : out    boolean);
alias hex_read is hread [ line, ufixed, boolean] ;

    procedure hwrite (l          : inout line;
                    value    : in    ufixed;
                    justified : in    side  := right;
                    field    : in    width := 0);
alias hex_write is hwrite [ line, ufixed, side, width] ;

-- sfixed
    procedure read(l     : inout line;
                value : out sfixed);
alias bread is read [ line, sfixed] ;
alias binary_read is read [ line, sfixed] ;
```

```
procedure read(l      : inout line;
               value : out    sfixed;
               good  : out    boolean);
alias bread is read [ line, sfixed, boolean] ;
alias binary_read is read [ line, sfixed, boolean] ;

procedure write (l          : inout line;
                 value     : in     sfixed;
                 justified : in     side  := right;
                 field     : in     width := 0);
alias bwrite is write [ line, sfixed, side, width] ;
alias binary_write is write [ line, sfixed, side, width] ;

procedure oread(l      : inout line;
                value : out    ufixed);
alias octal_read is oread [ line, sfixed] ;

procedure oread(l      : inout line;
                value : out    ufixed;
                good  : out    boolean);
alias octal_read is oread [ line, sfixed, boolean] ;

procedure owrite (l          : inout line;
                  value     : in     sfixed;
                  justified : in     side  := right;
                  field     : in     width := 0);
alias octal_write is owrite [ line, sfixed, side, width] ;

procedure hread(l      : inout line;
                value : out    sfixed);
alias hex_read is hread [ line, sfixed] ;

procedure hread(l      : inout line;
                value : out    sfixed;
                good  : out    boolean);
alias hex_read is hread [ line, sfixed, boolean] ;

procedure hwrite (l          : inout line;
                  value     : in     sfixed;
                  justified : in     side  := right;
                  field     : in     width := 0);
alias hex_write is hwrite [ line, sfixed, side, width] ;

--------------------------------------------------
-- string functions

-- ufixed
function to_string (value : ufixed) return string;
alias to_bstring is to_string [ ufixed return string] ;
alias to_binary_string is to_string [ ufixed return string] ;

function to_ostring (value : ufixed) return string;
alias to_octal_string is to_ostring [ ufixed return string] ;
```

```
function to_hstring (value : ufixed) return string;
alias to_hex_string is to_hstring [ufixed return string];

function from_string (bstring : string;
  left_index  : integer;
  right_index : integer)
  return ufixed;
alias from_bstring is
  from_string [ string, integer, integer return ufixed];
alias from_binary_string is
  from_string [ string, integer, integer return ufixed];

function from_string (bstring : string;
  size_res : ufixed)
  return ufixed;
alias from_bstring is
  from_string [ string, ufixed return ufixed];
alias from_binary_string is
  from_string [ string, ufixed return ufixed];

function from_string (bstring : string) return ufixed;
alias from_bstring is from_string [ string return ufixed];
alias from_binary_string is from_string [ string return ufixed];

function from_ostring (ostring : string;
  left_index  : integer;
  right_index : integer)
  return ufixed;
alias from_octal_string is
  from_ostring [ string, integer, integer return ufixed];

function from_ostring (ostring : string;
  size_res : ufixed)
  return ufixed;
alias from_octal_string is
  from_ostring [ string, ufixed return ufixed];

function from_ostring (ostring : string) return ufixed;
alias from_octal_string is from_ostring [ string return ufixed];

function from_hstring (hstring : string;
  left_index  : integer;
  right_index : integer)
  return ufixed;
alias from_hex_string is
  from_hstring [ string, integer, integer return ufixed];

function from_hstring (hstring : string;
  size_res : ufixed)
  return ufixed;
alias from_hex_string is
  from_hstring [ string, ufixed return ufixed];

function from_hstring (hstring : string) return ufixed;
```

```vhdl
alias from_hex_string is from_hstring [ string return ufixed] ;

--sfixed
function to_string (value : sfixed) return string;
alias to_bstring is to_string [ sfixed return string] ;
alias to_binary_string is to_string [ sfixed return string] ;

function to_ostring (value : sfixed) return string;
alias to_octal_string is to_ostring [ sfixed return string] ;

function to_hstring (value : sfixed) return string;
alias to_hex_string is to_hstring [ sfixed return string] ;

function from_string (bstring      : string;
  left_index  : integer;
  right_index : integer)
  return sfixed;
alias from_bstring is
  from_string [ string, integer, integer return sfixed] ;
alias from_binary_string is
  from_string [ string, integer, integer return sfixed] ;

function from_string (bstring  : string;
  size_res : sfixed)
  return sfixed;
alias from_bstring is
  from_string [ string, sfixed return sfixed] ;
alias from_binary_string is
  from_string [ string, sfixed return sfixed] ;

function from_string (bstring : string) return sfixed;
alias from_bstring is from_string [ string return sfixed] ;
alias from_binary_string is from_string [ string return sfixed] ;

function from_ostring (ostring : string;
  left_index  : integer;
  right_index : integer)
  return sfixed;
alias from_octal_string is
  from_ostring [ string, integer, integer return sfixed] ;

function from_ostring (ostring : string;
  size_res : sfixed)
  return sfixed;
alias from_octal_string is
  from_ostring [ string, sfixed return sfixed] ;

function from_ostring (ostring : string) return sfixed;
alias from_octal_string is from_ostring [ string return sfixed] ;

function from_hstring (hstring : string;
  left_index  : integer;
  right_index : integer)
  return sfixed;
```

```
    alias from_hex_string is
      from_hstring [ string, integer, integer return sfixed] ;

    function from_hstring (hstring : string;
      size_res : sfixed)
      return sfixed;
    alias from_hex_string is
      from_hstring [ string, sfixed return sfixed] ;

    function from_hstring (hstring : string) return sfixed;
    alias from_hex_string is from_hstring [ string return sfixed] ;

  end;
```

A.9   Package Float_Pkg

Package float_pkg listed here is the VHDL-1993 compatibility version of the VHDL-2008 instantiation of float_generic_pkg with the default generic parameters. It is therefore usable in both VHDL-1993 and VHDL-2008 systems without change.

```
    package float_pkg is

      ---------------------------------------------------

      constant float_exponent_width : natural    := 8;
      constant float_fraction_width : natural    := 23;
      constant float_round_style    : round_type := round_nearest;
      constant float_denormalize    : boolean    := true;
      constant float_check_error    : boolean    := true;
      constant float_guard_bits     : natural    := 3;
      constant no_warning           : boolean    := false;

      ---------------------------------------------------

      type unresolved_float is array (integer range <>) of std_ulogic;
      subtype u_float is unresolved_float;

      subtype float is unresolved_float;

      subtype unresolved_float32 is unresolved_float(8 downto -23);
      alias u_float32 is unresolved_float32;
      subtype float32 is float(8 downto -23);

      subtype unresolved_float64 is unresolved_float(11 downto -52);
      alias u_float64 is unresolved_float64;
      subtype float64 is float(11 downto -52);

      subtype unresolved_float128 is unresolved_float(15 downto -112);
      alias u_float128 is unresolved_float128;
      subtype float128 is float(15 downto -112);
```

```
    ----------------------------------------------------

    type valid_fpstate is (nan,
                           quiet_nan,
                           neg_inf,
                           neg_normal,
                           neg_denormal,
                           neg_zero,
                           pos_zero,
                           pos_denormal,
                           pos_normal,
                           pos_inf,
                           isx);


    function classfp (x : float; check_error : boolean := true)
      return valid_fpstate;


    ----------------------------------------------------

    function "abs" (arg : float) return float;
    function "-"   (arg : float) return float;
    function "+"   (l, r : float) return float;
    function "-"   (l, r : float) return float;
    function "*"   (l, r : float) return float;
    function "/"   (l, r : float) return float;
    function "rem"(l, r : float) return float;
    function "mod"(l, r : float) return float;

    function add (l, r : float;
      round_style : round_type := round_nearest;
      guard       : natural    := 3;
      check_error : boolean    := true;
      denormalize : boolean    := true)
      return float;

    function subtract (l, r : float;
      round_style : round_type := round_nearest;
      guard       : natural    := 3;
      check_error : boolean    := true;
      denormalize : boolean    := true)
      return float;

    function multiply (l, r : float;
      round_style : round_type := round_nearest;
      guard       : natural    := 3;
      check_error : boolean    := true;
      denormalize : boolean    := true)
      return float;
    function divide (l, r : float;
      round_style : round_type := round_nearest;
      guard       : natural    := 3;
      check_error : boolean    := true;
      denormalize : boolean    := true)
      return float;
```

```
   function remainder (l, r : float;
     round_style : round_type := round_nearest;
     guard       : natural    := 3;
     check_error : boolean     := true;
     denormalize : boolean     := true)
     return float;

   function modulo (l, r : float;
     round_style : round_type := round_nearest;
     guard       : natural    := 3;
     check_error : boolean     := true;
     denormalize : boolean     := true)
     return float;

   function reciprocal (arg : float;
     round_style : round_type := round_nearest;
     guard       : natural    := 3;
     check_error : boolean     := true;
     denormalize : boolean     := true)
     return float;

   function dividebyp2 (l, r : float;
     round_style : round_type := round_nearest;
     guard       : natural    := 3;
     check_error : boolean     := true;
     denormalize : boolean     := true)
     return float;

   function mac (l, r, c : float;
     round_style : round_type := round_nearest;
     guard       : natural    := 3;
     check_error : boolean     := true;
     denormalize : boolean     := true)
     return float;

   function sqrt (arg : float;
     round_style : round_type := round_nearest;
     guard       : natural    := 3;
     check_error : boolean     := true;
     denormalize : boolean     := true)
     return float;

   function is_negative (arg : float) return boolean;

   ----------------------------------------------------

   function "="  (l, r : float) return boolean;
   function "/=" (l, r : float) return boolean;
   function ">=" (l, r : float) return boolean;
   function "<=" (l, r : float) return boolean;
   function ">"  (l, r : float) return boolean;
   function "<"  (l, r : float) return boolean;

   function eq (l, r : float;
     check_error : boolean := true;
```

```
    denormalize : boolean := true)
    return boolean;

function ne (l, r : float;
  check_error : boolean := true;
  denormalize : boolean := true)
  return boolean;

function lt (l, r : float;
  check_error : boolean := true;
  denormalize : boolean := true)
  return boolean;

function gt (l, r : float;
  check_error : boolean := true;
  denormalize : boolean := true)
  return boolean;

function le (l, r : float;
  check_error : boolean := true;
  denormalize : boolean := true)
  return boolean;

function ge (l, r : float;
  check_error : boolean := true;
  denormalize : boolean := true)
  return boolean;

function \?=\   (l, r : float) return std_ulogic;
function \?/=\  (l, r : float) return std_ulogic;
function \?>\   (l, r : float) return std_ulogic;
function \?>=\  (l, r : float) return std_ulogic;
function \?<\   (l, r : float) return std_ulogic;
function \?<=\  (l, r : float) return std_ulogic;

function std_match (l, r : float) return boolean;
function find_rightmost (arg : float; y : std_ulogic)
  return integer;
function find_leftmost (arg : float; y : std_ulogic)
  return integer;
function maximum (l, r : float) return float;
function minimum (l, r : float) return float;


------------------------------------------------------

function resize (arg : float;
  exponent_width : natural    := 8;
  fraction_width : natural    := 23;
  round_style    : round_type := round_nearest;
  check_error    : boolean    := true;
  denormalize_in : boolean    := true;
  denormalize    : boolean    := true)
  return float;
```

```
function resize (arg : float;
  size_res       : float;
  round_style    : round_type := round_nearest;
  check_error    : boolean    := true;
  denormalize_in : boolean    := true;
  denormalize    : boolean    := true)
  return float;

function to_float32 (arg : float;
  round_style    : round_type := round_nearest;
  check_error    : boolean    := true;
  denormalize_in : boolean    := true;
  denormalize    : boolean    := true)
  return float32;

function to_float64 (arg : float;
  round_style    : round_type := round_nearest;
  check_error    : boolean    := true;
  denormalize_in : boolean    := true;
  denormalize    : boolean    := true)
  return float64;

function to_float128 (arg : float;
  round_style    : round_type := round_nearest;
  check_error    : boolean    := true;
  denormalize_in : boolean    := true;
  denormalize    : boolean    := true)
  return float128;


  ----------------------------------------------------

function to_slv (arg : float) return std_logic_vector;
alias to_stdlogicvector is
  to_slv [ float return std_logic_vector] ;
alias to_std_logic_vector is
  to_slv [ float return std_logic_vector] ;

function to_sulv (arg : float) return std_ulogic_vector;
alias to_stdulogicvector is
  to_sulv [ float return std_ulogic_vector] ;
alias to_std_ulogic_vector is
  to_sulv [ float return std_ulogic_vector] ;

function to_float (arg : std_ulogic_vector;
  exponent_width : natural := 8;
  fraction_width : natural := 23)
  return float;

function to_float (arg : integer;
  exponent_width : natural     := 8;
  fraction_width : natural     := 23;
  round_style    : round_type := round_nearest)
  return float;
```

```
function to_float (arg : real;
  exponent_width : natural    := 8;
  fraction_width : natural    := 23;
  round_style    : round_type := round_nearest;
  denormalize    : boolean    := true)
  return float;

function to_float (arg : unsigned;
  exponent_width : natural    := 8;
  fraction_width : natural    := 23;
  round_style    : round_type := round_nearest)
  return float;

function to_float (arg : signed;
  exponent_width : natural    := 8;
  fraction_width : natural    := 23;
  round_style    : round_type := round_nearest)
  return float;

function to_float (arg : ufixed;
  exponent_width : natural    := 8;
  fraction_width : natural    := 23;
  round_style    : round_type := round_nearest;
  denormalize    : boolean    := true)
  return float;

function to_float (arg : sfixed;
  exponent_width : natural    := 8;
  fraction_width : natural    := 23;
  round_style    : round_type := round_nearest;
  denormalize    : boolean    := true)
  return float;

function to_float (arg : integer;
  size_res    : float;
  round_style : round_type := round_nearest)
  return float;

function to_float (arg : real;
  size_res    : float;
  round_style : round_type := round_nearest;
  denormalize : boolean    := true)
  return float;

function to_float (arg : unsigned;
  size_res    : float;
  round_style : round_type := round_nearest)
  return float;

function to_float (arg : signed;
  size_res    : float;
  round_style : round_type := round_nearest)
  return float;
function to_float (arg : std_ulogic_vector;
  size_res : float)
```

```
    return float;

function to_float (arg : ufixed;
  size_res    : float;
  round_style : round_type := round_nearest;
  denormalize : boolean    := true)
  return float;

function to_float (arg : sfixed;
  size_res    : float;
  round_style : round_type := round_nearest;
  denormalize : boolean    := true)
  return float;

function to_float (arg : std_logic_vector;
  exponent_width : natural := 8;
  fraction_width : natural := 23)
  return float;

function to_float (arg : std_logic_vector;
  size_res : float)
  return float;

function to_unsigned (arg : float;
  size        : natural;
  round_style : round_type := round_nearest;
  check_error : boolean    := true)
  return unsigned;

function to_signed (arg : float;
  size        : natural;
  round_style : round_type := round_nearest;
  check_error : boolean    := true)
  return signed;

function to_ufixed (arg : float;
  left_index     : integer;
  right_index    : integer;
  overflow_style : fixed_overflow_style_type := fixed_saturate;
  round_style    : fixed_round_style_type    := fixed_round;
  check_error    : boolean                   := true;
  denormalize    : boolean                   := true)
  return ufixed;

function to_sfixed (arg : float;
  left_index     : integer;
  right_index    : integer;
  overflow_style : fixed_overflow_style_type := fixed_saturate;
  round_style    : fixed_round_style_type    := fixed_round;
  check_error    : boolean                   := true;
  denormalize    : boolean                   := true)
  return sfixed;

function to_unsigned (arg : float;
  size_res    : unsigned;
```

```
    round_style : round_type := round_nearest;
    check_error : boolean     := true)
    return unsigned;

function to_signed (arg : float;
  size_res    : signed;
  round_style : round_type := round_nearest;
  check_error : boolean     := true)
  return signed;

function to_ufixed (arg : float;
  size_res      : ufixed;
  overflow_style : fixed_overflow_style_type  := fixed_saturate;
  round_style    : fixed_round_style_type      := fixed_round;
  check_error    : boolean                      := true;
  denormalize    : boolean                      := true)
  return ufixed;

function to_sfixed (arg : float;
  size_res      : sfixed;
  overflow_style : fixed_overflow_style_type  := fixed_saturate;
  round_style    : fixed_round_style_type      := fixed_round;
  check_error    : boolean                      := true;
  denormalize    : boolean                      := true)
  return sfixed;

function to_real (arg : float;
  check_error : boolean := true;
  denormalize : boolean := true)
  return real;

function to_integer (arg : float;
  round_style : round_type := round_nearest;
  check_error : boolean     := true)
  return integer;

    -----------------------------------------------------

function realtobits (arg : real) return std_ulogic_vector;
function bitstoreal (arg : std_ulogic_vector) return real;
function realtobits (arg : real) return std_logic_vector;
function bitstoreal (arg : std_logic_vector) return real;

function to_01 (arg : float; xmap : std_logic := '0')
  return float;
function is_x (arg    : float) return boolean;
function to_x01 (arg : float) return float;
function to_x01z (arg : float) return float;
function to_ux01 (arg : float) return float;

    -----------------------------------------------------

procedure break_number (
  arg          : in  float;
```

```
    denormalize : in  boolean := true;
    check_error : in  boolean := true;
    fract       : out unsigned;
    expon       : out signed;
    sign        : out std_ulogic);

procedure break_number (
    arg         : in  float;
    denormalize : in  boolean := true;
    check_error : in  boolean := true;
    fract       : out ufixed;
    expon       : out signed;
    sign        : out std_ulogic);

function normalize (
    fract           : unsigned;
    expon           : signed;
    sign            : std_ulogic;
    sticky          : std_ulogic := '0';
    exponent_width  : natural    := 8;
    fraction_width  : natural    := 23;
    round_style     : round_type := round_nearest;
    denormalize     : boolean    := true;
    nguard          : natural    := 3)
    return float;

function normalize (
    fract           : ufixed;
    expon           : signed;
    sign            : std_ulogic;
    sticky          : std_ulogic := '0';
    exponent_width  : natural    := 8;
    fraction_width  : natural    := 23;
    round_style     : round_type := round_nearest;
    denormalize     : boolean    := true;
    nguard          : natural    := 3)
    return float;

function normalize (
    fract        : unsigned;
    expon        : signed;
    sign         : std_ulogic;
    sticky       : std_ulogic := '0';
    size_res     : float;
    round_style  : round_type := round_nearest;
    denormalize  : boolean    := true;
    nguard       : natural    := 3)
    return float;

function normalize (
    fract        : ufixed;
    expon        : signed;
    sign         : std_ulogic;
    sticky       : std_ulogic := '0';
```

```
    size_res    : float;
    round_style : round_type := round_nearest;
    denormalize : boolean     := true;
    nguard      : natural     := 3)
    return float;


  ----------------------------------------------------

  function "+"   (l : float; r : real) return float;
  function "+"   (l : real;  r : float) return float;
  function "+"   (l : float; r : integer) return float;
  function "+"   (l : integer; r : float) return float;
  function "-"   (l : float; r : real) return float;
  function "-"   (l : real;  r : float) return float;
  function "-"   (l : float; r : integer) return float;
  function "-"   (l : integer; r : float) return float;
  function "*"   (l : float; r : real) return float;
  function "*"   (l : real;  r : float) return float;
  function "*"   (l : float; r : integer) return float;
  function "*"   (l : integer; r : float) return float;
  function "/"   (l : float; r : real) return float;
  function "/"   (l : real;  r : float) return float;
  function "/"   (l : float; r : integer) return float;
  function "/"   (l : integer; r : float) return float;
  function "rem" (l : float; r : real) return float;
  function "rem" (l : real;  r : float) return float;
  function "rem" (l : float; r : integer) return float;
  function "rem" (l : integer; r : float) return float;
  function "mod" (l : float; r : real) return float;
  function "mod" (l : real;  r : float) return float;
  function "mod" (l : float; r : integer) return float;
  function "mod" (l : integer; r : float) return float;

  function "="   (l : float; r : real) return boolean;
  function "/="  (l : float; r : real) return boolean;
  function ">="  (l : float; r : real) return boolean;
  function "<="  (l : float; r : real) return boolean;
  function ">"   (l : float; r : real) return boolean;
  function "<"   (l : float; r : real) return boolean;
  function "="   (l : real;  r : float) return boolean;
  function "/="  (l : real;  r : float) return boolean;
  function ">="  (l : real;  r : float) return boolean;
  function "<="  (l : real;  r : float) return boolean;
  function ">"   (l : real;  r : float) return boolean;
  function "<"   (l : real;  r : float) return boolean;
  function "="   (l : float; r : integer) return boolean;
  function "/="  (l : float; r : integer) return boolean;
  function ">="  (l : float; r : integer) return boolean;
  function "<="  (l : float; r : integer) return boolean;
  function ">"   (l : float; r : integer) return boolean;
  function "<"   (l : float; r : integer) return boolean;
  function "="   (l : integer; r : float) return boolean;
  function "/="  (l : integer; r : float) return boolean;
  function ">="  (l : integer; r : float) return boolean;
```

```
function "<="  (l : integer; r : float) return boolean;
function ">"   (l : integer; r : float) return boolean;
function "<"   (l : integer; r : float) return boolean;
function \?=\  (l : float; r : real) return std_ulogic;
function \?/=\ (l : float; r : real) return std_ulogic;
function \?>\  (l : float; r : real) return std_ulogic;
function \?>=\ (l : float; r : real) return std_ulogic;
function \?<\  (l : float; r : real) return std_ulogic;
function \?<=\ (l : float; r : real) return std_ulogic;
function \?=\  (l : real;  r : float) return std_ulogic;
function \?/=\ (l : real;  r : float) return std_ulogic;
function \?>\  (l : real;  r : float) return std_ulogic;
function \?>=\ (l : real;  r : float) return std_ulogic;
function \?<\  (l : real;  r : float) return std_ulogic;
function \?<=\ (l : real;  r : float) return std_ulogic;
function \?=\  (l : float; r : integer) return std_ulogic;
function \?/=\ (l : float; r : integer) return std_ulogic;
function \?>\  (l : float; r : integer) return std_ulogic;
function \?>=\ (l : float; r : integer) return std_ulogic;
function \?<\  (l : float; r : integer) return std_ulogic;
function \?<=\ (l : float; r : integer) return std_ulogic;
function \?=\  (l : integer; r : float) return std_ulogic;
function \?/=\ (l : integer; r : float) return std_ulogic;
function \?>\  (l : integer; r : float) return std_ulogic;
function \?>=\ (l : integer; r : float) return std_ulogic;
function \?<\  (l : integer; r : float) return std_ulogic;
function \?<=\ (l : integer; r : float) return std_ulogic;

function maximum (l : float; r : real) return float;
function minimum (l : float; r : real) return float;
function maximum (l : real;  r : float) return float;
function minimum (l : real;  r : float) return float;
function maximum (l : float; r : integer) return float;
function minimum (l : float; r : integer) return float;
function maximum (l : integer; r : float) return float;
function minimum (l : integer; r : float) return float;


-------------------------------------------------------


function "not"  (l    : float) return float;
function "and"  (l, r : float) return float;
function "or"   (l, r : float) return float;
function "nand" (l, r : float) return float;
function "nor"  (l, r : float) return float;
function "xor"  (l, r : float) return float;
function "xnor" (l, r : float) return float;

function "and" (l : std_ulogic; r : float) return float;
function "and" (l : float; r : std_ulogic) return float;
function "or" (l : std_ulogic; r : float) return float;
function "or" (l : float; r : std_ulogic) return float;
function "nand" (l : std_ulogic; r : float) return float;
function "nand" (l : float; r : std_ulogic) return float;
function "nor" (l : std_ulogic; r : float) return float;
```

```
function "nor" (l : float; r : std_ulogic) return float;
function "xor" (l : std_ulogic; r : float) return float;
function "xor" (l : float; r : std_ulogic) return float;
function "xnor" (l : std_ulogic; r : float) return float;
function "xnor" (l : float; r : std_ulogic) return float;

function and_reduce  (l : float) return std_ulogic;
function nand_reduce (l : float) return std_ulogic;
function or_reduce   (l : float) return std_ulogic;
function nor_reduce  (l : float) return std_ulogic;
function xor_reduce  (l : float) return std_ulogic;
function xnor_reduce (l : float) return std_ulogic;


  ------------------------------------------------------


function copysign (x, y : float) return float;

function scalb (y : float; n : integer;
  round_style : round_type := round_nearest;
  check_error : boolean    := true;
  denormalize : boolean    := true)
  return float;

function scalb (y : float; n : signed;
  round_style : round_type := round_nearest;
  check_error : boolean    := true;
  denormalize : boolean    := true)
  return float;

function logb (x : float) return integer;
function logb (x : float) return signed;

function nextafter (x, y : float;
  check_error : boolean := true;
  denormalize : boolean := true)
  return float;

function unordered (x, y : float) return boolean;
function finite (x : float) return boolean;
function isnan (x : float) return boolean;

function zerofp (
  exponent_width : natural := 8;
  fraction_width : natural := 23)
  return float;

function nanfp (
  exponent_width : natural := 8;
  fraction_width : natural := 23)
  return float;

function qnanfp (
  exponent_width : natural := 8;
  fraction_width : natural := 23)
  return float;
```

```
function pos_inffp (
  exponent_width : natural := 8;
  fraction_width : natural := 23)
  return float;

function neg_inffp (
  exponent_width : natural := 8;
  fraction_width : natural := 23)
  return float;

function neg_zerofp (
  exponent_width : natural := 8;
  fraction_width : natural := 23)
  return float;

function zerofp (size_res : float) return float;
function nanfp (size_res : float) return float;
function qnanfp (size_res : float) return float;
function pos_inffp (size_res : float) return float;
function neg_inffp (size_res : float) return float;
function neg_zerofp (size_res : float) return float;


---------------------------------------------------


procedure read (l     : inout line;
                value : out float);
alias bread is read [ line, float];
alias binary_read is read [ line, float, boolean];

procedure read (l     : inout line;
                value : out float;
                good  : out    boolean);
alias bread is read [ line, float, boolean];
alias binary_read is read [ line, float];

procedure write (l          : inout line;
                 value      : in    float;
                 justified  : in    side := right;
                 field      : in    width := 0);
alias bwrite is write [ line, float, side, width];
alias binary_write is write [ line, float, side, width];

procedure oread (l     : inout line;
                 value : out float);
alias octal_read is oread [ line, float];

procedure oread (l     : inout line;
                 value : out float;
                 good  : out    boolean);
alias octal_read is oread [ line, float, boolean];

procedure owrite (l          : inout line;
                  value      : in    float;
                  justified  : in    side := right;
                  field      : in    width := 0);
```

```
alias octal_write is owrite [line, float, side, width];

procedure hread (l     : inout line;
                 value : out float);
alias hex_read is hread [line, float];

procedure hread (l     : inout line;
                 value : out float;
                 good  : out   boolean);
alias hex_read is hread [line, float, boolean];

procedure hwrite (l         : inout line;
                  value     : in    float;
                  justified : in    side := right;
                  field     : in    width := 0);
alias hex_write is hwrite [line, float, side, width];

-----------------------------------------------------

function to_string (value : float) return string;
alias to_bstring is to_string [float return string];
alias to_binary_string is to_string [float return string];

function to_hstring (value : float) return string;
alias to_hex_string is to_hstring [float return string];

function to_ostring (value : float) return string;
alias to_octal_string is to_ostring [float return string];

function from_string (bstring : string;
  exponent_width : natural := 8;
  fraction_width : natural := 23)
  return float;
alias from_bstring is
  from_string [string, natural, natural return float];
alias from_binary_string is
  from_string [string, natural, natural return float];

function from_ostring (ostring : string;
  exponent_width : natural := 8;
  fraction_width : natural := 23)
  return float;
alias from_octal_string is
  from_ostring [string, natural, natural return float];

function from_hstring (hstring : string;
  exponent_width : natural := 8;
  fraction_width : natural := 23)
  return float;
alias from_hex_string is
  from_hstring [string, natural, natural return float];

function from_string (bstring : string;
  size_res : float)
```

```
      return float;
   alias from_bstring is
      from_string [ string, float return float] ;
   alias from_binary_string is
      from_string [ string, float return float] ;

   function from_ostring (ostring : string;
      size_res : float)
      return float;
   alias from_octal_string is
      from_ostring [ string, float return float] ;

   function from_hstring (hstring : string;
      size_res : float)
      return float;
   alias from_hex_string is
      from_hstring [ string, float return float] ;


   -----------------------------------------------------

end;
```

A.10   Package TextIO

Package `textio` defines the basic I/O provided by VHDL. This is the VHDL-1993 version and is supplemented by package `standard_textio_additions` (see Appendix A.11). In VHDL-2008, the additions are merged into package `textio`.

```
package textio is

   type line is access string;
   function "="(l, r: line) return boolean;
   function "/="(l, r: line) return boolean;

   type text is file of string;
   procedure file_open (file f: text;
                        external_name; in string;
                        open_kind: in file_open_kind := read_mode);
   procedure file_open (status: out file_open_status;
                        file f: text;
                        external_name: in string;
                        open_kind: in file_open_kind := read_mode);
   procedure file_close (file f: text);
   procedure read (file f: text; value: out string);
   procedure write (file f: text; value: in string);
   function endfile (file f: text) return boolean;

   type side is (right, left);
   function "="(l, r: side) return boolean;
   function "/="(l, r: side) return boolean;
   function "<"(l, r: side)  return boolean;
```

```
function "<="(l, r: side) return boolean;
function ">"(l, r: side) return boolean;
function ">="(l, r: side) return boolean;

subtype width is natural;

file input:text open read_mode is "std_input";
file output:text open write_mode is "std_output";

procedure readline (file f: text; l: inout line);

procedure read (l: inout line;value: out bit;
                good: out boolean);
procedure read (l: inout line;value: out bit);

procedure read (l: inout line;value: out bit_vector;
                good: out boolean);
procedure read (l: inout line;value: out bit_vector);

procedure read (l: inout line;value: out boolean;
                good: out boolean);
procedure read (l: inout line;value: out boolean);

procedure read (l: inout line;value: out character;
                good: out boolean);
procedure read (l: inout line;value: out character);

procedure read (l: inout line;value: out integer;
                good: out boolean);
procedure read (l: inout line;value: out integer);

procedure read (l: inout line;value: out real;
                good: out boolean);
procedure read (l: inout line;value: out real);

procedure read (l: inout line;value: out string;
                good: out boolean);
procedure read (l: inout line;value: out string);

procedure read (l: inout line;value: out time;
                good: out boolean);
procedure read (l: inout line;value: out time);

procedure writeline (file f: text; l: inout line);
procedure write (l: inout line;
                 value: in bit;
                 justified: in side:= right;
                 field: in width := 0);
procedure write (l: inout line;
                 value: in bit_vector;
                 justified: in side:= right;
                 field: in width := 0);
procedure write (l: inout line;
                 value: in boolean;
```

```
                          justified: in side:= right;
                          field: in width := 0);
     procedure write (l: inout line;
                          value: in character;
                          justified: in side:= right;
                          field: in width := 0);
     procedure write (l: inout line;
                          value: in integer;
                          justified: in side:= right;
                          field: in width := 0);
     procedure write (l: inout line;
                          value: in real;
                          justified: in side:= right;
                          field: in width := 0;
                          digits: in natural:= 0);
     procedure write (l: inout line;
                          value: in string;
                          justified: in side:= right;
                          field: in width := 0);
     procedure write (l: inout line;
                          value: in time;
                          justified: in side:= right;
                          field: in width := 0;
                          unit: in time:= ns);
end;
```

A.11   Package Standard_Textio_Additions

Package `standard_textio_additions` is the VHDL-1993 supplement to package `textio` (see Appendix A.10). In VHDL-2008, the additions are merged into package `textio` and so this package is empty.

```
   package standard_textio_additions is

     procedure deallocate (p : inout line);
     procedure flush (file f : text);

     function minimum (l, r : side) return side;
     function maximum (l, r : side) return side;

     function to_string (value : side) return string;

     function justify (value : string;
                          justified : side := right;
                          field : width := 0)
        return string;

     procedure sread (l : inout line;
                          value : out string;
                          strlen : out natural);
```

```
     alias string_read is sread [ line, string, natural] ;
     alias bread is read [ line, bit_vector, boolean] ;
     alias bread is read [ line, bit_vector] ;
     alias binary_read is read [ line, bit_vector, boolean] ;
     alias binary_read is read [ line, bit_vector] ;

     procedure oread (l : inout line;
                      value : out bit_vector;
                      good : out boolean);
     procedure oread (l : inout line;
                      value : out bit_vector);
     alias octal_read is oread [ line, bit_vector, boolean] ;
     alias octal_read is oread [ line, bit_vector] ;

     procedure hread (l : inout line;
                      value : out bit_vector;
                      good : out boolean);
     procedure hread (l : inout line;
                      value : out bit_vector);
     alias hex_read is hread [ line, bit_vector, boolean] ;
     alias hex_read is hread [ line, bit_vector] ;

     procedure tee (file f : text; l : inout line);

     procedure write (l : inout line;
                      value : in real;
                      format : in    string);
     alias swrite is write [ line, string, side, width] ;
     alias string_write is write [ line, string, side, width] ;
     alias bwrite is write [ line, bit_vector, side, width] ;
     alias binary_write is write [ line, bit_vector, side, width] ;

     procedure owrite (l : inout line;
                       value : in bit_vector;
                       justified : in    side := right;
                       field : in width := 0);
     alias octal_write is owrite [ line, bit_vector, side, width] ;

     procedure hwrite (l : inout line;
                       value : in bit_vector;
                       justified : in    side := right;
                       field : in width := 0);
     alias hex_write is hwrite [ line, bit_vector, side, width] ;

   end;
```

A.12   Package Std_Logic_Arith

Package std_logic_arith is the old arbitrary-precision integer package from Synopsys, before the standardised package numeric_std became the preferred package for this functionality. It's use is now deprecated and it is only listed here for reference when working on changes to old designs.

```
--------------------------------------------------------------
--
-- Copyright (c) 1990,1991,1992 by Synopsys, inc.
-- all rights reserved.
--
-- this source file may be used and distributed without restriction
-- provided that this copyright statement is not removed from the
-- file and that any derivative work contains this copyright notice
--
--------------------------------------------------------------

library ieee;
use ieee.std_logic_1164.all;
package std_logic_arith is

    type unsigned is array (natural range <>) of std_logic;
    type signed is array (natural range <>) of std_logic;
    subtype small_int is integer range 0 to 1;

    function "+"(l: unsigned; r: unsigned) return unsigned;
    function "+"(l: signed; r: signed) return signed;
    function "+"(l: unsigned; r: signed) return signed;
    function "+"(l: signed; r: unsigned) return signed;
    function "+"(l: unsigned; r: integer) return unsigned;
    function "+"(l: integer; r: unsigned) return unsigned;
    function "+"(l: signed; r: integer) return signed;
    function "+"(l: integer; r: signed) return signed;
    function "+"(l: unsigned; r: std_ulogic) return unsigned;
    function "+"(l: std_ulogic; r: unsigned) return unsigned;
    function "+"(l: signed; r: std_ulogic) return signed;
    function "+"(l: std_ulogic; r: signed) return signed;

    function "+"(l: unsigned; r: unsigned) return std_logic_vector;
    function "+"(l: signed; r: signed) return std_logic_vector;
    function "+"(l: unsigned; r: signed) return std_logic_vector;
    function "+"(l: signed; r: unsigned) return std_logic_vector;
    function "+"(l: unsigned; r: integer) return std_logic_vector;
    function "+"(l: integer; r: unsigned) return std_logic_vector;
    function "+"(l: signed; r: integer) return std_logic_vector;
    function "+"(l: integer; r: signed) return std_logic_vector;
    function "+"(l: unsigned; r: std_ulogic) return std_logic_vector;
    function "+"(l: std_ulogic; r: unsigned) return std_logic_vector;
    function "+"(l: signed; r: std_ulogic) return std_logic_vector;
    function "+"(l: std_ulogic; r: signed) return std_logic_vector;

    function "-"(l: unsigned; r: unsigned) return unsigned;
    function "-"(l: signed; r: signed) return signed;
    function "-"(l: unsigned; r: signed) return signed;
    function "-"(l: signed; r: unsigned) return signed;
    function "-"(l: unsigned; r: integer) return unsigned;
    function "-"(l: integer; r: unsigned) return unsigned;
    function "-"(l: signed; r: integer) return signed;
    function "-"(l: integer; r: signed) return signed;
    function "-"(l: unsigned; r: std_ulogic) return unsigned;
    function "-"(l: std_ulogic; r: unsigned) return unsigned;
```

```
function "-"(l: signed; r: std_ulogic) return signed;
function "-"(l: std_ulogic; r: signed) return signed;

function "-"(l: unsigned; r: unsigned) return std_logic_vector;
function "-"(l: signed; r: signed) return std_logic_vector;
function "-"(l: unsigned; r: signed) return std_logic_vector;
function "-"(l: signed; r: unsigned) return std_logic_vector;
function "-"(l: unsigned; r: integer) return std_logic_vector;
function "-"(l: integer; r: unsigned) return std_logic_vector;
function "-"(l: signed; r: integer) return std_logic_vector;
function "-"(l: integer; r: signed) return std_logic_vector;
function "-"(l: unsigned; r: std_ulogic) return std_logic_vector;
function "-"(l: std_ulogic; r: unsigned) return std_logic_vector;
function "-"(l: signed; r: std_ulogic) return std_logic_vector;
function "-"(l: std_ulogic; r: signed) return std_logic_vector;

function "+"(l: unsigned) return unsigned;
function "+"(l: signed) return signed;
function "-"(l: signed) return signed;
function "abs"(l: signed) return signed;

function "+"(l: unsigned) return std_logic_vector;
function "+"(l: signed) return std_logic_vector;
function "-"(l: signed) return std_logic_vector;
function "abs"(l: signed) return std_logic_vector;

function "*"(l: unsigned; r: unsigned) return unsigned;
function "*"(l: signed; r: signed) return signed;
function "*"(l: signed; r: unsigned) return signed;
function "*"(l: unsigned; r: signed) return signed;

function "*"(l: unsigned; r: unsigned) return std_logic_vector;
function "*"(l: signed; r: signed) return std_logic_vector;
function "*"(l: signed; r: unsigned) return std_logic_vector;
function "*"(l: unsigned; r: signed) return std_logic_vector;

function "<"(l: unsigned; r: unsigned) return boolean;
function "<"(l: signed; r: signed) return boolean;
function "<"(l: unsigned; r: signed) return boolean;
function "<"(l: signed; r: unsigned) return boolean;
function "<"(l: unsigned; r: integer) return boolean;
function "<"(l: integer; r: unsigned) return boolean;
function "<"(l: signed; r: integer) return boolean;
function "<"(l: integer; r: signed) return boolean;

function "<="(l: unsigned; r: unsigned) return boolean;
function "<="(l: signed; r: signed) return boolean;
function "<="(l: unsigned; r: signed) return boolean;
function "<="(l: signed; r: unsigned) return boolean;
function "<="(l: unsigned; r: integer) return boolean;
function "<="(l: integer; r: unsigned) return boolean;
function "<="(l: signed; r: integer) return boolean;
function "<="(l: integer; r: signed) return boolean;
```

```
function ">"(l: unsigned; r: unsigned) return boolean;
function ">"(l: signed; r: signed) return boolean;
function ">"(l: unsigned; r: signed) return boolean;
function ">"(l: signed; r: unsigned) return boolean;
function ">"(l: unsigned; r: integer) return boolean;
function ">"(l: integer; r: unsigned) return boolean;
function ">"(l: signed; r: integer) return boolean;
function ">"(l: integer; r: signed) return boolean;

function ">="(l: unsigned; r: unsigned) return boolean;
function ">="(l: signed; r: signed) return boolean;
function ">="(l: unsigned; r: signed) return boolean;
function ">="(l: signed; r: unsigned) return boolean;
function ">="(l: unsigned; r: integer) return boolean;
function ">="(l: integer; r: unsigned) return boolean;
function ">="(l: signed; r: integer) return boolean;
function ">="(l: integer; r: signed) return boolean;

function "="(l: unsigned; r: unsigned) return boolean;
function "="(l: signed; r: signed) return boolean;
function "="(l: unsigned; r: signed) return boolean;
function "="(l: signed; r: unsigned) return boolean;
function "="(l: unsigned; r: integer) return boolean;
function "="(l: integer; r: unsigned) return boolean;
function "="(l: signed; r: integer) return boolean;
function "="(l: integer; r: signed) return boolean;

function "/="(l: unsigned; r: unsigned) return boolean;
function "/="(l: signed; r: signed) return boolean;
function "/="(l: unsigned; r: signed) return boolean;
function "/="(l: signed; r: unsigned) return boolean;
function "/="(l: unsigned; r: integer) return boolean;
function "/="(l: integer; r: unsigned) return boolean;
function "/="(l: signed; r: integer) return boolean;
function "/="(l: integer; r: signed) return boolean;

function shl(arg: unsigned; count: unsigned) return unsigned;
function shl(arg: signed; count: unsigned) return signed;
function shr(arg: unsigned; count: unsigned) return unsigned;
function shr(arg: signed; count: unsigned) return signed;

function conv_integer(arg: integer)
   return integer;
function conv_integer(arg: unsigned)
   return integer;
function conv_integer(arg: signed)
   return integer;
function conv_integer(arg: std_ulogic)
   return small_int;

function conv_unsigned(arg: integer; size: integer)
   return unsigned;
function conv_unsigned(arg: unsigned; size: integer)
   return unsigned;
```

```
    function conv_unsigned(arg: signed; size: integer)
      return unsigned;
    function conv_unsigned(arg: std_ulogic; size: integer)
      return unsigned;

    function conv_signed(arg: integer; size: integer)
      return signed;
    function conv_signed(arg: unsigned; size: integer)
      return signed;
    function conv_signed(arg: signed; size: integer)
      return signed;
    function conv_signed(arg: std_ulogic; size: integer)
      return signed;

    function conv_std_logic_vector(arg: integer; size: integer)
      return std_logic_vector;
    function conv_std_logic_vector(arg: unsigned; size: integer)
      return std_logic_vector;
    function conv_std_logic_vector(arg: signed; size: integer)
      return std_logic_vector;
    function conv_std_logic_vector(arg: std_ulogic; size: integer)
      return std_logic_vector;

    function ext(arg: std_logic_vector; size: integer)
      return std_logic_vector;

    function sxt(arg: std_logic_vector; size: integer)
      return std_logic_vector;

  end;
```

A.13   Package Math_Real

Package math_real is completely unsynthesisable. It is included here because it is useful for writing test benches, as demonstrated by the case study in Chapter 15.

```
    package math_real is

      constant math_e              : real := 2.71828_18284_59045_23536;
      constant math_1_over_e       : real := 0.36787_94411_71442_32160;
      constant math_pi             : real := 3.14159_26535_89793_23846;
      constant math_2_pi           : real := 6.28318_53071_79586_47693;
      constant math_1_over_pi      : real := 0.31830_98861_83790_67154;
      constant math_pi_over_2      : real := 1.57079_63267_94896_61923;
      constant math_pi_over_3      : real := 1.04719_75511_96597_74615;
      constant math_pi_over_4      : real := 0.78539_81633_97448_30962;
      constant math_3_pi_over_2    : real := 4.71238_89803_84689_85769;
      constant math_log_of_2       : real := 0.69314_71805_59945_30942;
      constant math_log_of_10      : real := 2.30258_50929_94045_68402;
      constant math_log2_of_e      : real := 1.44269_50408_88963_4074;
      constant math_log10_of_e     : real := 0.43429_44819_03251_82765;
      constant math_sqrt_2         : real := 1.41421_35623_73095_04880;
```

```
      constant math_1_over_sqrt_2 : real := 0.70710_67811_86547_52440;
      constant math_sqrt_pi        : real := 1.77245_38509_05516_02730;
      constant math_deg_to_rad     : real := 0.01745_32925_19943_29577;
      constant math_rad_to_deg     : real := 57.29577_95130_82320_87680;

      function sign (x : in real) return real;
      function ceil (x : in real) return real;
      function floor (x : in real) return real;
      function round (x : in real) return real;
      function trunc (x : in real) return real;
      function "mod" (x, y : in real) return real;
      function realmax (x, y : in real) return real;
      function realmin (x, y : in real) return real;

      procedure uniform(variable seed1, seed2 : inout positive;
                        variable x : out real);

      function sqrt (x : in real) return real;
      function cbrt (x : in real) return real;
      function "**" (x : in integer; y : in real) return real;
      function "**" (x : in real; y : in real) return real;
      function exp (x : in real) return real;
      function log (x : in real) return real;
      function log2 (x : in real) return real;
      function log (x : in real; base : in real) return real;
      function sin (x : in real) return real;
      function cos (x : in real) return real;
      function tan (x : in real) return real;
      function arcsin (x : in real) return real;
      function arccos (x : in real) return real;
      function arctan (y : in real) return real;
      function arctan (y : in real; x : in real) return real;
      function sinh (x : in real) return real;
      function cosh (x : in real) return real;
      function tanh (x : in real) return real;
      function arcsinh (x : in real) return real;
      function arccosh (x : in real) return real;
      function arctanh (x : in real) return real;
   end;
```

Appendix B

Syntax Reference

This section gives the syntax of the main synthesis structures introduced throughout the book. It only covers the synthesis subset, not the whole language, and it excludes structures used for test benches.

B.1 Keywords

The following are the keywords in VHDL. Since these are reserved words, they cannot be used as names of signals, variables, functions or design units. In addition to this set of keywords, you should not use the name `work` as the name of a library.

```
abs access after alias all and architecture array assert assume
assume_guarantee attribute
begin block body buffer bus
case component configuration constant context cover
default disconnect downto
else elsif end entity exit
fairness file for force function
generate generic group guarded
if impure in inertial inout is
label library linkage literal loop
map mod
nand new next nor not null
of on open or others out
package parameter port postponed procedure process property pro-
tected pure
range record register reject release rem report restrict
restrict_guarantee return rol ror
select sequence severity shared signal sla sll sra srl strong
subtype then to transport type
unaffected units until use
variable vmode vprop vunit
wait when while with
xnor xor
```

VHDL for Logic Synthesis, Third Edition. Andrew Rushton.
© 2011 John Wiley & Sons, Ltd. Published 2011 by John Wiley & Sons, Ltd.

B.2   Design Units

All design units can be preceded by context items:

```
context ::= { use_clause | library_clause | context_clause }
use_clause ::= use selected_name { , selected_name } ;
library_clause ::= library identifier { , identifier } ;
context_clause ::= context selected_name { , selected_name } ;
```

Context clauses are only available in VHDL-2008.

B.2.1   Entity

```
entity ::=
     context
     entity identifier is
        [ generic ( generic_list ); ]
        [ port ( port_list ); ]
         declarations
     [ begin
           concurrent_statements ]
     end;
generic_list ::=
     constant_interface_declaration
     { ; constant_interface_declaration}
port_list ::=
     signal_interface_declaration
     { ; signal_interface_declaration}
```

The declarations are limited to subprograms, types and subtypes, constants and signals.
   The concurrent statements must be passive – that is, they must not update any signals.

B.2.2   Architecture

```
context
architecture identifier of identifier is
     declarations
begin
     concurrent_statements
end;
```

The *declarations* are limited to subprograms, types and subtypes, constants, signals, components and configuration specifications.

B.2.3   Package

```
context
package identifier is
     declarations
end;
```

The *declarations* are limited to subprogram declarations (but not bodies), types and subtypes, constants, signals and components.

B.2.4 Package Body

```
context
package body identifier is
    declarations
end;
```

The *declarations* are limited to subprograms, types and subtypes and constants. All subprograms declared in the package must have bodies in the package body.

B.2.5 Context Declaration

This is available in VHDL-2008 only.

```
context identifier is
    context
end;
```

B.3 Concurrent Statements

```
concurrent_statements ::= { concurrent_statement ; }
concurrent_statement ::=
    block_statement |
    process_statement |
    concurrent_procedure_call |
    concurrent_assertion |
    concurrent_signal_assignment |
    component_instance |
    generate_statement
block_statement ::=
    label : block
        declarations
    begin
        concurrent_statements
    end block
```

The set of declarations and statements allowed in a block statement is the same as for an architecture.

```
process_statement ::=
    [ label : ] process [ ( sensitivity_list ) ]
        declarations
    begin
        sequential_statements
    end process
```

The declarations are limited to subprograms, types and subtypes, constants and variables. A process with a sensitivity list cannot contain wait statements. A process with no sensitivity list must contain one wait statement.

```
sensitivity_list ::= name { , name }
```

A sensitivity list can only contain signal names, slices and elements.

```
concurrent_procedure_call ::= [ label : ] procedure_call
concurrent_assertion ::= [ label : ] assertion
```

The procedure call and assertion are sequential statements.

```
concurrent_signal_assignment ::=
    [ label : ] conditional_signal_assignment |
    [ label : ] selected_signal_assignment
conditional_signal_assignment ::=
    target <= { expression when expression else } expression
```

The source expressions must match the type of the target. The when expressions must be boolean. A simple signal assignment is simply the minimal form of a conditional signal assignment.

```
selected_signal_assignment ::=
    with expression select
        target <= { expression when choices , }
                  expression when choices
```

The choices must match the type of the selection expression. The source expressions must match the type of the target.

```
component_instance ::=
    label : name
        [ generic map ( association_list ) ]
        [ port map ( association_list ) ]
generate_statement ::= for_generate | if_generate
for_generate ::=
    label : for identifier in discrete_range generate
        concurrent_statements
    end generate
```

The discrete range must be constant,

```
if_generate ::=
    label : if expression generate
        concurrent_statements
    end generate
```

The expression must be constant and boolean.

B.4   Sequential Statements

```
sequential_statements ::= { sequential_statement ; }
sequential_statement ::=
    wait_statement |
    assertion |
    signal_assignment |
    variable_assignment |
    procedure_call |
    if_statement |
    case_statement |
    for_loop |
    next_statement |
    exit_statement |
    return_statement |
    null_statement
wait_statement ::= wait [ on sensitivity_list ] [ until expression ]
```

The expression must be boolean.

```
assertion ::=
    assert expression [ report expression ] [ severity expression ]
```

The assertion expression must be boolean, the report expression must be string and the severity expression must be severity_type.

```
signal_assignment ::= target <= expression
variable_assignment ::= target := expression
procedure_call ::= identifier [ ( association_list ) ]
if_statement ::=
    if expression then
       sequential_statements
    { elsif expression then
       sequential_statements }
    [ else
       sequential_statements ]
    end if
```

The expressions must be boolean.

```
case_statement ::=
    case expression is
      when choices =>
         sequential_statements
      { when choices =>
         sequential_statements }
    end case
```

The choices must match the type of the case expression.

```
for_loop ::=
    [ label : ] for identifier in discrete_range loop
         sequential_statements
    end loop
```

The discrete range must be constant.

```
next_statement ::= next [ label ] [ when expression ]
exit_statement ::= exit [ label ] [ when expression ]
```

The expressions must be boolean. Next and exit statements can only be used within a loop.
The label determines which loop to next or exit. No label means exit or next the innermost loop.

```
return_statement ::= function_return &verbar; procedure_return
function_return ::= return expression
procedure_return ::= return
```

The return expression must match the return type of the function. A procedure return can only
be used in a procedure and a function return can only be used in a function.

```
null_statement ::= null
```


B.5   Expressions

```
expression ::=
      relation { and relation }
      relation { or relation }
      relation { xor relation }
      relation { xnor relation }
      relation [ nand relation ]
      relation [ nor relation ]
relation ::=
      shift_expression [ relational_operator shift_expression ]
relational_operator ::=
      = | /= | < | <= | >| >=
shift_expression ::=
      simple_expression [ shift_operator simple_expression ]
shift_operator ::=
      sll | srl | sla | sra | rol | ror
simple_expression ::=
      [ sign ] term { adding_operator  term }
sign ::=
      + | -
adding_operator ::=
      + | - | &
term ::=
      factor { multiplying_operator factor }
multiplying_operator ::=
      * | / | mod | rem
factor ::=
      primary [ ** primary ] | abs primary | not primary
primary ::=
      name | literal | aggregate | function_call |
      qualified_expression | type_conversion | ( expression )
name ::=
      identifier | operator_symbol | selected_name |
      indexed_name | slice_name | attribute_name
```

```
operator_symbol ::= string_literal
selected_name ::= prefix . suffix
prefix ::= name | function_call
suffix ::= identifier | character_literal | operator_symbol | all
indexed_name ::= prefix ( expression { , expression } )
slice_name ::= prefix ( discrete_range )
attribute_name ::= prefix ' identifier [ ( expression ) ]
aggregate ::=
     ( [ choices => ] expression { , [ choices => ] expression } )
function_call ::= function_name [ ( association_list ) ]
function_name ::= identifier | operator_symbol
qualified_expression ::=
     identifier ' ( expression ) | identifier ' aggregate
type_conversion ::= identifier ( expression )
choices ::= choice { '|' choice }
choice ::= identifier | simple_expression | discrete_range | others
```

Choices must be locally static expressions – that is, constants. Thus, a simple expression is allowed just so that a choice of, for example, -1 can be used.

```
association_list ::= association_element { , association_element }
association_element ::= [ name => ] actual_part
actual_part ::= expression | open
discrete_range ::= subtype_indication | range
subtype_indication ::= identifier [ constraint ]
constraint ::= range_constraint | index_constraint
range_constraint ::= range range
index_constraint ::= ( discrete_range { , discrete_range } )
range ::=
     attribute_name |
     simple_expression direction simple_expression
direction ::= to | downto
target ::= name | aggregate
```

B.6   Declarations

```
declarations ::= { declaration ; }
declaration ::=
     function_declaration |
     function_body |
     procedure_declaration |
     procedure_body |
     type_declaration |
     subtype_declaration |
     constant_declaration |
     variable_declaration |
     signal_declaration |
     component_declaration |
     configuration_specification
function_declaration ::=
     function_designator [ ( interface_list ) ] return identifier
```

```
function_designator ::= identifier | operator_symbol
function_body ::=
    function_designator [ ( interface_list ) ] return identifier is
        declarations
    begin
        sequential_statements
    end;
procedure_declaration ::=
    identifier [ ( interface_list ) ]
procedure_body ::=
    identifier [ ( interface_list ) ] is
        declarations
    begin
        sequential_statements
    end;
interface_list ::= interface_declaration { ; interface_declaration }
interface_declaration ::=
    constant_interface_declaration |
    variable_interface_declaration |
    signal_interface_declaration
constant_interface_declaration ::=
    [ constant ] identifier_list : [ mode ] subtype_indication
    [ := expression ]
variable_interface_declaration ::=
    [ variable ] identifier_list : [ mode ] subtype_indication
    [ := expression ]
signal_interface_declaration ::=
    [ signal ] identifier_list : [ mode ] subtype_indication
    [ := expression ]
identifier_list ::= identifier { , identifier }
mode ::= in | out | inout | buffer
```

The keyword constant, variable or signal can be omitted if it is the default. The default for entity ports is signal, the default for subprogram in parameters and generics is constant and the default for other subprogram parameters is variable. The mode can be omitted if it is the default – the default mode is always in. Mode buffer only applies to entity ports and is deprecated.

```
type_declaration ::= type identifier is type_definition
type_definition ::=
    enumeration_type | integer_type | array_type | record_type
enumeration_type ::=
    ( enumeration_literal [ , enumeration_literal ] )
enumeration_literal ::= identifier | character_literal
integer_type ::= range range
array_type ::= array ( array_constraint ) of subtype_indication
array_constraint ::= identifier range <> | discrete_range
record_type ::=
    record
        identifier_list : subtype_indication ;
        { identifier_list : subtype_indication ;}
    end record
subtype_declaration ::= subtype identifier is subtype_indication
```

```
constant_declaration ::=
    constant identifier_list : subtype_indication := expression
variable_declaration ::=
    variable identifier_list : subtype_indication [ := expression ]
signal_declaration ::=
    signal identifier_list : subtype_indication [ := expression ]
component_declaration ::=
    component identifier
       [ generic ( generic_list ); ]
       [ port ( port_list ); ]
    end component
configuration_specification ::=
    for instances : identifier
       use entity selected_name [ ( identifier ) ]
instances ::= identifier { , identifier } | all | others
```

References

Ashenden, P. and Lewis, J. (2008) *VHDL-2008 – Just the New Stuff*, Morgan Kaufman Publishers, Burlington, MA, USA, ISBN 978-0-12-374249-0.

EDA Industry Working Groups (2009) *Fixed-Point and Floating Point Packages*, [Online] Available from http://www.eda.org/fphdl/, [Accessed: 19 July 2010].

Horowitz, P. and Hill, W. (1989) *The Art of Electronics*, 2nd edn, Cambridge University Press, Cambridge, UK, ISBN 978-0-521370950.

IEEE Design Automation Standards Committee (2008) Std 1076-2008, *IEEE Standard VHDL Language Reference Manual*, IEEE, New York, NY, USA, ISBN 978 0 7381 5800 6.

IEEE Design Automation Standards Committee (1997) Std 1076.3-1997, *IEEE Standard VHDL Synthesis Packages*, IEEE, New York, NY, USA, ISBN 1-55937-923-5.

IEEE Design Automation Standards Committee (2004) Std 1076.6-2004, *IEEE Standard VHDL Register Transfer-Level (RTL) Synthesis*, IEEE, New York, NY, USA, ISBN 0-7381-4064-3.

IEEE Design Automation Standards Committee (1993) Std 1164-1993, *IEEE Standard Multivalue Logic System for VHDL Model Interoperability*, IEEE, New York, NY, USA, ISBN 1-55937-299-0.

IEEE Microprocessor Standards Committee (2008) Std 754-2008, *IEEE Standard for Floating-Point Arithmetic*, IEEE, New York, NY, USA, ISBN 978-0-7381-5753-5.

Open Cores (2010) *Open-source Hardware Library*, [Online] Available from http://opencores.org/, [Accessed: 19 July 2010].

Robin, I. (2005) *Digital Signal Processing*, [Online] Available from http://www.dsptutor.freeuk.com/, [Accessed: 19 July 2010].

Index

VHDL for Logic Synthesis, Third Edition. Andrew Rushton.
© 2011 John Wiley & Sons, Ltd. Published 2011 by John Wiley & Sons, Ltd.

Printed and bound by CPI Group (UK) Ltd, Croydon, CR0 4YY

27/10/2024

14580148-0005